Zoology: Principles and Concepts

Zoology: Principles and Concepts

Edited by Simon Benson

SYRAWOOD
PUBLISHING HOUSE

New York

Published by Syrawood Publishing House,
750 Third Avenue, 9th Floor,
New York, NY 10017, USA
www.syrawoodpublishinghouse.com

Zoology: Principles and Concepts
Edited by Simon Benson

International Standard Book Number: 978-1-68286-834-8 (Hardback)

Cataloging-in-Publication Data

Zoology : principles and concepts / edited by Simon Benson.
    p. cm.
Includes bibliographical references and index.
ISBN 978-1-68286-834-8
1. Zoology. 2. Animals. I. Benson, Simon.
QL45.2 .Z66 2019
590--dc23

TABLE OF CONTENTS

Permissions

List of Contributors

Index

PREFACE

Zoology is a branch of biology that studies the animal kingdom. This encompasses the study of structure, classification, embryology, habit and distribution of animals, whether extinct or living, and their interactions with their environment. Structural studies in zoology are approached from the domains of cell biology and anatomy. This enables the study of unicellular and multicellular organisms from the perspective of the structure and function of cells, organs and organ systems. Physiology studies how the different structures of a living organism function as a whole. The origin and descent of species, and their change over time is tackled in evolutionary biology. The grouping and categorization of organisms based on their genus or species allows ease of study and research and is known as scientific classification. It is considered an important aspect of zoology. This book discusses the fundamentals as well as modern approaches of zoology. It unravels the recent studies in this field. It will prove to be immensely beneficial to students and researchers in this field.

This book is the end result of constructive efforts and intensive research done by experts in this field. The aim of this book is to enlighten the readers with recent information in this area of research. The information provided in this profound book would serve as a valuable reference to students and researchers in this field.

At the end, I would like to thank all the authors for devoting their precious time and providing their valuable contribution to this book. I would also like to express my gratitude to my fellow colleagues who encouraged me throughout the process.

Editor

Behavioral event occurrence differs between behavioral states in *Sotalia guianensis* (Cetarctiodactyla: Delphinidae) dolphins: a multivariate approach

Rodrigo H. Tardin[1,2,4], Míriam P. Pinto[2,3], Maria Alice S. Alves[2] & Sheila M. Simão[1]

[1] *Laboratório de Bioacústica e Ecologia de Cetáceos, Departamento de Ciências Ambientais, Instituto de Florestas, Universidade Federal Rural do Rio de Janeiro. Rodovia BR 465 km 7, 23890-000 Seropédica, RJ, Brazil.*
[2] *Departamento de Ecologia, IBRAG, Universidade do Estado do Rio de Janeiro. Rua São Francisco Xavier 524, Maracanã, 20550-011 Rio de Janeiro, RJ, Brazil.*
[3] *Departamento de Botânica, Ecologia e Zoologia, Centro de Biociências, Universidade Federal do Rio Grande do Norte. Rodovia BR-101, Campus Universitário, Lagoa Nova, 59072-970 Natal, RN, Brasil.*
[4] *Corresponding author. Email: rhtardin@gmail.com*

ABSTRACT. Difficulties in quantifying behavioral events can cause loss of information about cetacean behavior, especially behaviors whose functions are still debated. The lack of knowledge is greater for South American species such as *Sotalia guianensis* (Van Benédén, 1864). Our objective was to contextualize the behavioral events inside behavioral states using a Permutational Multivariate Analysis of Variance (MANOVA). Three events occurred in the Feeding, Socio-Sexual and Travelling states (Porpoising, Side flop, Tail out dive), and five events occurred in the Feeding and Travelling states (Back flop, Horizontal jump, Lobtail, Spy-hop, Partial flop ahead). Three events (Belly exposure, Club, and Heading) occurred exclusively in the Socio-sexual state. Partial Back flop and Head flop occurred exclusively in the Feeding state. For the events that occurred in multiple states, we observed that some events occurred more frequently in one of the states (p < 0.001), such as Lobtail, Tail out dive horizontal Jump, Partial flop ahead and Side flop. Our multivariate analysis, which separated Socio-sexual behavior from Feeding and Travelling, showed that the abundance of behavioral events differs between states. This differentiation indicates that some events are associated with specific behavioral states. Almost 40% of the events observed were exclusively performed in one state, which indicates a high specialization for some events. Proper discrimination and contextualization of behavioral events may be efficient tools to better understand dolphin behaviors. Similar studies in other habitats and with other species, will help build a broader scenario to aid our understanding of the functions of dolphin behavioral events.

KEY WORDS. Aerial behavior; behavioral analysis; Guiana dolphin; surface behavior.

One of the major problems of cetacean behavioral studies is the clustering of different behaviors into only one category, such as jumps (Lusseau 2006, Azevedo *et al.* 2009). Altmann (1974) described two classes of behaviors: long duration behaviors in which the most common measure is the duration of this behavior and behavioral events, short duration behaviors in which the most common measure is the occurrence of each event. Under this classification, several events may occur inside a single behavioral state. Although most cetacean behavioral studies investigate behavioral state patterns (Shane 1990, Geise *et al.* 1999, Karczmarski *et al.* 2000, Daura-Jorge *et al.* 2007, Nery *et al.* 2010), behavioral events may also provide important information.

Difficulties associated with the quantification of behavioral events in the wild have resulted in a scarcity of studies in this area and limited information about cetacean behavior (Lusseau 2006, Vaughn *et al.* 2011). Cetaceans only spend a fraction of their time on the surface, and oftentimes the lack of underwater visibility makes full understanding of their behavior difficult. Thus, the investigation of which behavioral events occur in the different behavioral states will provide new knowledge in this area.

This contextualization is better understood for some cetacean species such as the bottlenose dolphin, *Tursiops truncatus* (Montagu, 1821) (Acevedo-Gutierrez 1999, Lusseau 2006, Miller *et al.* 2010). However, for "data deficient" species, such as the Guiana dolphin, *Sotalia guianensis* (Van Benédén, 1864) (Iucn 2013) there are no available data. Although some studies have analyzed behavioral events in *S. guianensis*'s repertoire (e.g., Araujo *et al.* 2008, Nascimento *et al.* 2008), quantification and proper discrimination between states and events is still lacking.

The Guiana dolphin, *S. guianensis*, is a small delphinid that inhabits estuaries and bays from northern Honduras (Carr & Bonde 2000) to southern Brazil (Simões-Lopes 1988).

Our general purpose was to investigate and quantify the behavioral events that occur within behavioral states. Our hypothesis is that some behavioral events are associated to one behavioral state or are more frequent in just one. If this is true, it is possible to clearly define groups of behavioral states in a multivariate analysis from behavioral event occurrences. Our hypothesis derives from findings on related species (e.g., *T. truncatus*) in which some events are associated to one behavioral state (e.g., LUSSEAU 2006). The investigation of the association between behavioral events and states aids the understanding of the general behavior of *S. guianensis*, since oftentimes only some conspicuous events (such as jumps) can be observed, which provides little information about the behavior of the group.

MATERIAL AND METHODS

Ilha Grande Bay (23°8'26"S, 44°14'50"W) is a large estuarine system on the southern coast of Rio de Janeiro, southeastern Brazil (SIGNORINI 1980). Marine habitats of this bay act as transition areas between the land and the sea by receiving organic matter from river drainage and mangrove production (NOGARA 2000). The bay receives deep nutrients from the sea derived from the South Atlantic Central Waters – SACW (SIGNORINI 1980).

Our sampling occurred in the western part of the bay, which includes shallow areas (<10 m) inhabited by a *S. guianensis* population (LODI 2003). *Sotalia guianensis* inhabits this part of the bay throughout the year, and 63.9% of the population presents some kind of residency (ESPÉCIE *et al.* 2010). Groups include up to 18 individuals and may include offspring (TARDIN *et al.* 2013a, b). The Ilha Grande Bay population, the largest population of *S. guianensis*, has an estimated 1,311 individuals (CI 95%: 1,232-1,389). (Mariana de Assis Espécie, pers. comm.)

We used continuous samplings using a digital handycam SONY DCR-TRV 120® (ALTMANN 1974) on board a 7.5 m vessel from May 2007 to March 2010. Two different approaches were used: focal group for behavioral states recording and all-occurrence sampling for behavioral events (MANN 1999). We followed random routes to search for dolphins, which maximized coverage of the sampling area. When a group of dolphins was sighted, we reduced the boat's velocity and filmed the group from a distance of 15 m. A group consisted of individuals within 10m from one another, according to the chain rule definition (SMOLKER *et al.* 1992).

In our study we used ALTMANN's (1974) categorization of long duration behaviors, or behavioral states, and short duration behaviors, or events. The behavioral states used were: Feeding, Travelling, and Socio-Sexual. Definitions and group sizes for each behavioral state are given in Table I. In Ilha Grande Bay, all occurrences of Feeding state were performed in groups (TARDIN *et al.* 2011). Other behavioral states like rest and play are not listed because they were not observed in our study.

We used videoclips as sampling units. Videoclips were restricted to a maximum of 300 seconds, since clips larger than 300 seconds were rare. Clips larger than 300 seconds were divided into several 80 second clips, and two of these were randomly selected. The 80 seconds interval duration was chosen to standardize our dataset since this represents the mean duration of the videoclips.

Every time we spotted a group of dolphins, we recorded following the above cited protocol. We first identified the behavioral state of the group and then analyzed all the behavioral events performed by all individuals inside that group in each videoclip by counting the number of times each event was performed. The slow frame-to-frame analysis of videoclips allowed a proper identification from both states and events. Each performance of a behavioral event was considered as one occurrence.

We observed 12 behavioral events. Seven of these events were also observed by LUSSEAU (2006), and we thus used his terms to ease comparisons: Horizontal Jump, Lobtail, Side flop, Spy-hop, Tail out dive, Back flop and Head flop. For a description of these events see LUSSEAU (2006). Belly up and Forward slap definitions were taken from SHANE (1990) and Porpoising from WEIHS (2002). Other two events we defined as: Heading, when an individual dolphin hits another with its beak/head; and Club, when a dolphin stands vertically in the water with body partially out of the water and re-enters the water hitting another dolphin.

We constructed an abundance matrix of the different events in each sampling unit (video clip). Then we counted the relative proportion of each event in the different states. We performed a Multi Dimensional Scaling ordination (MDS) using Bray-Curtis dissimilarity index to investigate the co-occurrences of the different events (LEGENDRE & LEGENDRE 1998). In the MDS plot we used behavioral states as a grouping variable to investigate whether some events or combinations of events were more associated with one or more other states.

We used a one-way non-parametric MANOVA permutation test (5,000 permutations, ANDERSON 2001) to test whether the multivariate mean of the abundance of events varied among the three behavioral states using the software R (Package: vegan, OKSANEN *et al.* 2013). This test was chosen because our data did not meet parametric assumptions (ANDERSON 2001). Post-hoc tests were done using separate permutation tests across the pair of groups being compared (ANDERSON 2001).

RESULTS

We performed 28 boat trips, in which we spent 100.5 hours at the sea and 42.1 hours (41.9%) directly observing *S. guianensis*. The behavioral state we observed most frequently was Feeding (338.3 minutes, 65.1%), followed by Travelling (160.5 minutes, 30.9%) and Socio-sexual (20.6 minutes, 4%).

Summary statistics for group size related to each behavioral state is presented in Table I. From these hours of observation we created 678 videos. The duration of the 462 videos (68.1%) in which we did not observe behavioral events was: Feeding: 44.6s ± 37.1s; and Travelling: 50.1s ± 42.0s. The duration of the 216 videos (31.9%) in which we observed at least one event: Feeding: 88.5s ± 102.1s; Travelling: 87.0s ± 104.4s; and Socio-Sexual: 38.7s ± 31.2s.Videos with events were the only ones used in the multivariate analyses.

The proportion of each event in the three different behavioral states is shown in Fig. 1. Three events occurred in all three states (Porpoising, Side flop, Tail-out-dive), five of them occurred in two states (Back flop, Horizontal jump, Lobtail, Spy-hop, Partial flop ahead). Three events (Belly exposure, Club, and Heading) occurred exclusively during the Socio-sexual state. Partial back flop and Head flop occurred exclusively during the Feeding state. For the events that occurred in more than one state, we observed that some events occurred more frequently in one of them, such as Lobtail (66% in Feeding), Tail out dive (80% in Feeding), Horizontal jump (88% in Feeding), Partial flop ahead (86% in Feeding) and Side flop (61% in Feeding) (Table II).

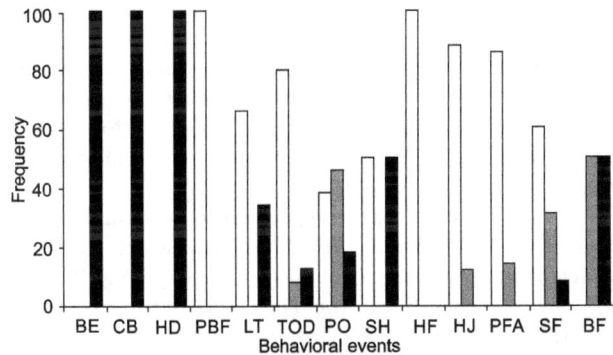

Figure 1. Frequency of occurrence for each event in the three behavioral states. (HJ) Horizontal jump, (PFA) partial flop ahead, (PO) porpoising, (SF) side flop, (PBF) partial back flop, (LT) Lobtail, (TOD) tail out dive, (SH) spy-hopping, (CB) club, (HD) heading, (HF) head flop, (BE) belly exposure, (BF) back flop, (□) feeding, (■) travelling, (■) socio-sexual.

A strong separation between Socio-sexual and the other Feeding and Travelling events was found through MDS ordination, which explained 91.04% of the variance (Fig. 2). Behavioral events abundance differed among the behavioral states (MANOVA: F = 17.8, d.f. = 2, p < 0.01). The post-hoc pairwise comparisons were significant between Feeding and Travelling events (F = 2.5, d.f. = 1, p = 0.03), Feeding and Socio-sexual (F = 28.2, d.f. = 1, p < 0.01) and Travelling and Socio-sexual events (F = 28.2, d.f. = 1, p < 0.01).

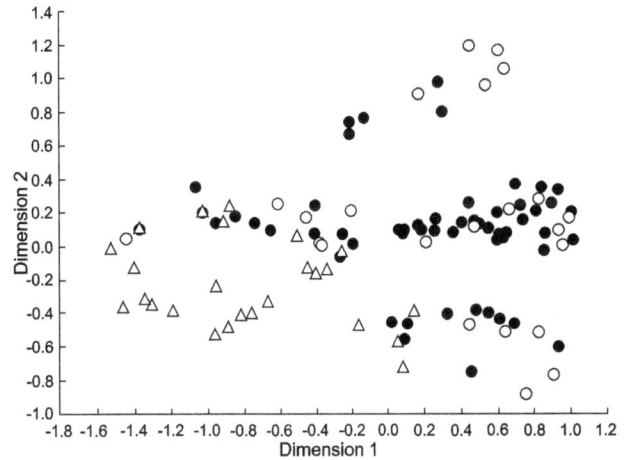

Figure 2. MDS ordination using occurrences of behavioral events grouped by behavioral states. Stress: 0.14. (●) feeding, (○) travelling, (▲) socio-sexual.

Table I. Definitions and group size for each behavioral state observed in Guiana dolphins' population in Ilha Grande bay, Rio de Janeiro, Brazil.

	Definition	Reference	Group size
Feeding	When individuals did not show directional movements and dove frequently in asynchronous fashion	Karczmarski et al. (2000)	21.4 ± 29.2 (3 to 200)
Travelling	Directional and persistent movements	Karczmarski et al. (2000)	20.9 ± 29.1 (4 to 150)
Socializing	Socio-sexual behavioral states occurred when individuals focused on each other, and the belly-to-belly position was frequently observed	Slooten (1994)	6.2 ± 2.8 (2 to 15)

Table II. Frequency of behavioral events in the different behavioral states of Sotalia guianensis at Ilha Grande Bay, Brazil. In parenthesis, there is the proportion of each event in relation to the total occurrences.

Behavioral events	Feeding	Travelling	Socio-sexual
Horizontal jump	337 (0.88)	44 (0.12)	0 (0.0)
Partial flop ahead	214 (0.86)	34 (0.14)	0
Porpoising	24 (0.38)	29 (0.46)	10 (0.18)
Side flop	8 (0.61)	4 (0.31)	1 (0.08)
Partial back flop	1 (1.0)	0	0
Lobtail	33 (0.66)	0 (0.0)	17 (0.34)
Tail out dive	32 (0.8)	3 (0.075)	5 (0.125)
Spy-hopping	4 (0.5)	0 (0.0)	4 (0.5)
Club	0	0	3 (1.0)
Heading	0	0	9 (1.0)
Head flop	6 (1.0)	0	0
Belly exposure	0	0	45 (1.0)
Back flop	0	1 (0.5)	1 (0.5)

DISCUSSION

Our data rejected the null hypothesis that some events would co-occur with similar frequencies in some behavioral states. The abundance of behavioral events is clearly different between states, as revealed by our multivariate analysis results, which separated mainly Socio-sexual behavior from Feeding and Traveling. This quantitatively supports the strong association of some events to specific behavioral states. Our proposed analysis tried to exclude the subjectivity of observers while at the same time maintaining flexibility to assess regional adaptations. This is the first study that quantitatively contextualized behavioral events for the lesser known *S. guianensis*.

Three of these events were performed in the Socio-sexual context. Belly exposure was observed 45 times only during the Socio-sexual state which indicates a strong relationship with this state. In fact, Belly exposure was frequently associated with movements of an individual (presumably males) directing its belly to another individual's belly (presumably females). This movement is considered to indicate courtship, since the genital regions of males and females are located in the posterior part of the belly. However, instead of engaging in sexual intercourse the individual on top turned its belly to the surface and avoided the copulation attempt by placing the genital region out of reach of the male. This may occur when females do not want to copulate with the specific male for many reasons. Interestingly, in one situation that lasted about thirty minutes we could observe twenty individuals copulating in belly-to-belly behavior without any sign of Belly exposure. This situation reported in our study and quantitatively assigned to the Socio-sexual state may be an indication of female active choice of males, a subject that remains unclear for dolphins (WHITEHEAD & MANN 2000), especially for *S. guianensis*. Another hypothesis that we can draw from our data is that these females may be feeding their calves and therefore are not in an estrous cycle. Most dolphins display a polyestrous cycle, which allows females to copulate several times in a year. Forced copulations may occur during the estrous cycle and females may actively avoid these situations (WHITEHEAD & MANN 2000) by exposing their belly to the surface, out of the water. Sometimes males force females to enter into an estrous cycle by killing their offspring. When their offspring die, the female may be able to copulate and generate other offspring (WHITEHEAD & MANN 2000). Although it is rare, an infanticide case was already reported for an adjacent population of *S. guianensis* in Sepetiba Bay in which males were harassing one female, which resulted in the death of her calf (NERY & SIMÃO 2009). This kind of interaction has also been documented in other dolphins, such as the common bottlenose dolphin in northeast Scotland (PATTERSON *et al.* 1998). In order to understand the proximate function of this behavior, the investigation of male rejection frequency by females with and without offspring in the population may unravel interesting questions about female active mate selection in dolphins.

The other two events exclusively performed in the Socio-sexual state (Club and Heading) seem to be related to aggression cases since they involve hitting other individuals. These events may be performed by males harassing females for sexual purposes or by females avoiding aggressive bachelor males. *Tursiops aduncus* (Ehrenberg, 1833) males form alliances to aggressively harass females at Shark Bay (Australia) (CONNOR *et al.* 2001). Aggressive interactions establish dominance hierarchies and can be an important driving force in the regulation of population density and dynamics.

The existence of some events exclusively performed in one state may have some evolutionary benefits. For instance, if some individuals recognize that some events are particularly performed for courtship purposes, the non-verbal communication displayed with the event would be better transmitted and, therefore, a lower energetic cost would be associated when individuals are seeking mates. This might also be true for aggressive interactions. When two or more individuals are displaying specific events that indicate agonistic interactions, nearby individuals that can understand the signs will be able to act, whether escaping or engaging the conflict. For instance, in common bottlenose dolphins aggression is expressed through posture, movement and sound, in which dolphins slam, ram and bites each other (CONNOR *et al.* 2000). Therefore it is important for the individuals to understand the signals to adopt a better strategy to minimize the energy expenditure and maximize their survival or breeding success. Nevertheless, for future studies, the recognition that some behavioral events are aggressive and context-dependent may help to quantify agonistic interactions in the future and to understand and test hypotheses about aggression in *S. guianensis*.

Our data suggest that dolphins may use a single behavioral repertoire for different functions. Aerial events may have different implications when used in Feeding, Travelling or Socio-sexual states. In Feeding states, aerial events may facilitate feeding by driving fish towards other dolphins and increasing capture success. Jumps associated with coordinated feeding tactics were documented in an Ilha Grande Bay population of *S. guianensis* (TARDIN *et al.* 2011). These jumps may serve as a complementary strategy to herd and increase prey capture. In a *T. truncatus* population from Isla del Coco (Costa Rica), jumps were used to aid in capturing prey (ACEVEDO-GUTIERREZ 1999). Our data showed that Horizontal Jump, Side Flop and Partial Flop Ahead occurred most frequently in the Feeding state. Jumps in which dolphins entirely leave the water (Horizontal jump and Side flop) may produce the largest percussive effect in the water, which disrupts schools of fish. However, in the Travelling state, leaps may be used to increase swimming speed since friction is lower in the air and turbulence forces are lower (PURCELL 1977). The aerial events observed in the Socio-sexual state were Back flops and Side flops. Leaps may have a communicative function as well. These movements may function as visual cues for dolphins to aggregate (NORRIS

& Dohl 1980), as a type of communication which would avoid competition between the groups (Lusseau 2006). Knowledge about why dolphins jump in the wild may be important in understanding their behavior. This event is conspicuous and therefore easy to visualize in the wild.

Comprehending the context of conspicuous events may help scientists to more easily understand the nature of dolphins' behavior and therefore provide tools for their conservation. For example, if we understand that some jumps are directly related to the Feeding behavioral state, we can prioritize conservation areas that are used for feeding purposes. This seems to be especially true for the Ilha Grande Bay population since fisheries and tourist activities are continually increasing. The presence of an oil maritime station, two nuclear power plants, and a large shipyard currently threaten the condition of the environmental reserve. Since this area supports the largest documented population of *S. guianensis*, these threats may harm the genetic diversity of this species in a global context.

Tail-out-dive, mostly observed during Feeding, suggests a diving behavior to capture demersal prey such as *Porichtys porosissimus* (Cuvier, 1829), which is one of the principal prey items of Guiana dolphins in the study area (Bernardes *et al.* 1989). The western part of Ilha Grande Bay has a sandy/muddy bottom with cryptic prey species, and dolphins may have to use echolocalization clicks more often to find prey in this region. Since the study area is shallow, dolphins may forage on the bottom for a longer time, which increases their chances of finding and capturing prey.

Lobtails can also be used as percussive events to drive fish in the Feeding state, the state with which they are most associated. This event was used by *T. aduncus* in Shark Bay (Australia) to induce an alarm reaction in fish and facilitate their detection (Connor *et al.* 2000).This event was also associated with *T. truncatus* in shallow seagrass waters in Sarasota Bay (Florida, USA) (Weiss 2006). Although depth was not measured in the present study, the western part of Ilha Grande Bay is also a shallow habitat (<10 m). In the Socio-sexual state, Lobtails can be used in agonistic interactions, especially male-female ones. Lobtails are used in a non-vocal communicative fashion by *Stenella frontalis* (Cuvier, 1829) in the Bahamas (Herzing 2000). These actions are used particularly to gain the attention of other dolphins, which will observe the direction that the animal is heading.

Spy-hopping occurred in similar proportions in Feeding and Socio-sexual states. In Feeding, this event could occur as cues to orientation, and dolphins might use it to aggregate to feed (Shane 1990). In socializing, this event could have sexual purposes (Slooten 1994, Lusseau 2006).

The Porpoising event occurred in similar proportions for Travelling and Feeding, with a slightly higher occurrence in the Travelling state. In this context, Porpoising may be used to increase speed to reach a destination. Dolphins hold their dorsal region out of the water, which reduces turbulence forces,

and their ventral region in the water, which provides benefits from the lower gravity forces (Purcell 1977). In the Travelling state this event was performed by all and/or almost all of the individuals of the group and several times by each one. In the Feeding state, Porpoising seems to be shorter and involves fewer individuals, and it might be used to chase prey, especially individually.

The exclusivity of some events and multi-purpose of others can indicate that Guiana dolphins repertoire may be broad and only a subset of these events may be displayed by different population along its distribution. Nevertheless it is important to have in mind that some specific behavioral events can have a pattern to be compared among populations, but exclusivity and sharing of events can be good proxies to understand the diversity of Guiana dolphin behavior. An interesting subject that we also have to consider is that some events could have been more frequent in our study because Guiana dolphins used Ilha Grande bay primary as a feeding place (more than 60% of groups behavioral budget was feeding). It is possible that other regions that Guiana dolphins use more frequently for other purposes than feeding may present a different subset of events.

The present study creates a methodological framework to analyze Guiana dolphins' behavior in a multidimensional perspective. At the behavioral and ecological aspect, this study is a valuable contribution to investigate *S. guianensis* behavioral diversity and the comparison among different populations may highlight regional differences along its distribution. Moreover, the validation and replication of this framework for other Guiana dolphins' population will allow using behavioral events as an important conservation tool.

ACKNOWLEDGMENTS

We thank Sergio C. Moreira, Dona Elza, Tico, Gilberto and students of the LBEC for their support. We also thank the solicitude and availability from two anonymous reviewers. Rodrigo H. Tardin is in Programa de Pós-Graduação em Ecologia e Evolução, Departamento de Ecologia, IBRAG, Universidade do Estado do Rio de Janeiro (UERJ). Personnel for this study were partially supported by the Fundação de Amparo à Pesquisa do Estado do Rio de Janeiro (FAPERJ) (R.H.O. Tardin, Grant number E-26/151.047/2007 and Grant number E-26/100.866/ 2011); The Coordenação de Aperfeiçoamento de Pessoal de Nível Superior (CAPES) (R.H.O. Tardin) and Cetacean Society International. M.A.S.Alves received fellowship and research grant associated from CNPq – process 308792/2009-2.

REFERENCES

Acevedo-Gutierrez, A. 1999. Aerial behavior is not a social facilitator in bottlenose dolphins hunting in small groups. **Journal of Mammalogy** 80 (3): 768-776.

ANDERSON, M.J. 2001. A new method for non-parametric multivariate analysis of variance. **Austral Ecology 26**: 32-46.

ALTMANN, J. 1974. Observational study of behavior: sampling methods. **Behavior 49** (3-4): 227-267.

ARAUJO, J.P.; A. SOUTO; L. GEISE & M.E. ARAUJO. 2008. The behavior of *Sotalia guianensis* (Van Benédén) in Pernambuco coastal waters, and a further analysis of its reaction to boat traffic. **Revista Brasileira de Zoologia 25** (1): 1-9.

AZEVEDO, A. F.; T.L. BISI; M. VAN SLUYS; P.R. DORNELES & L. BRITO JR. 2009. Comportamento do boto-cinza *Sotalia guianensis* (Cetacea: Delphinidae): Amostragem, termos e definições. **Oecologia Brasiliensis 13** (1): 192-200.

BERNARDES, A.T.; A.B. MACHADO & A.B. RYLAND. 1989. **Fauna brasileira ameaçada de extinção**. Belo Horizonte, Fundação Biodiversitas.

CARR, T. & R.K. BONDE. 2000. Tucuxi (*Sotalia fluviatilis*) occurs in Nicaragua, 800km north of its previously known range. **Marine Mammal Science 16** (2): 447-452. doi: 10.1111/j.1748-7692.2000.tb00936.x.

CONNOR, R.C.; J. MANN; P. TYACK & H. WHITEHEAD. 2000. The social lives of whales and dolphins, p.1-6. *In*: J. MANN; R.C. CONNOR; P. TYACK & H. WHITEHEAD (Eds). **Cetacean Societies: Field studies of dolphins and whales.** Chicago, The University of Chicago Press, 433p.

CONNOR, R.C.; M. HEITHAUS & L. BARRÉ. 2001. Complex social structure, alliance stability and mating access in a bottlenose dolphin "supper-alliance". **Proceedings of Royal Society B: Biological Sciences 268** (1464): 263-267. doi: 10.1098/rspb.2000.1357.

DAURA-JORGE, F.G.; M.R. ROSSI-SANTOS; L.L. WEDEKIN & P.C. SIMÕES-LOPES. 2007. Behavioral patterns and movement intensity of *Sotalia guianensis* (P.J. Van Benéden) (Cetacea, Deplphinidae) in two different areas in Brazilian coast. **Revista Brasileira de Zoologia 24** (2): 265-270.

ESPÉCIE, M.A.; R.H.O. TARDIN & S.M. SIMÃO. 2010. Degrees of residence of Guiana dolphins (*Sotalia guianensis*) in Ilha Grande Bay, south-eastern Brazil: a preliminary assessment. **Journal of Marine Biological Association of United Kingdom 90** (8): 1633-1639. doi:10.1017/S0025315410001256.

GEISE, L.; N. GOMES & R. CERQUEIRA. 1999. Behaviour, habitat use and population size of Sotalia fluviatilis (Gervais, 1853) (Cetacea: Delphinidae) in the Cananéia estuary region, SP, Brazil. **Revista Brasileira de Biologia 59** (2): 183-194.

HERZING, D. L. 2000. Acoustics and social behaviour of wild dolphins: implications for a sound society, p. 224-271. *In*: W.L. AU; A.N. POPPER & R.R. FAY (Eds.). **Hearing by Whales and Dolphins.** New York, Springer, 487p.

IUCN. 2013. Red List of threatened species. Accessed in; 14/06/2013

KARCZMARSKI, L.; V.C. COCKCROFT & A. MCLACHLAN. 2000. Habitat use and preferences of Indo-Pacific humpback dolphins *Sousa chinensis* in Algoa Bay, South Africa. **Marine Mammal Science 16** (1): 65-79. doi: 10.1111/j.1748-7692.2000.tb00904.x.

LEGENDRE, P & L. LEGENDRE. 1998. **Numerical Ecology.** Amsterdam, Elsevier, 990p.

LODI, L. & B. HETZEL. 1998. Grandes agregações do boto-cinza (*Sotalia fluviatilis*) na Baía da Ilha Grande, Rio de Janeiro. **Bioikos 12** (2): 26-30.

LUSSEAU, D. 2006. Why do dolphins jump? Interpreting the behavioural repertoire of bottlenose dolphins (*Tursiops* sp.) in Doubtful Sound, New Zealand. **Behavioural Processes 73**: 257-265. doi:10.1016/j.beproc.2006.06.006.

MANN, J. 1999. Behavioral Sampling Methods For Cetaceans: A Review And Critique. **Marine Mammal Science 15** (1): 102-122.

MILLER, L.J.; M. SOLANGI & S.A. KUCZAJ. 2010. Seasonal and diurnal patterns of behavior exhibited by Atlantic bottlenose dolphins (*Tursiops truncatus*) in the Mississippi Sound. **Ethology 116** (12): 1127-1137. doi: 10.1111/j.1439-0310.2010.01824.x.

NASCIMENTO, L.F.; P.I.A.P. MEDEIROS & M.E. YAMAMOTO. 2008. Descrição do Comportamento de Superfície do Boto Cinza, *Sotalia guianensis*, na Praia de Pipa – RN. **Psicologia: reflexão e crítica 21** (3): 509-517.

NERY, M.F. & S.M. SIMÃO. 2009. Sexual coercion and aggression towards a newborn calf of marine tucuxi dolphins (*Sotalia guianensis*). **Marine Mammal Science 25** (2): 450-454. doi: 10.1111/j.1748-7692.2008.00275.x.

NERY, M.F.; S.M. SIMÃO & T. PEREIRA. 2010. Ecology and behavior of the estuarine dolphin, *Sotalia guianensis* (Cetacea: Delphinidae) in Sepetiba Bay, South-eastern **Brazilian Journal of Ecology and Natural Environment 2** (9): 194-200.

NOGARA, P.J. 2000. **Caracterização dos ambientes marinhos da Área de Proteção Ambiental de Cairuçu – Município de Paraty – RJ.** Technical report, Fundação SOS Mata Atlântica. Rio de Janeiro, 83p.

NORRIS, K.S. & T.P. DOHL. 1980. Behavior of the Hawaiian Spinner Dolphin *S. longirostris*. **Fishery Bulletin 77** (4): 821-849.

OKSANEN, J.; F.G. BLANCHET; R. KINDT; P. LEGENDRE; P.R. MINCHIN; R.B. O'HARA; G.L. SIMPSON; P. SOLYMOS; M.H.H. STEVENS & H. WAGNER. 2012: vegan: Community Ecology Package. R package version 2.0-3. Available online at: http://CRAN.R-project.org/package=vegan vegan [Accessed: 14/VI/2013].

PATTERSON, I.A.P.; R. J. REID; B. WILSON; K. GRELLIER; H.M. ROSS & P.M. THOMPSON. 1999. Evidence for infanticide in bottlenose dolphins: an explanation for violent interactions with harbour porpoises? **Proceeding of the Royal Society B: Biological Sciences 265** (1402): 1167-1170.

PURCELL, E.M. 1977. Life at low Reynolds number. **American Journal of Physics 45**: 3-11.

SHANE, S.H. 1990. Behavior and ecology of the bottlenose dolphin at Sanibel Island, Florida, p. 245-265. *In*: S. LEATHERWOOD & R.R. REEVES (Eds). **The Bottlenose Dolphin.** San Diego, Academic press.

SIGNORINI, S.R. 1980. A study of the circulation in bay of Ilha Grande and bay of Sepetiba. Part I, an assessment to the tidally and wind-driven circulation using a finite element numerical model. **Boletim do Instituto Oceanográfico 29** (1): 41-55.

SIMÕES-LOPES, P.C. 1988. Ocorrência de uma população de *Sotalia fluviatilis* Gervais, 1853, (Cetacea: Delphinidae) no limite sul da sua distribuição, Santa Catarina, Brasil. **Biotemas 1** (1): 57-62.

SLOOTEN, E. 1994. Behaviour of Hector's dolphin – classifying behaviour by sequence analysis. **Journal of Mammalogy 75** (4): 956-964.

SMOLKER, R.A.; A.F. RICHARDS; R.C. CONNOR & J.W. PEPPER. 1992. Sex differences in patterns of association among Indian Ocean bottlenose dolphins. **Behavior 123** (1-2): 38-69.

TARDIN, R.H.O.; M.A. ESPÉCIE; F.T. D'AZEREDO; M.F. NERY & S.M. SIMÃO. 2011. Coordinated feeding tactics of the Guiana dolphin, *Sotalia guianensis* (Cetacea: Delphinidae), in Ilha Grande Bay, Rio de Janeiro, Brazil. **Zoologia 28** (3): 291-296. doi: 10.1590/S1984-46702011000300002.

TARDIN, R.; C. GALVÃO; M.A. ESPÉCIE & S.M. SIMÃO. 2013a. Group structure of Guiana dolphins, Sotalia guianensis (Cetacea, Delphinidae) in Ilha Grande Bay, Rio de Janeiro, southeastern Brazil. **Latin American Journal of Aquatic Research 41** (2): 313-322. doi: 10.3856/vol41-issue2-fulltext-10.

TARDIN, R.H.O.; M.A. ESPÉCIE; L. LODI & S.M. SIMÃO. 2013b. Parental care behavior in the Guiana dolphin, Sotalia guianensis (Cetacea: Delphinidae), in Ilha Grande Bay, southeastern Brazil. **Zoologia 30** (1): 15-23. doi: 10.1590/S1984-46702013000100002.

VAUGHN, R.L.; E. MUZI; J.L. RICHARDSON & B. WÜRSIG. 2011. Dolphin bait-balling behaviors in relation to prey ball escape behaviors. **Ethology 117** (10): 859-871.

WEISS, J. 2006. Foraging habitats and associated preferential foraging specializations of bottlenose dolphin (*Tursiops truncatus*) mother-calf pairs. **Aquatic Mammals 32** (1): 9-19.

WHITEHEAD, H. & J. MANN. 2000. Female reproductive strategies of cetaceans: Life histories and calf care, p. 219-246. *In*: J. MANN, R.C. CONNOR, P. TYACK & H. WHITEHEAD. (Eds). **Cetacean Societies: Field studies of dolphins and whales.** Chicago, The University of Chicago Press, 433p.

WEIHS, D. 2002. Dynamics of Dolphin Porpoising Revisited. **Integrative and Comparative Biology 42** (5): 1071-1078.

The surface morphology of the ctenidia of *Spondylus spinosus* (Mollusca: Bivalvia) from Antalya Bay, Turkey

Deniz Aksit[1,2], Beria Falakalı Mutaf[1] & Ahmet Balcı[1]

[1] Akdeniz University, Faculty of Aquatic Sciences and Fisheries, Antalya, Turkey.
[2] Corresponding author. E-mail: denizaksit@akdeniz.edu.tr

ABSTRACT. The surface morphology of the ctenidia of *Spondylus spinosus* Schreibers, 1793 was studied with light and scanning electron microscopy for comparison with the gill structures of other bivalves. The demibranch of *S. spinosus* is heterorhabdic, with the principal filaments at the descending lamellae and ordinary filaments at the ascending lamellae. The gill lamellae have a prominent gauze-like structure at their distal part, with numerous groups of eight ordinary filaments. They bear ciliary arrays on their frontal surfaces and ostia at their latero-frontal surfaces. Frequent cirral plates form regular interfilamentary junctions. The description of the gill structure of *S. spinosus* presented here can be used to derive implications for the correlations among the structure, habitat and mode of life of this species. At a particular stage of its adult life, *Spondylus spinosus* could be used as a subject for biomonitoring studies in natural and experimental environments.

KEY WORDS. Ciliary structures; ctenidia; morphology; scanning electron microscopy (SEM); spiny oyster.

Species of *Spondylus* Linnaeus, 1758, also known as spiny oysters, are the only bivalve molluscs of the family Spondylidae Gray, 1826. *Spondylus spinosus* Schreibers, 1793 is highly abundant on Mediterranean shores (ZENETOS *et al.* 2003), a habitat where these bivalves live attached to rocky substrates. As a result of evolutionary processes, the gills of *Spondylus* spp. perform many functions, collecting food particles and facilitating the dispersal of gametes in addition to their role in respiration, which includes establishing a water current in the mantle cavity as part of the circulatory system (GOSLING 2003). In brief, the gills consist of two ctenidia, each containing two demibranchs; each V-shaped demibranch has inner and outer components consisting of two lamellae with or without marginal grooves at their free end (GOSLING 2003).

The use of bivalve mollusks to assess national marine resources and their biological status has been the subject of many extensive studies (DOMOUHTSIDOU & DIMITRIADIS 2000, DAVID & FONTANETTI 2005, USHEVA *et al.* 2006). In species of Bivalvia, the gills and digestive glands are the first organs exposed to water and its pollutants (DOMOUHTSIDOU & DIMITRIADIS 2000). Despite the abundance of the genus in tropical and subtropical seawaters worldwide, very little is known about the basic biology and ecology of *Spondylus* spp. Although, some research on the reproductive biology of several species has been performed (VILLALEJO-FUERTE *et al.* 2002), very little information is available on the morphology of the organs of this bivalve (YOUNGE 1973).

The main purpose of this study was to use light and scanning electron microscopy to provide detailed information on the structure of the gill filaments of *S. spinosus*, as we believe that an understanding of this structure will serve as a tool of fundamental importance in biomonitoring studies.

MATERIAL AND METHODS

Spondylus spinosus individuals were collected from the littoral zone of the rocky shores of Antalya Bay (between 36°36′25.22″-36°53′04.26″N and 30°42′03.62″-31°46′31.30″E) on the southwestern coast of Turkey, a site where dense aggregations were found attached to the substrate. The individuals were extracted from rocks that were detached from the substrate by breakage using special sledgehammers during scuba diving excursions. All of the specimens were transported to the laboratory at the Faculty of Aquatic Sciences and Fisheries at Akdeniz University. Four specimens were washed, opened by breaking their shells and washed again in sea water.

The general morphology of the ctenidia of all the individuals was observed using stereo-light microscopy (LM). Undamaged areas of the gills were excised, and the dissected specimens were examined by LM and scanning electron microscopy (SEM). The gill lamellae were fixed in Bouin's solution for LM and in 2.5% glutaraldehyde in Sorensen's buffer (0.1 M, pH 7.3) for SEM. For LM, the specimens were cleared in xylene after rinsing in a graded-ethanol series and embedded in wax. Sections of 4-7 μm were stained with haematoxylin and eosin (H-E) (DRURY & WALLINGTON 1973). For SEM, the dehydrated and air-dried lamellae were mounted flat on aluminium stubs and coated with gold-palladium (AKSIT & FALAKALI

Mutaf 2011). We examined the lamellar morphology using a Zeiss Leo 1430 Scanning Electron Microscope at the Akdeniz University Medical School EM Unit (TEMGA).

The following abbreviations are used in this report: (Af.bv) afferent blood vessel, (al) ascending lamella, (C.a) ctenidial axis, (cd) ciliary disc, (ce) ciliary epithelium, (dl) descending lamella, (Ef.bv.) efferent blood vessel, (H-E) haematoxylin and eosin, (mg) marginal gutter, (nce) non-ciliated epithelium, (of) ordinary filaments, (pf) principal filaments, (wt) water tubule, (lf) latero-frontal cilia, (fc) frontal cilia, (◄) ostia.

RESULTS

The ctenidia of *S. spinosus* represent the second largest gross anatomical structure of the visceral parts (Fig. 1), exceeded in size only by the single adductor muscle. Each of the ctenidia of *S. spinosus* is formed by two V-shaped demibranchs attached to a thick ctenidial axis, with the ascending lamella one-third shorter in length than the descending lamella (Fig. 2). The branchial haemolymph vessel was prominent (Fig. 3). There was no marginal food groove at the free distal end.

The filaments presented different appearances on the frontal surfaces of the ascending (al) and descending (dl) lamellae. The descending lamellae were thicker than the ascending, and the ascending lamellae differed from the descending by the presence of minor lamellar subunits and transparent water tubules (Fig. 3). The filaments of this fillibranch gill were heterorhabdic in appearance (Figs 1-6).

On the descending lamella, thick principal filaments were present, consisting of two different components: a straight outer part and spirally wound cord-like inner pieces (Figs 4-6). The straight branch was covered with a thick ciliary sheet, and the wound portion contained more than two pieced folds, which appeared to fold backwards, assuming a thick pocket-like appearance (Fig. 6). No interlamellar junctions were observed.

Ordinary filaments were the only components of the ascending lamellae located at the ventral part of each demibranch. Eight ordinary filaments formed the continuation of every principal filament formed by branching at the lower part of the demibranch. The ordinary filaments, resembling translucent channels, extended parallel and caused a fan-like expansion (Figs 6 and 7). Many ordinary filaments formed a group, and numerous groups were found in the terminal portion of the dorsal region with a flattened undulating edge, which was shaped into a marginal gutter by outwards folding.

Frequent ciliary discs connected each of the eight ordinary filaments to each other at regular intervals and extended in two directions (Figs 7 and 8). These interfilamentary connections showed a complex structure (Figs 9-11), with each connective disc consisting either of 5-7 cirral or microvillar plates, interlocked at the top with each other (Fig. 11). Their location on the surface epithelium was clearly visible under light and electron microscopy (Figs 8-11).

The frontal surface of the ordinary filaments showed a uniform array of cilia characterized by ciliary rootlets, whereas the abfrontal surfaces were covered with a non-ciliated epithe-

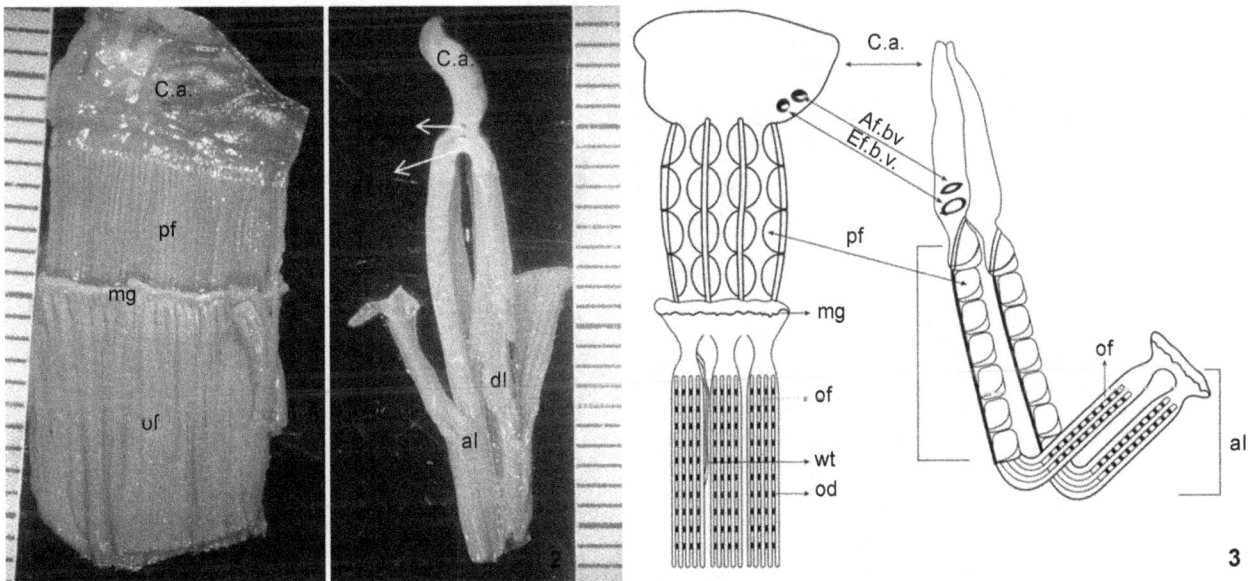

Figures 1-3. General appearance of the ctenidia of *Spondylus spinosus*: (1) ascending and descending lamellae attached at the ctenidial axis, forming a W shape; (2) frontal view of the principal and ordinary filaments; (3) schematic drawings of the ctenidia, with frontal (left) and side (right) views of the gill filaments. Scale: 1-2 = a ruler in mm, 3 = the drawing is merely to specify the structural definition and is not the actual size.

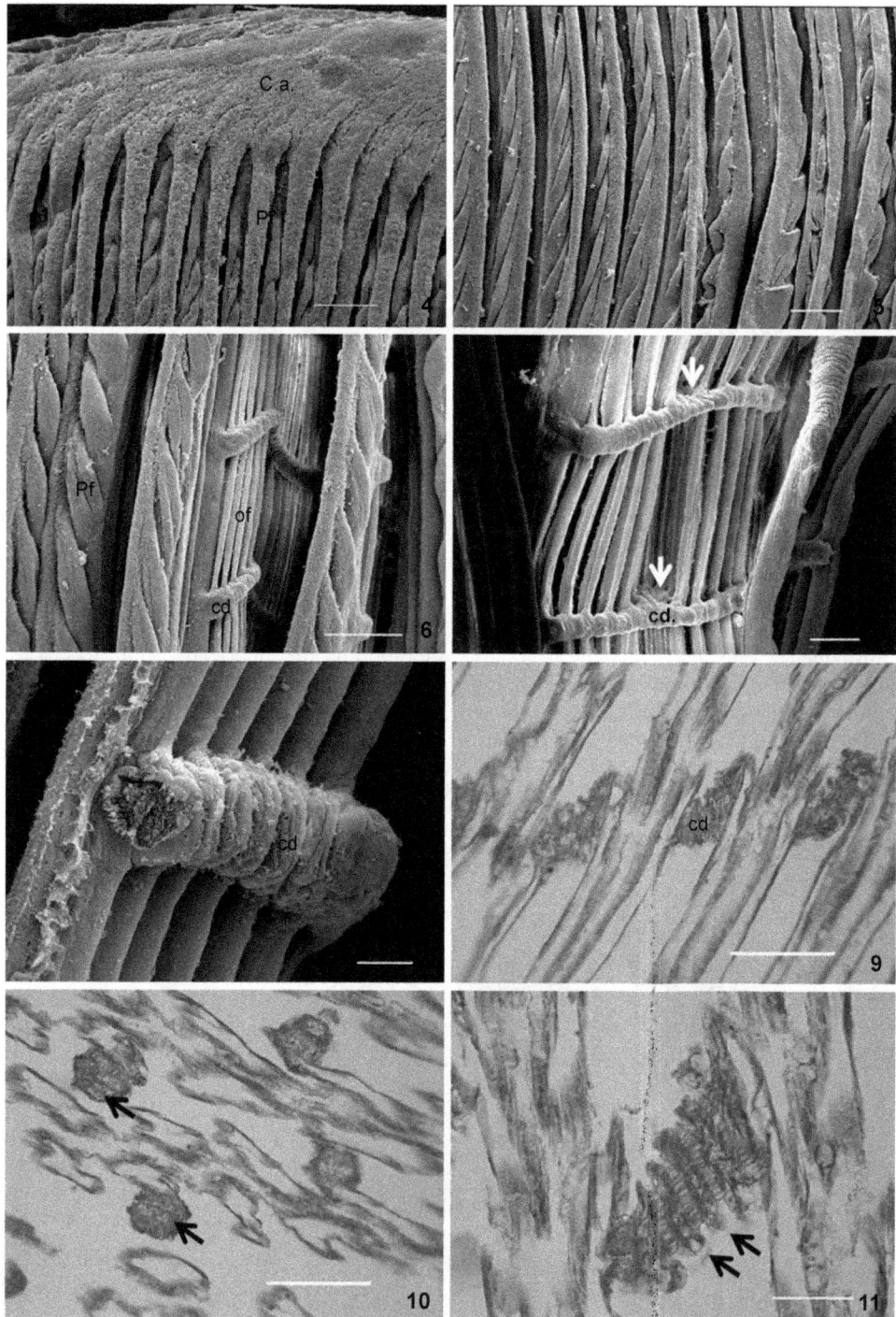

Figures 4-11. SEM and LM of the ctenidia of *Spondylus spinosus*: (4) proximal region of a demibranch, with principal filaments extending from the axis; (5) view of the principal filaments at different axes; (6) ordinary filaments located behind the principal filaments at the distal region of the demibranch; (7) ordinary filaments attached by continual ciliary junctions, with connective discs at regular intervals and extended inwardly (arrows); (8) ciliary discs interlocked with each other; (9) ciliary junctions between ordinary filaments (LM/H-E); (10) cross-section at the connection area (arrow) (LM/H-E-H); (11) ciliary junction and connective disc consisting of 5-7 cirral plates (black arrows) interlocked with each other (LM/H-E). Scale bars: 4-6 = 200 μm, 7 = 60 μm, 8 = 30 μm, 9-10 = 100 μm, 11 = 50 μm.

lium consisting of microvilli (Figs 12 and 13). The latter cells were characterized by large nuclei. A frontal view of the ordinary filaments showed abundant ciliary cover. The composite microcilia with an elongated latero-frontal array were clearly evident when the filaments were inflated and expanded into the interfilamentary spaces (Fig. 14). No ciliar differentiation was detected. All the cilia appeared to be simple cilia. Next to the ciliary band, an irregular number of ostia were observed on the lateral surfaces (Fig. 15). No mucus strings or layers or any lesional morphology that could be associated with environmental stress were observed in any part of the gill lamellae.

DISCUSSION

Light microscopy revealed that the general structure of the gills in *S. spinosus* was similar, in part, to that described for other bivalves, such as Mytilidae and Pectinidae (BENNINGER & ST JEAN 1997, GREGORY & GEORGE 2000, BENNINGER & DECOTTIGNES 2008). In brief, the ctenidia of *Spondylus* spp. were described as *Pecten* type (Type B1b), without a marginal groove (ATKINS 1937a,b). The general morphology observed in the present study was in accordance with previous descriptions. However, the ascending lamella of *S. spinosus* was found to be two-thirds

the size of the descending lamella, which is shorter than that described for *S. americanus* (YOUNGE 1973).

In many bivalve species, ciliation covers both the frontal surface and the lateral surfaces of the gill, and scattered cilia were described on the abfrontal surface of the ordinary filaments (DUFOUR & BENINGER 2001, DAVID & FONTANETTI 2005). Only the frontal surface showed a ciliary cover on the ordinary filaments of the gills of *S. spinosus*. AIELLO & SLEIGH (1972) suggested that the absence of ciliation provided a space for the eu-latero-frontal cilia to perform their movement, particularly in *Mytilus edulis* Linnaeus, 1758, whereas others have suggested that the lack of ciliation served to increase the surface available for the absorptive cells to obtain materials from the environment, as in *Mytella falcata* (Orbigny, 1846) (DAVID *et al.* 2008). Retention of the food particles in filter-feeding bivalves is performed either by the beating of cilia or the muscular control of ostia (GOSLING 2003). *Spondylus spinosus* appears to employ both mechanisms, exhibiting a sieve in many ostia next to the ciliary array. Oralwards currents could occur primarily in the dorsal channels, as described by YONGE (1973), and the prominent marginal gutter in *S. spinosus* could be proposed as an effective channel for particle movement because no marginal groove exists.

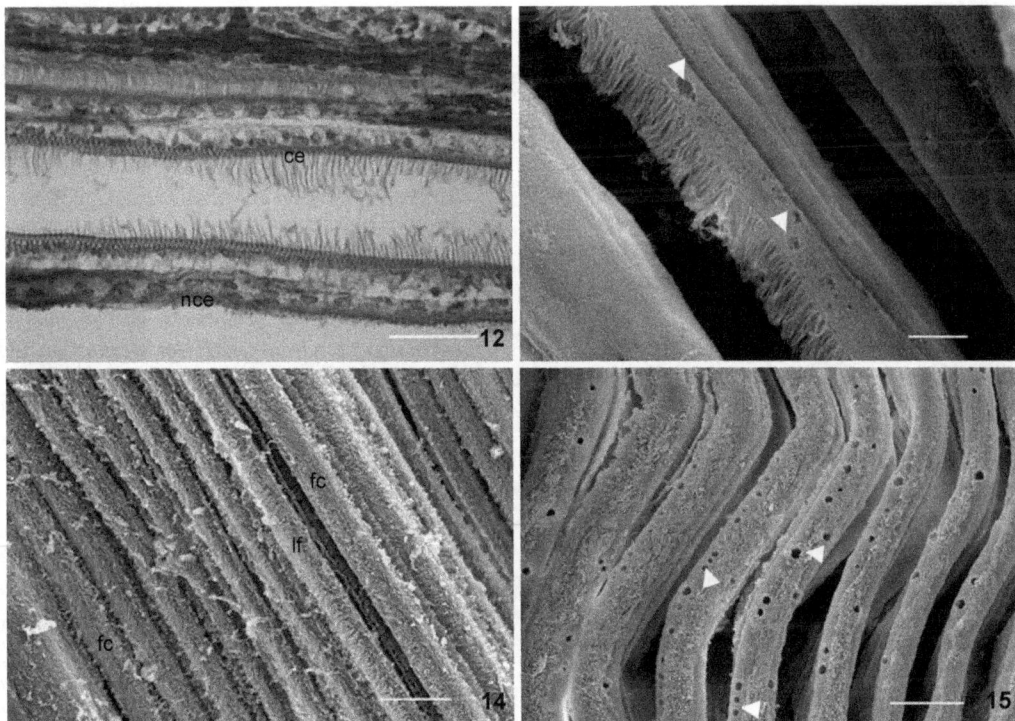

Figures 12-15. SEM and LM of the gill filaments of *Spondylus spinosus*: (12) general view showing the cellular organisation of the ordinary filaments, ciliated epithelium at only the frontal surface and prominent ciliary bases; (13) detail of a single filament with lateral cilia (the arrow indicates ostia); (14) detail of the ordinary filaments showing the ciliary cover; (15) frontal ciliary row on the inflated filaments and scattered ostia (◄) at the lateral surface. Scale bars: 12, 14-15 = 30 μm, 13 = 4 μm.

The regular distribution of ciliary junctions was similar to that found in certain species of Mytilidae, such as *Perna perna* Linnaeus, 1758 and *M. falcata* (Gregory & George 2000, David & Fontanetti 2005), but several differences were also noted. The ciliary discs were located on either side of the ordinary filaments and extended laterally at eight filament intervals in *Spondylus spinosus*. The ciliary discs contained a smaller number of cirral plates and were, therefore, much more complex than those reported to occur in certain mytilid and pectinid species, such as *M. falcata* (David & Fontanetti 2005), *Modiolus barbatus* Linnaeus, 1758 (Falakali Mutaf *et al.* 2009) and *Placopecten magellanicus* (Gmelin, 1791) (Morse & Zardus 1997), species in which only a single ciliary plate forms the disc. The distribution of the ciliary discs appears to serve predominantly to protect the filaments from hydrodynamic forces and to regulate the flow of water through the interfilamentary spaces, thus facilitating currents in the lower part of the demibranch in addition to those occurring in the dorsal channels.

Bivalve mollusks have been considered for monitoring aquatic habitats in many countries of the Northern and Southern Hemispheres, particularly well-settled species and those described as sentinel organisms in cases of pollution (Gregory *et al.* 2002). In Turkey, only a few pilot projects of this nature have been conducted to establish a local assessment programme for marine pollution or to contribute to international surveys based on the study of sedentary species, such as *S. spinosus*. Indeed, gill morphology has been extensively used as an indicator organ of pollution (Domouhtsidou & Dimitriadis 2000, Gregory & George 2000, David & Fontanetti 2005, Koehler *et al.* 2008).

The results presented here provide data on the general structure of the gills of *S. spinosus*, with no pollution effect, for comparison in histopathological studies. Although increased mucus production on gill surfaces has been suggested to result from the accumulation of metallic or microorganismal pollutants (Gregory *et al.* 2002, Aksit *et al.* 2008), no trace of mucus was observed on the specimens studied.

In summary, *S. spinosus* have gills with heterorhabdic filaments, with the principal ones showing a complex structure. The ascending lamellae of the ordinary filaments, consisting of many minor lamellar subunits, are shorter than the thicker descending lamellae. Every eight ordinary filaments, connected by regular ciliary discs of 5-7 cirral plates, ended with a marginal gutter at the dorsal end. No marginal food groove at the free distal end was evident.

In conclusion, the external morphology of the ctenidia of.*S. spinosus* can provide many useful features for evolutionary studies and can be the subject of biomonitoring studies within both natural and experimental contexts.

ACKNOWLEDGEMENTS

We thank M. Gokoglu for providing us with the specimens of *S. spinosus*. This work was funded by the Akdeniz University Research Projects Coordination Unit.

REFERENCES

Aiello, E. & M.A. Sleigh. 1972. The metachronal wave of lateral cilia of *Mytilus edulis*. The **Journal of Cell Biology 54**: 493-506.

Aksit, D.; B. Falakali- Mutaf; B. Gocmen & G.Gurelli. 2008. A preliminary observations on *Trichodina* sp. (Ciliophora: Pertricha) on the gills of limpets (*Patella* spp.) in Antalya (Turkey). **North-Western Journal of Zoology 4**: 295-298.

Aksit, D.& B. Falakali-Mutaf. 2011. The external morphology of the gill of *Patella caerulea* L. (Mollusca: Gastropoda). **Turkish Journal of Zoology 35**: 603-606.

Atkins, D. 1937a. On the ciliary mechanisms and interelationships of lamellibranchs. Part II. Sorting devices on the gills. **Quarterly Journal of Microscopical Science 78**: 339-373.

Atkins, D. 1937b. On the ciliary mechanisms and interelationships of lamellibranchs. Part III. Types of lamellibranch gills and their food currents. **Quarterly Journal of Microscopical Science 79**: 375-421.

Beninger P.G. & S.D. St-Jean. 1997. The role of mucus in particle processing by suspension-feeding marine bivalves: unifying principles. **Marine Biology 129**: 389-397.

Beninger, P.G. & P. Decottignes. 2008. Worth a second look: gill structure in *Hemipecten forbesianus* (Adams & Reeve, 1849) and taxonomic implications for Pectinidae. *Journal* of *Molluscan* Studies **74**: 137-142. doi: 10.1093/mollus/eyn001

David J.A.O. & C.S. Fontanetti. 2005. Surface morphology of *Mytella falcata* gill filaments from three regions of the santos estuary. **Brazillian Journal of Morphological Science 22**: 203-210.

David, J.A.O.; R.B. Salaroli. & C.S. Fontanetti. 2008. Fine structure of *Mytella falcata* (Bivalvia) gill filaments. **Micron 39**: 329-336. doi: 10.1016/j.micron.2007.06.002

Domouhtsidou, G.P. & V.K. Dimitriadis. 2000. Ultrastructural localization of heavy metals (Hg, Ag, Pb and Cu) in gills and digestive gland of mussels, *Mytilus galloprovincialis* (L). **Archives of *Environmental Contamination* and *Toxicology* 38**: 472-478. doi: 10.1007/s002449910062

Dufour, S.C. & P.G. Beninger. 2001. A functional interpretation of cilia and mucocyte distributions on the abfrontal surface of bivalve gills. **Marine Biology 138**: 295-309.

Drury, R.A. & E.A.Wallington. 1973. **Carleton's Histological Technique**. London, Oxford University Press, 4th ed., X+432p.

Falakali Mutaf, B.; D. Aksit & M. Gokoglu. 2009. Scanning electron microscopy of the gill morphology of *Modiolus barbatus* (Bivalvia: Mollusca). *In*: **XV National Symposium of Aquatic Sciences**, Rize, Turkey, p. 331.

Gosling, E. 2003. **Bivalve Molluscs. Biology and Culture**. Oxford, Fishing New Books, Blackwell Scientific Publications, X+443p.

Gregory, M.A. & R.C. George. 2000. The structure and surface morphology of gill filaments in the Brown mussel *Perna perna*. **Africa Zoology 35**: 121-129.

GREGORY, M.A.; D.J. MARSHALL; R.C. GEORGE; ANANDRAJ A; T.P. MCCLURG. 2002. Correlation between metal uptake in the soft tissue of *Perna perna* and gill filament pathology after exposure to mercury. **Marine *Pollution* Bulletin 45**: 114-125.

KOEHLER, A.; U. MARX; K.BROEG; S. BAHNS & J.BRESSLING. 2008. Effects of nanoparticles in *Mytilus edulis* gills and hepatopancreas – A new threat to marine life? **Marine Environmental Research 66**: 12-14.

MORSE, M.P. & J.D. ZARDUS. 1997. Bivalvia. *In*: F.W. HARRISON & A.J. KOHN (Eds). "**Microscopic Anatomy of Invertebrates: Mollusca II**. New York, Wiley-Liss Inc., vol. 6A, p. 7-118.

VILLALEJO-FUERTE, M.; M ARELLALNO-MARTINEZ; B.P. CEBALLOS-VAZQUEZ & F. GARCIA-DOMINGUEZ. 2002. Reproductive cycle of *Spondylus calcifer* Carpenter, 1857 (Bivalvia: Spondyludae) in the "Bahai de Loreto" National Park, Gulf of California. **The *Journal* of *Shellfish* Research 21**: 103-108.

ZENETOS, A.; S. GOFAS; G. RUSSO & J. TEMPLADO. 2003. **CIESM Atlas of Exotic Species in the Mediterranean**. Monaco, CIESM Pub., vol. 3, 376p.

YONGE, C.M. 1973. Functional morphology with particular reference to hinge and ligament in Spondylus and Plicatula and a discussion on relations with superfamily Pectinicea (Mollusca:Bivalvia). **Philosophical Transactions of the Royal Society of London B 267** (883): 173-208.

3

Molecular evidence for the polyphyly of *Bostryx* (Gastropoda: Bulimulidae) and genetic diversity of *Bostryx aguilari*

Jorge L. Ramirez[1, 2] & Rina Ramírez[1]

[1] *Departamento de Malacología y Carcinología, Museo de Historia Natural, Universidad Nacional Mayor de San Marcos, Apartado 14-0434, Lima-14, Perú.*
[2] *Coresponding author. E-mail: jolobio@hotmail.com*

ABSTRACT. *Bostryx* is largely distributed in Andean Valleys and Lomas formations along the coast of Peru and Chile. One species, *Bostryx aguilari*, is restricted to Lomas formations located in the Department of Lima (Peru). The use of genetic information has become essential in phylogenetic and population studies with conservation purposes. Considering the rapid degradation of desert ecosystems, which threatens the survival of vulnerable species, the aim of this study was, first, to resolve evolutionary relationships within *Bostryx* and to determine the position of *Bostryx* within the Bulimulidae, and second, to survey the genetic diversity of *Bostryx aguilari*, a species considered rare. Sequences of the mitochondrial 16S rRNA and nuclear rRNA regions were obtained for 12 and 11 species of Bulimulidae, respectively, including seven species of *Bostryx*. Sequences of the 16S rRNA gene were obtained for 14 individuals (from four different populations) of *Bostryx aguilari*. Phylogenetic reconstructions were carried out using Neighbor-Joining, Maximum Parsimony, Maximum Likelihood and Bayesian Inference methods. The monophyly of *Bostryx* was not supported. In our results, *B. solutus* (type species of *Bostryx*) grouped only with *B. aguilari*, *B. conspersus*, *B. modestus*, *B. scalariformis* and *B. sordidus*, forming a monophyletic group that is strongly supported in all analyses. In case the taxonomy of *Bostryx* is reviewed in the future, this group should keep the generic name. *Bostryx aguilari* was found to have both low genetic diversity and small population size. We recommend that conservation efforts should be increased in Lomas ecosystems to ensure the survival of *B. aguilari*, and a large number of other rare species restricted to Lomas.

KEY WORDS. Land snails; Lomas; molecular systematic; Orthalicoidea; rRNA.

Among Neotropical land snails, Bulimulidae is one of the most diverse (BREURE 1979, RAMÍREZ *et al*. 2003a). The phylogenetic relationships among its members, however, are still problematic. Genera such as *Bostryx* and *Scutalus* are distributed in desert ecosystems and are adapted to survive under some of the harshest climatic conditions (AGUILAR & ARRARTE 1974, RAMÍREZ *et al*. 2003b). *Bostryx* is found in Argentina, Bolivia, Chile, Peru, Ecuador, and possibly in Venezuela (BREURE 1979). It is spread throughout Peru, but is more prevalent in the Pacific coastal desert and the western Andean slopes (RAMÍREZ *et al*. 2003a). Among the *Bostryx* species, *B. aguilari* Weyrauch, 1967 (Fig. 2) is of particular interest, due to its vulnerable status and lack of information on the genetic diversity of its populations. It is a species associated with bushy Lomas and is found at elevations of 200 to 600 m. *Bostryx aguilari* was originally reported for the Lomas of Amancaes, Atocongo and Pachacamac, but there is also a record of an unknown locality near the city of Junín, in the Peruvian Andes (WEYRAUCH 1967). To date, this species has been reported for at least 12 Lomas in the Department of Lima, and is distributed from the Lomas of Lachay, in the north, to the Lomas of Pacta, in the south (R. Ramírez unpublished data). *Bostryx aguilari*, unlike other gastropod species of Lomas, is very

difficult to find, not only alive, but also as shell remnant. An exception is the Lomas of Atocongo, where *B. aguilari* can be found more easily. The Lomas formations are seasonal ecosystems occurring along the coast of Peru and Chile, between 8° and 30° SL (RUNDEL *et al*. 1990), where the main source of humidity are fogs brought from the Pacific Ocean during the winter months (DILLON *et al*. 2003). Periodically (every few years), the El Niño-Southern Oscillation (ENSO) alters the seasonality of the Lomas, causing summer drizzles that promote the development of out of season vegetation. The steady growth of cities is threatening the biodiversity in desert ecosystems, and particularly the Lomas, which are still poorly known and described. They are beginning to disappear at a fast pace, and with them, their endemic species. Our objectives are to resolve evolutionary relationships within *Bostryx* to clarify the position of the genus among the Bulimulidae, and to survey the genetic diversity of *B. aguilari*, a rare species threatened by loss of habitat and human pressure. Because the maintenance of genetic diversity is vital to the survival of populations and species, this information will be crucial to the establishment of guidelines for the conservation of *B. aguilari* and for the Lomas ecosystems they inhabit.

MATERIAL AND METHODS

Species of *Bostryx* were collected from several Peruvian localities comprising Lomas, Inter-Andean valleys and tropical forests (Table I). We also included species of *Scutalus*, *Drymaeus*, *Naesiotus* and *Neopetraeus* as outgroups (Table I and Figs 1-12). Samples were fixed in 96% ethanol and deposited in the collection at Department of Malacology and Carcinology, Museum of Natural History, San Marcos University. Individuals of *B. aguilari* were obtained from seven Lomas, all located in the Department of Lima in the central coast of Peru (Amancaes, Atocongo, Iguanil, Lúcumo, Manzano, Paraíso and Picapiedra), although live specimens were only found in three locations (Amancaes, Atocongo and Iguanil).

DNA was isolated using a modified CTAB method (Doyle & Doyle 1987) from 1-2 mm³ of tissue from the snail foot. The tissue sample was digested in 300 µL of extraction buffer (100 mM Tris/HCl, 1.4 M NaCl, 20 mM EDTA, 2% CTAB, 2% PVP and 0.2% of β-mercaptoethanol) with 0.05 mg Proteinase K and incubated at 60°C for approximately two hours. Proteins were removed twice with 310 µL of chloroform/isoamyl alcohol (24:1), centrifugation was at 13,000 rpm for 15 minutes before removal of the aqueous phase. The DNA was precipitated using 600 µL of cold absolute ethanol and 25 µL of 3M ammonium acetate and incubated at -20°C for at least 30 minutes, then centrifuged at 13,000 rpm for 15 min. The pellet obtained was washed twice in 1 mL of 70% ethanol and centri-fuged at 13,000 rpm for 15 min. Finally, the pellet was dried at room temperature for 24 hours, resuspended in 50 µL of double-distilled water at 37°C, and stored at -20°C.

Using total genomic DNA, we amplified and sequenced the 16S rRNA gene and the rRNA gene-cluster. Amplifications were carried out using the polymerase chain reaction (PCR) (Saiki *et al.* 1988). For the amplification of the 16S rRNA gene, we used primers developed by (R. Ramírez unpublished data): 16SF-104 (5'-GACTGTGCTAAGGTAGCATAAT-3') and 16SR-472 (5'-TCGTAGTCCAACATCGAGGTCA-3'). To obtain the nuclear rRNA gene-cluster, including the 3'-end of the 5.8S rRNA gene, the complete internal transcribed spacer 2 (ITS-2) region, and the 5'-end of the large subunit (28S rRNA) gene, we used primers LSU1 and LSU3 developed for mollusks by Wade & Mordan (2000).

For the 16S rRNA, PCR amplification were performed in a final volume of 30 µL, containing 1 U of *Taq* DNA polymerase (Fermentas Inc., Maryland, US), 1.5 mM MgCl$_2$, 0.2 mM dNTP and 0.2 iM of each primer, 1X buffer, and 3 µL of DNA template. Amplifications consisted of 35 cycles of denaturation at 94°C for 30s, annealing at 48°C for 30s, and extension at 72°C for 60s. PCR reagents used for the amplification of nuclear markers were the same as above; amplifications consisted of 35 cycles of 96°C for 60s, 50-55°C for 30s and 72°C for 60s. Amplicons were electrophoresed on 1% agarose gels to verify the amplification. PCR products were purified and sequenced for both strands using the commercial services at Macrogen USA.

Figures 1-12. Species of Bulimulidae analyzed in this work: (1) *Bostryx solutes*, MUSM 5515-82G1; (2) *B. aguilari* MUSM 5501-42A3; (3) *B. conspersus* MUSM 5505-23F1; (4) *B. modestus* MUSM 5507-74F1; (5) *B. sordidus* MUSM 5511-14A15; (6) *B. scalariformis* MUSM 5510-75.3; (7) *B. turritus* MUSM 5514-1F1; (8) *Scutalus proteus* MUSM 5519-35G1; (9) *S. versicolor* MUSM 5518-11.8; (10) *Drymaeus arcuatostriatus* MUSM 5516-59Eu; (11) *Neopetraeus tessellates* MUSM 4020-62E1; (12) *Naesiotus geophilus* MUSM 5517-18G1. Photographs: 1 and 8 by D. Maldonado; 2, 10 and 11 by J. Ramirez; 3-7 by A. Chumbe; and 12 by V. Borda. Escale bars: 1, 3, 4, 6, 7, 12 = 2 mm; 2, 5, 8-11 = 5 mm.

Table I. Voucher information and GenBank accession numbers for individuals included in the analyses. Sequences generated for this study are in bold. MUSM: Museum of Natural History, San Marcos University.

Species	Population	Voucher MUSM	GenBank accession 16S	GenBank accession LSU 1-3
Bostryx aguilari Weyrauch, 1967	Lima: Amancaes[1]	**MUSM 5501-42A3**	**HQ225813**	HM116230
	Lima: Amancaes[1]	**MUSM 5501-43A6**	**HQ225814**	
	Lima: Amancaes[1]	**MUSM 5500-53.10**	**HQ225815**	
	Lima: Amancaes[4]	MUSM 5041-Ama3	JQ669492	
	Lima: Iguanil[1]	**MUSM 5504-29A**	**HQ225820**	JQ669461
	Lima: Iguanil[1]	**MUSM 5504-31A**	**HQ225821**	
	Lima: Iguanil[1]	**MUSM 5504-26A**	**HQ225819**	
	Lima: Iguanil[1]	**MUSM 5504-32A**	**HQ225822**	
	Lima: Atocongo[1]	**MUSM 5503-25F1**	**HQ225816**	
	Lima: Atocongo[1]	**MUSM 5502-17F5**	**HQ225817**	
	Lima: Atocongo[1]	**MUSM 5502-19F8**	**HQ225818**	
	Lima: Atocongo[1]	**MUSM 5505-23F1**	**HM057172**	JQ669462
	Lima: Atocongo[4]	MUSM 5042-Atoc39	JQ669493	
	Lima: Lachay[4]	MUSM 5043-Lach.u	JQ669494	
Bostryx conspersus (Sowerby, 1833)	Lima: Atocongo[1]	**MUSM 5505-23F1**	**HM057173**	
	Lima: Iguanil[1]	**MUSM 5036-Ig5**		JQ669463
	Lima: Lachay[1]	**MUSM 5506-51G3**	**JQ669456**	JQ669464
Bostryx modestus (Broderip, in Broderip & Sowerby 1832)	Lima: Atocongo[1]	**MUSM 5507-74F1**	**HM057174**	
	Lima: Paraiso[1]	**MUSM 5508-6F1**	**JQ669457**	JQ669465
Bostryx scalariformis (Broderip, in Broderip & Sowerby 1832)	Lima: Pasamayo[1]	**MUSM 5510-75.3**	**HM057181.1**	JQ669466
	Lima: N Pan American Hwy Km 115[1]	**MUSM 5509-83.a**	**FJ969796.1**	
	Lima: N Pan American Hwy Km 115[1]	**MUSM 5509-84.b**		JQ669467
Bostryx solutus (Troschel, 1847)	Lima: Infiernillo[2]	**MUSM 5515-82G1**	**JQ669458**	JQ669468
	Lima: Infiernillo[2]	**MUSM 5515-80G6**	**HQ225824**	
Bostryx sordidus (Lesson, 1826)	Lima: Iguanil[1]	**MUSM 5511-14A15**	**HM057176.1**	
	Lima: Lupin[1]	**MUSM 5512-62.12**	**FJ969797.1**	
	Lima: Santa Eulalia[2]	**MUSM 5513-77E5**	**JQ669459**	JQ669469
Bostryx turritus (Broderip, in Broderip & Sowerby 1832)	Lima: Santa Eulalia[2]	**MUSM 5514-1F1**	**HM057175**	JQ669470
	Lima: Santa Eulalia[2]	**MUSM 5514-4F4**	**JQ669460**	JQ669471
Bostryx bilineatus (Sowerby, 1833)	Ecuador[5]			HM027501
Bostryx strobeli (Parodiz, 1956)	Argentina[5]			HM027498
Bulimulus guadalupensis (Bruguière, 1789)	Puerto Rico[6]			AY841298
Bulimulus tenuissimus (Férussac, 1832)	Brazil[5]			HM027507
Bulimulus sporadicus (d'Orbigny, 1835)	Brazil[6]			AY841299
Clessinia pagoda Hylton Scott, 1967	Argentina[5]			HM027497
Drymaeus discrepans (Sowerby, 1833)	Guatemala[6]			AY841300
Drymaeus inusitatus (Fulton, 1900)	Costa Rica[5]			HM027503
Dryamaues laticinctus (Guppy, 1868)	Dominica[5]			HM027492
Drymaeus serratus (Pfeiffer, 1855)	Peru[5]			HM027499
Drymaeus arcuatostriatus (Pfeiffer, 1855)	San Martin: Juan Guerra[3]	**MUSM 5516-59Eu**	**HM057178**	JQ669472
Naesiotus quitensis (Pfeiffer, 1848)	Ecuador[5]			HM027510
Naesiotus stenogyroides (Guppy, 1868)	Dominica[5]			HM027494
Naesiotus geophilus Weyrauch, 1967	San Martin: Juan Guerra[3]	**MUSM 5517-18G1**	**HM057180**	
Neopetraeus tessellatus (Shuttleworth, 1852)	Ancash: nr. Pontó[2]	**MUSM 4020-62E1**	**HM057179**	JQ669473
Plagiodontes multiplicatus Döring, 1874	Argentina[5]			HM027496
Scutalus proteus (Broderip, in Broderip & Sowerby 1832)	Lima: Santa Eulalia[2]	**MUSM 5519-35G1**	**HQ225823**	JQ669474
Scutalus versicolor (Broderip, in Broderip & Sowerby 1832)	Lima: Mongón[1]	**MUSM 5518-11.8**	**FJ969798**	JQ669475
Spixia popana Döring, 1876	Argentina[5]			HM027502
Placostylus bivaricosus (Gascoin, 1885)	Lord Howe Island[7]			AY165846
Placostylus bivaricosus (Gascoin, 1885)	Lord Howe Island[7]			AY165850

[1]Lomas, [2]Andean region, [3]Tropical forest, [4]R. Ramírez (unpublished data), [5]BREURE *et al.* (2010), [6]WADE *et al.* (2006); [7]PONDER *et al.* (2003).

Sequences of the mitochondrial 16S rRNA and nuclear rRNA regions were obtained for 12 and 11 species of Bulimulidae, respectively, including seven species of *Bostryx*. Eleven sequences of the partial 16S rRNA gene were obtained from different populations of *B. aguilari*; three samples were sequenced with the LSU1/LSU3 primer pair. Nineteen sequences were retrieved from GenBank. Voucher information and GenBank accession numbers are given in Table I.

Sequences were edited with Chromas (McCarthy 1996), assembled with CAP3WIN (Huang & Madan 1999), aligned with ClustalX 2.0 (Larkin *et al.* 2007) and adjusted manually in BioEdit v7.0.9 (Hall 1999). Gaps were treated as a fifth character. For the phylogenetic analyses we used, in addition to our data, seven sequences of the nuclear marker retrieved from GenBank (Table I). We were very careful when aligning the 16S rRNA marker, because it has a high mutation rate and indels are extremely common. In order to get a better hypothesis of homology, we used the secondary structure of the 16S rRNA of *Albinaria caerulea* (Lydeard *et al.* 2000, Ramirez & Ramírez 2010) as a template for the alignment.

Different phylogenetic analyses were performed. The cladogram for all taxa was constructed using Neighbor-Joining (NJ) (Saitou & Nei 1987) as implemented in PAUP* 4.0b10 (Swofford 2003). Tree searching was heuristic, with tree-bisection-reconnection branch swapping. Branch support was evaluated using bootstrap resampling (Felsenstein 1985) with 1,000 replicates. Maximum Parsimony (MP) was implemented using PAUP* 4.0b10 (Swofford 2003), initial heuristic searches were conducted with random stepwise addition, Tree-Bisection-Reconnection (TBR) branch swapping, and bootstrap with 1,000 replicates. Maximum Likelihood (ML) analyses were conducted using heuristic search, the initial tree was obtained by stepwise addition and TBR in PAUP* 4.0b10. Support for nodes was estimated with 1,000 bootstrap replicates. The nucleotide substitution model, base frequencies, proportion of invariant sites and shape parameter of the gamma distribution were estimated based on Akaike criterion using JModeltest (Posada 2008). Bayesian inference (BI) was performed using MrBayes 3.1.2 (Ronquist & Huelsenbeck 2003); four chains of a Markov Chain Monte Carlo algorithm were run simultaneously for 10 million generations, sampled every 1,000 generations, and burn-in of 9,000 generations. A consensus tree and final posterior probabilities were calculated using the remaining trees. The tree based on 16S rRNA was rooted using *Placostylus* (Placostylidae). For the nuclear rRNA gene-cluster, trees were rooted using species belonging Odontostomidae, which is sister to Bulimulidae according to Breure *et al.* (2010).

Sequences of *Bostryx aguilari* were evaluated in DAMBE v5.0.8 (Xia & Xie 2001). We calculated nucleotide frequencies, percentage of CpG islands, percentage of CG and the extent of saturation, by plotting pairwise genetic distances against the distribution of transitions and transversions. Values of genetic diversity, such as haplotype diversity (*h*) and nucleotide diver-

sity (π) were obtained using DnaSP v5.10 (Librado & Rozas 2009). Pairwise distances were obtained in MEGA v4.02 (Kumar *et al.* 2008) including all positions and using a Maximum Composite Likelihood method. Relationships among haplotypes of the 16S rRNA marker were evaluated using the Median Joining algorithm obtained in Network 4.5.1.0 (Bandelt *et al.* 1999). *Fst* statistics was calculated using Arlequin v3.11. In order to estimate the time to most recent common ancestor (TMRCA) for *B. aguilari*, we calibrated a Linearized NJ tree for a conservative rate for terrestrial mollusks (0.06 substitutions per site per million years) for the 16S rRNA, using MEGA (Excoffier *et al.* 2005).

RESULTS

Interspecific phylogeny

The alignment generated for the phylogenetic reconstruction of the partial 16S rRNA gene consisted of 26 sequences (only the four haplotypes of *B. aguilari* were used) corresponding to 13 species of Bulimulidae. This alignment had 382 positions, with 202 variable sites (of which 179 were informative), 170 conserved sites, and 22 singletons. The nucleotide substitution model selected was TPM1uf+G. For the nuclear rRNA, the alignment of 29 sequences resulted in 868 sites, 656 of which were conserved and 199 were variable sites (158 informative), and 41 were singletons. The nucleotide substitution model selected was GTR+G.

Phylogenetic reconstructions based on the partial 16S rRNA gene using NJ, MP, ML and BI resulted in trees with similar topologies (Fig. 13). The group of species known as the "*Bostryx modestus* species complex", which includes *B. modestus*, *B. sordidus* and *B. scalariformis* (R. Ramírez unpublished data), was strongly supported in our analyses. It grouped along with *B. solutus*, *B. aguilari*, and *B. conspersus* with weak to strong support. However, *B. turritus* did not cluster with any other species of *Bostryx*. Regarding the phylogenetic analyses using the nuclear rRNA marker, again the four phylogenetic methods used yielded trees with similar topologies (Fig. 14). Bulimulidae was supported by maximum values. The sequences of *B. modestus*, *B. scalariformis*, and *B. sordidus* (*B. modestus* species complex) grouped with strong support. The *B. modestus* species complex, along with *B. solutus*, *B. conspersus*, and *B. aguilari* grouped together with strong support. *Neopetraeus* and *Drymaeus* formed a monophyletic group with good to strong support. *Naesiotus quitensis* and *Bostryx strobeli* clustered with strong support and, in our data, formed a strongly supported monophyletic group with *Bulimulus*.

The trees obtained with the two markers have similar topologies. Both trees grouped *B. solutus* with *B. modestus* species complex, *B. aguilari* and *B. conspersus*.

Genetic diversity of *Bostryx aguilari*

The alignment of 14 sequences of the partial 16S rRNA gene of *B. aguilari* resulted in 345 sites without indels. There were three variable sites, which were informative. By compar-

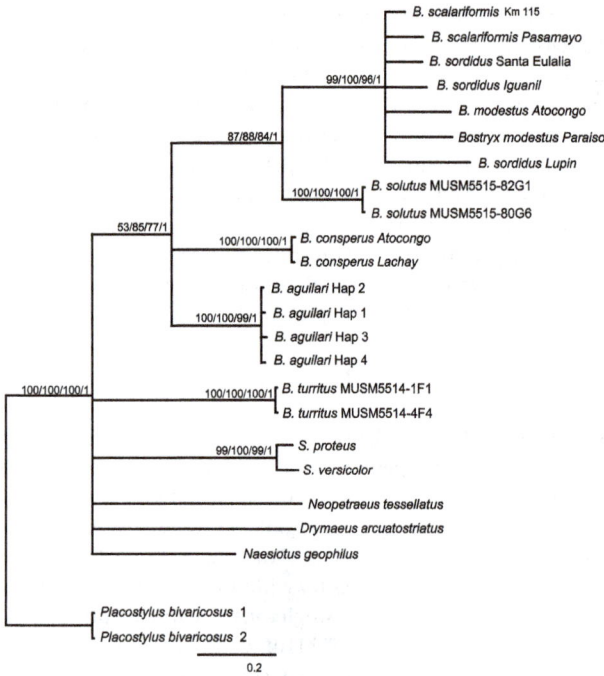

Figure 13. Phylogenetics relationships based on the 16S rRNA. Numbers correspond to bootstrap values for Neighbor-Joining, Maximum Parsimony and Maximum Likelihood, respectively, and posterior probabilities for Bayesian inference. Only nodes with bootstrap values greater than 50% and posterior probabilities of 0.9 are represented.

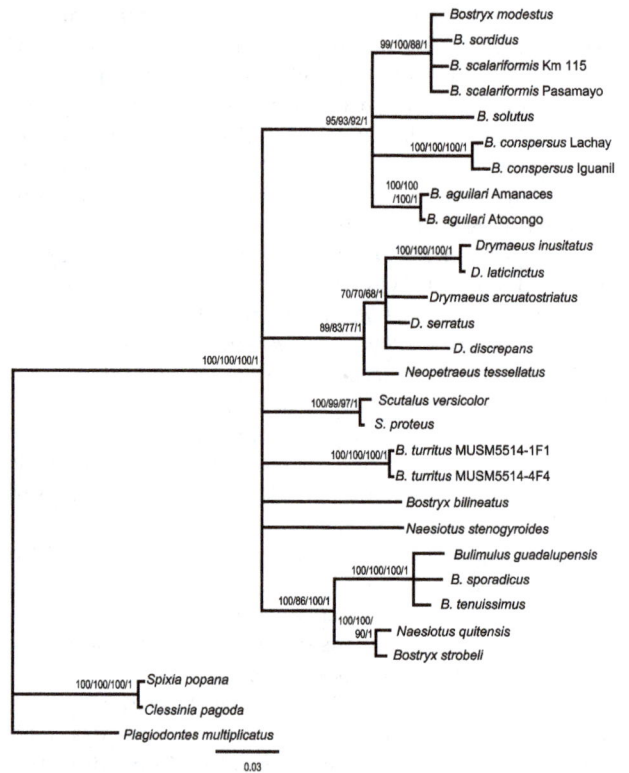

Figure 14. Phylogenetics relationships based on the nuclear rRNA gene cluster. Numbers represent bootstrap values for Neighbor-Joining, Maximum Parsimony and Maximum Likelihood, respectively, and posterior probabilities for Bayesian inference. Only nodes with bootstrap values greater than 50% and posterior probabilities of 0.9 are represented.

ing these sequences with other Bulimulids, we were able to observe the presence of several indels up to 4 bp long. This region of the mitochondrial genome of *B. aguilari* is larger than that found in other *Bostryx* from Lomas, as well as in other genera of Bulimulidae evaluated so far (4 to 23 bp difference). The nucleotide composition showed a predominance of AT (71.66%) over GC (28.34%). Sequences obtained for the nuclear rRNA were 826 bp long. The three individuals had the same haplotype. The percentage of GC (55.83%) was slightly higher than that of AT.

The 14 16S rRNA sequences collapsed into four haplotypes. The haplotype diversity (h) was 0.7802 and π was 0.00347. By comparing these results with values found for other species of *Bostryx* from Lomas (R. Ramírez unpublished data), we observed that *B. aguilari* has the lowest values of haplotype diversity. The haplotype network in Figure 15 shows a correlation between haplotypes and the geographic distribution of *B. aguilari*, revealing the Atocongo population as the only one with unique haplotypes. The Amancaes population showed only one haplotype, which was shared with an individual from Iguanil, in spite of the geographic distance (70 km) and the apparent absence of intermediate populations between the two Lomas (no live individuals or shells recorded). The individual

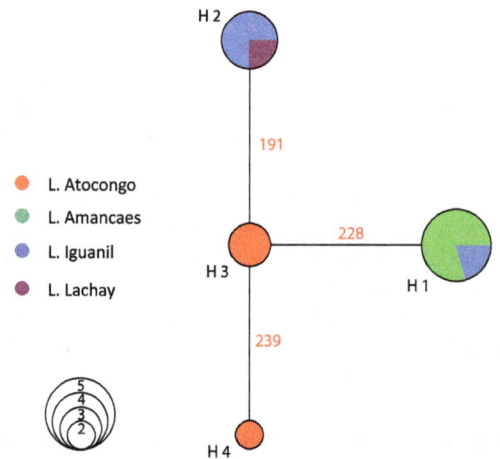

Figure 15. Haplotype Network based on 16S rRNA of *B. aguilari*. Circles are proportional to frequencies. Colors indicate locality of samples. There is only one mutation between haplotypes. Numbers indicate the position of mutation in the alignment.

from Lachay shared its haplotype with individuals from Iguanil. These haplotypes are differentiated by a single mutational step between them. The *Fst* analyses showed significant values only between Atocongo and the remaining populations. The TMRCA for *B. aguilari* was estimated in 38,565 years.

DISCUSSION

The polyphyly of *Bostryx*

BREURE (1979) conducted a study on evolutionary relationships and geographic distribution of the genera in Bulimulinae. More recently, molecular studies have shed new light on the diversity and the relationships within the land snails (WADE & MORDAN 2000, WADE *et al.* 2001, 2006) and within Orthalicoidea (PONDER *et al.* 2003, PARENT & CRESPI 2006, HERBERT & MITCHELL 2009, RAMÍREZ *et al.* 2009, BREURE *et al.* 2010, BUCKLEY *et al.* 2011). BREURE *et al.* (2010), after revisiting the phylogeny of Orthalicoidea, found that Orthalicidae and Amphibulimulidae are the most basal families, whereas Placostylidae is basal to the clade consisting of Odontostomidae and Bulimulidae.

Bostryx belongs to Orthalicidae, after BOUCHET & ROCROI (2005), which is composed of several subfamilies, including Bulimulinae. This subfamily had been considered as a separate family by several authors (VAUGHT 1989). In this study, we considered *Bostryx* as a member of Bulimulidae, following BREURE *et al.* (2010). In our analyses based on the nuclear rRNA, we added more genera and species to the set of taxa analyzed by BREURE *et al.* (2010), and confirmed that Bulimulidae is clearly a monophyletic group, and that *Bostryx* is a member of this family. Our results, obtained with both nuclear and mitochondrial markers, show that the *B. modestus* species complex, *B. aguilari* and *B. conspersus*, is related to *B. solutus*, a land snail that lives at 3300 m in the Western Andes. The position of *B. turritus*, a Peruvian species found in Inter-Andean valleys, was not resolved, showing low support for any relationships with the other *Bostryx* species analyzed. The monophyly of *Bostryx* was not supported by the different analyses. It is important to note that *Bostryx* was described using *Bostryx solutus* as the type species (BREURE 1979). *Bostryx solutus* was recovered in a strongly supported group together with *B. aguilari*, *B. conspersus*, *B. modestus*, *B. scalariformis*, and *B. sordidus*. These results suggest that only this group should be considered as *Bostryx*. More studies are needed to establish the position of *B. turritus*, as well as the other two species of *Bostryx* (*B. bilineatus* and *B. strobeli*) included in the nuclear analyses.

Genetic diversity of *Bostryx aguilari*

Bostryx aguilari had the lowest value of haplotype diversity compared to other species of *Bostryx* from Lomas. This may be due to the small size of the populations (suggested by the extreme difficulty in locating live individuals), and their possible recent origin. To compare other values of genetic diversity such as π, we examined the work of P. Romero (unpublished data), where a value of 0.04028 for π was found for popula-

tions of *B. scalariformis*. This is about 10 times higher than what was observed for *B. aguilari*. P. Romero found intraspecific distance values up to 0.0608 for *B. scalariformis*, while the maximum value found for *B. aguilari* was 0.006.

The distribution of the 16S rRNA haplotypes of *B. aguilari* and *Fst* values are consistent with results reported by R. Ramírez (unpublished data) regarding the variation of shells, and revealed the population of Atocongo as the most differentiated. The fact that an individual from Amancaes shared its haplotype with individuals from Iguanil, two distant Lomas and without intermediate populations of *B. aguilari*, suggests a recent origin of these populations from a common ancestor. The possibility that this distribution is due to an event of recent geographic expansion after a genetic bottleneck (from refuge) cannot be discarded. The single individual of *B. aguilari* from Lachay shared the same haplotype with individuals from Iguanil. Coupled with the proximity of these Lomas (17 km), we propose a likely phenomenon for the historic gene flow between them. The occurrence of ENSO events could allow the establishment of corridors connecting Lomas that are considered islands of vegetation (RAMÍREZ *et al.* 2003b).

Several ENSO events have left their marks on the genetic structure of populations of land snails from Lomas. ENSO events of greater magnitude have changed dramatically the landscape of the desert, generating larger Lomas and even connecting adjacent Lomas, whereas in dry periods and ENSO of low intensity, Lomas would become a refuge for these species (RAMÍREZ *et al.* 2003b). Both TUDHOPE *et al.* (2001) and LA TORRE *et al.* (2002) reported a strong ENSO about 40 thousand years, which agrees with the estimated date for the geographical expansion of *B. aguilari*.

Implications for Conservation

The Lomas are unique ecosystems in the world. They harbor endemic species whose restricted distribution has been caused by different historical processes (drastic climatic changes, population expansion, bottlenecks, isolation of populations by physical barriers, etc.). Unfortunately, humans have started to invade and occupy different Lomas, threatening the local biodiversity. For instance, cities are an almost insurmountable physical barrier to desert species, generating a new type of isolation that cannot be overcome by periodic favorable conditions of the ENSO. In most localities where *B. aguilari* has been reported, live individuals could not be found, and in those where they were found alive, their numbers were low. Atocongo was an exception to this rule, as it had a larger number of individuals, greater variation in shells, and a more differentiated population with exclusive haplotypes. Major conservation efforts should be applied to this area, which is currently threatened by the expansion of shanty towns, and which has been temporarily put under the custody of a cement factory performing work in the area; the company has surrounded the place with a concrete fence to prevent imminent invasions by the surrounding shanty towns. A worrying situation is found in the Lomas of Amancaes, *B. aguilari* was originally reported for this Loma from 200 to 600 m, and at the present time the Loma is virtually occupied by urbanization up to the 400 meters, being restricted to a fraction of the original size. Due to the damage that these incursions cause, as well as the scarce conservation efforts, it is not difficult to imagine the immediate future of this ecosystem. A different picture is seen in Lachay and Iguanil; Lachay is a National Reserve of great extension (5070 ha.) and Iguanil is far from the city and surrounded by farming communities. Both Lomas guarantee the conservation of part of the low diversity of *B. aguilari*, whose populations are the most distinct besides Atocongo. *Bostryx aguilari* is considered a rare species that has low genetic diversity and small populations. Therefore, there is an urgent need to increase conservation efforts, which should focus on stopping the degradation of its habitat.

ACKNOWLEDGMENTS

We thank A. Chumbe, C. Congrains, D. Fernandez, J. Chirinos, N, Medina, P. Matos and P. Romero for assistance during laboratory and field work. We thank A. Chumbe, D. Maldonado and V. Borda for provided photographs of species. We also thank to M. Arakaki for helping to improve a former version of the manuscript. This study was sponsored by the Instituto de Investigación en Ciencias Biológicas Antonio Raimondi (ICBAR), and funded by Consejo Superior de Investigaciones (CSI) of Vicerrectorado Académico, UNMSM (061001071, 071001221).

REFERENCES

Aguilar, P. & J. Arrarte. 1974. Moluscos de las lomas costeras del Perú. **Anales Científicos, Universidad Nacional Agraria, (Perú) 12** (3-4): 93-98.

Bandelt, H.; P. Forster & A. Röhl. 1999. Median-joining networks for inferring intraspecific phylogenies. **Molecular Biology and Evolution 16**: 37-48.

Breure, A. 1979. Systematics, phylogeny and zoogeography of Bulimulinae (Mollusca). **Zoologische Verhandelingen 168**: 1-215.

Breure, A.; D. Groenenberg & M. Schilthuizen. 2010. New insights in the phylogenetic relations within the Orthalicoidea (Gastropoda, Stylommatophora) based on 28S sequence data. **Basteria 74** (1-3): 25-32.

Bouchet, P. & J. Rocroi. 2005. Classification and Nomenclator of Gastropod Families. **Malacologia 47**: 1-2.

Buckley, T.R.; I. Stringer; D. Gleeson; R. Howitt; D. Attanayake; R. Parrish; G. Sherley & M. Rohan. 2011. A revision of the New Zealand *Placostylus* land snails using mitochondrial DNA and shell morphometric analyses, with implications for conservation. **New Zealand Journal of Zoology 38** (1): 55-81.

Dillon, M.; M. Nakazawa & S. Leiva. 2003. The *Lomas* formations of coastal Peru: Composition and biogeographic history. **Fieldiana Botany 43**: 1-9.

Doyle, J.J. & J.L. Doyle. 1987. A rapid DNA isolation procedure for small amounts of fresh leaf tissue. **Phytochemical Bulletin 19**: 11-15.

Excoffier, L.; G. Laval & S. Schneider. 2005. Arlequin ver. 3.0: An integrated software package for population genetics data analysis. **Evolutionary Bioinformatics Online 1**: 47-50.

Felsenstein, J. 1985. Confidence limits on phylogenies: an approach using the bootstrap. **Evolution 39**: 783-791.

Hall, T. 1999. BioEdit: a user-friendly biological sequence alignment editor and analysis program for Windows 95/98/NT. **Nucleic Acids Symposium Series 41**: 95-98.

Herbert, D. & A. Mitchell. 2009. Phylogenetic relationships of the enigmatic land snail genus Prestonella – the missing African element in the Gondwanan superfamily Orthalicoidea (Mollusca: Stylommatophora). **Biological Journal of the Linnean Society 96**: 203-221.

Huang, X. & A. Madan. 1999. CAP3: A DNA Sequence Assembly Program. **Genome Research 9**: 868-877.

Kumar, S.; J. Dudley; M. Nei & K. Tamura. 2008. MEGA: A biologist-centric software for evolutionary analysis of DNA and protein sequences. **Briefings in Bioinformatics 9**: 299-306.

La Torre, C.; J. Betancourt; K. Rylander & J. Quade. 2002. Vegetation invasions into absolute desert: A 45 000 yr rodent midden record from the Calama-Salar de Atacama basins, northern Chile (lat 228-248S). **GSA Bulletin 114** (3): 349-366.

Larkin, M.; G. Blackshields; N. Brown; R. Chenna; P. Mcgettigan; H. Mcwilliam; F. Valentin; I. Wallace; A. Wilm; R. Lopez; J. Thompson; T. Gibson & D. Higgins. 2007. Clustal W and Clustal X version 2.0. **Bioinformatics 23**: 2947-2948.

Librado, P. & J. Rozas. 2009. DnaSP v5: A software for comprehensive analysis of DNA polymorphism data. **Bioinformatics 25**: 1451-1452.

Lydeard, C.; W. Holznagel; M. Schnare & R. Gutell. 2000. Phylogenetic analysis of molluscan mitochondrial LSU rDNA sequences and secondary structures. **Molecular Phylogenetics and Evolution 15**: 83-102.

Mccarthy, C. 1996. **Chromas: version 1.3.** Brisbane, Griffith University.

Parent, C.E. & B.J. Crespi. 2006. Sequential colonization and diversification of Galápagos endemic land snail genus *Bulimulus* (Gastropoda, Stylommatophora). **Evolution 60**: 2311-2328.

Ponder, W.; D. Colgan; D. Gleeson & G. Sherley. 2003. The relationships of *Placostylus* from Lord Howe Island. **Molluscan Research 23**: 159-178.

Posada, D. 2008. jModelTest: phylogenetic model averaging. **Molecular Biology and Evolution 25**: 1253-1256.

Ramirez, J. & R. Ramírez, 2010. Utility of secondary structure of mitochondrial LSU rRNA in the phylogenetic reconstruction for land snails (Orthalicidae: Gastropoda). **Revista Peruana de Biología 17** (1): 53-57.

RAMÍREZ, J.; R. RAMÍREZ; P. ROMERO; A. CHUMBE & P. RAMÍREZ. 2009. Posición evolutiva de caracoles terrestres peruanos (Orthalicidae) entre los Stylommatophora (Mollusca: Gastropoda). **Revista Peruana de Biología 16** (1): 51-56.

RAMÍREZ, R.; C. PAREDES & J. ARENAS. 2003a. Moluscos del Perú. **Revista de Biologia Tropical 51** (3): 225-284.

RAMÍREZ, R.; S. CÓRDOVA; K. CARO & J. DUÁREZ. 2003b. Response of a land snail species (*Bostryx conspersus*) in the Peruvian Central Coast *Lomas* Ecosystem to the 1982-1983 and 1997-1998 El Niño events. **Fieldiana, Botany 43**: 10-23.

RONQUIST, F. & J. HUELSENBECK. 2003. MrBayes 3: Bayesian phylogenetic inference under mixed models. **Bioinformatics 19** (12): 1572-1574.

RUNDEL, P.; M. DILLON; B. PALMA; H. MOONEY; S. GULMON & J.R. EHLERINGER. 1990. The Phytogeography and Ecology of the Coastal Atacama and Peruvian Deserts. **Aliso 13** (1): 1-50.

SAIKI, R.; D. GELFAND; S. STOFFEL; S. SCHARFJ; R. HIGUCHI; G. HORN; K. MULLIS & H. ERLICH. 1988. Primer-directed enzymatic amplification of DNA with a thermostable DNA polymerase. **Science 239**: 487-491.

SAITOU, N. & M. NEI. 1987. The neighbor-joining method: A new method for reconstructing phylogenetic trees. **Molecular Biology and Evolution 4**: 406-425.

SWOFFORD, D. L. 2003. **PAUP*. Phylogenetic Analysis Using Parsimony (*and Other Methods)**. Sunderland,Sinauer Associates, version 4.

TUDHOPE, A.; C. CHILCOTT; M. MCCULLOCH; E. COOK; J. CHAPPELL; R. ELLAM; D. LEA; J. LOUGH & G. SHIMMIELD. 2001. Variability in the El Niño-Southern Oscillation through a glacial-interglacial cycle. **Science 291**: 1511-1517.

VAUGHT, K. 1989. **A classification of the living Mollusca**. Melbourne, American Malacologist Inc.

XIA, X. & Z. XIE. 2001. DAMBE: Data analysis in molecular biology and evolution. **Journal of Heredity 92**: 371-373.

WADE, C. & P. MORDAN. 2000. Evolution within the gastropod mollusks; using the ribosomal RNA gene-cluster as an indicator of phylogenetic relationships. **Journal of Molluscan Studies 66**: 565-570.

WADE,C; P. MORDAN & B. CLARKE. 2001. A phylogeny of the land snails (Gastropoda: Pulmonata). **Proceedings of the Royal Society of London Series B 268**: 413-422.

WADE, C.; P. MORDAN & F. NAGGS. 2006. Evolutionary relationships among the Pulmonate land snails and slugs (Pulmonata, Stylommatophora). **Biological Journal of the Linnean Society 87**(4): 593-610.

WEYRAUCH, W. 1967. Treinta y Ocho Nuevos Gastropodos Terrestres de Perú. **Acta Zoológica Lilloana 21**: 349-351.

4

Anuran community composition along two large rivers in a tropical disturbed landscape

Mauricio Almeida-Gomes[1,3], Carlos Frederico Duarte Rocha[2] & Marcus Vinícius Vieira[1]

[1]*Departamento de Ecologia, Universidade Federal do Rio de Janeiro. Avenida Carlos Chagas Filho 373, Cidade Universitária, 21941-902 Rio de Janeiro, RJ, Brazil.*
[2]*Departamento de Ecologia, Universidade do Estado do Rio de Janeiro. Rua São Francisco Xavier 524, 20550-900 Rio de Janeiro, RJ, Brazil.*
[3]*Corresponding author. E-mail: almeida.gomes@yahoo.com.br*

ABSTRACT. In this study we evaluated how anuran species were distributed in riparian habitats along two large rivers. Sampling was carried out between January and March 2012 in the municipality of Cachoeiras de Macacu, state of Rio de Janeiro. We delimited 20 plots along each river, ten in portions inside the forest of the Reserva Ecológica de Guapiaçu (REGUA), and with comparatively greater amount of forest cover, and ten outside REGUA, with comparatively lesser forest cover surrounding the rivers. We recorded 70 individuals from 14 frog species in the Manoel Alexandre River and 63 individuals from 15 frog species in the Guapiaçu River. The most abundant species in both rivers was *Cycloramphus brasiliensis* (Steindachner, 1864), and it was more abundant in sections with greater amount of forest cover. This information, coupled with the occurrence of species that are more adapted to open and more disturbed habitats in river sections that harbor lesser riparian vegetation, help to explain differences in amphibian species composition between river sections with greater and lesser forest cover. The results of our study highlight the importance of preserving riparian vegetation associated with rivers in the Atlantic Forest for the conservation of amphibians.

KEY WORDS. Amphibian conservation; aquatic environments; connectivity; frog species; riparian forests.

Land-use change represents a serious threat to amphibians, and poses a challenge for their conservation (Hof et al. 2011). Local amphibian communities can be structured by several factors, some of which are affected by land use. Examples are the size of forest fragments (Bell & Donnelly 2006), the presence of suitable reproductive sites (Zimmerman & Bierregaard 1986), and the presence of riparian vegetation (Parris & McCarthy 1999). Riparian habitats can be important for maintaining the structural and physicochemical characteristics of aquatic environments, hence they are important factors determining community composition in these habitats (Gomi et al. 2006). In fact, anuran assemblages are frequently structured along a continuum of riparian habitats (e.g., Keller et al. 2009, Rodríguez-Mendoza & Pineda 2010), which can be crucial for vulnerable amphibian species in tropical forest areas. Besides, riparian habitats provide connectivity between forest patches and facilitate gene flow between amphibian populations (Richards-Zawacki 2009). Since many amphibian species depend on these environments for reproduction (Haddad & Prado 2005), and some of them live all their life cycles closely associated with water bodies that are near riparian vegetation, preservation of the latter may be critical to the viability of several species (Ficetola et al. 2009, Todd et al. 2009).

The Atlantic Forest has high biodiversity and endemism, being among the five most threatened biomes in the world (Mittermeier et al. 2005). However, despite the fact that human interference has reduced the Atlantic Forest to about 12% of its original area (Ribeiro et al. 2009), it still harbors more than 480 amphibian species. Most of these species have an aquatic larval phase (Becker et al. 2007). Since the mid 2010s there has been an intense discussion in Brazil regarding a review of the current Brazilian Forest Act (Metzger et al. 2010). Among the most controversial points of this review is the proposed reduction of the minimum width of riparian vegetation in each property, which can result in serious consequences for the persistence of many amphibian species. This situation is worsened by the fact that even under the current legislation the protection area of riparian vegetation has not been enforced (Becker et al. 2007). In this context, it is urgent to gather field data to evaluate the importance of riparian vegetation for amphibian species and local communities. It is still unknown how riparian forest loss affects anuran communities in Atlantic Forest rivers. The aim of this study was to compare anuran species composition in riparian habitats of two rivers with different amounts of forest cover (greater and lesser forest cover). We expected that community composition would differ between plots with greater and lesser riparian forest cover: those with more deforested plots were expected to harbor more generalist species than plots with greater forest cover, which were expected to be mostly occupied by anuran species that are strictly adapted to forests.

MATERIAL AND METHODS

We conducted this study between January and March 2012 in the Reserva Ecológica de Guapiaçu, REGUA (22°24'S, 42°44'W), in the municipality of Cachoeiras de Macacu, state of Rio de Janeiro, Brazil. REGUA is a private reserve comprising about 7,600 ha of Atlantic forest, most of which is continuous with the large forest patch of the Serra dos Órgãos mountain range (ALMEIDA-GOMES & ROCHA 2014). Sampling was carried out in two rivers, Manoel Alexandre and Guapiaçu (Figs. 1-3). Both rivers have a gradient of riparian forest cover: the portions of rivers inside REGUA have greater forest cover, which progressively reduces as the rivers leave REGUA. Rivers outside REGUA have little riparian vegetation surrounded mostly by pasture areas. A total of 20 plots, each 20 m in length were delimited along each river, 10 inside and 10 outside REGUA, totaling 400 m sampled in each river. Each plot ranged from one margin of the river to the other, and was separated from the nearest plot by at least 10 m. To limit altitudinal variation, the maximum difference in altitude between upstream and downstream plots was approximately 100 m at the Manoel Alexandre River(108-220 m), and 60 m at the Guapiaçu River (39-98 m).

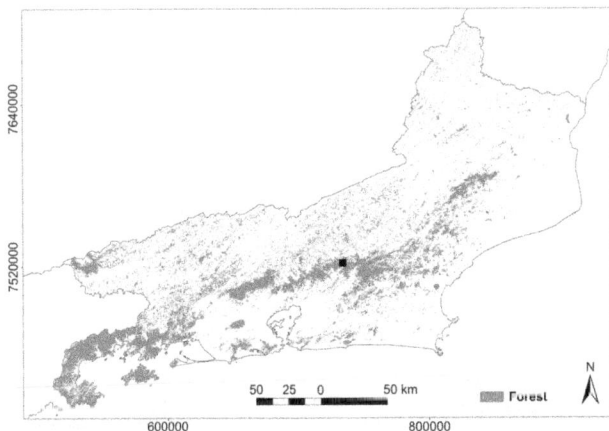

Figures 2-3. Study area indicating the location of plots along the Manoel Alexandre (2) and Guapiaçu (3) rivers, municipality of Cachoeiras de Macacu, state of Rio de Janeiro, Brazil. (squares) Plots with greater forest cover, (triangles) plots with lesser forest cover.

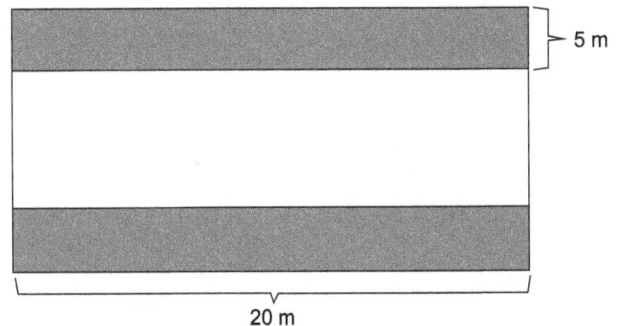

Figure 4. A river plot showing the total sampled area for anurans, including margins (gray) and river channel (white).

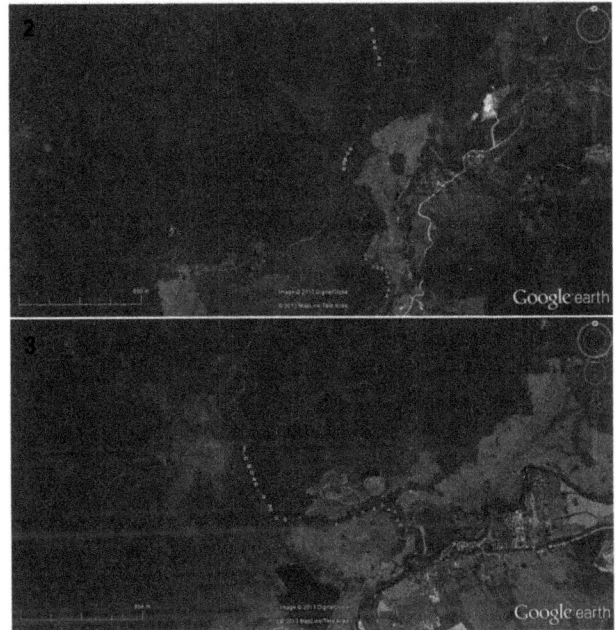

Figure 1. Location of Reserva Ecológica de Guapiaçu (REGUA, black square) in the state of Rio de Janeiro.

We conducted visual encounter surveys, VES (CRUMP & SCOTT 1994) at nighttime (19:00 to 23:00 h) using headlamps. We conducted VES within each plot up to five meters distance from the water and along the entire river channel (Fig. 4), with a sampling effort of two hours in each plot. The total sampling

effort at each river was 40 hours of nocturnal active search. Each individual was identified to species and then released at the same point it was originally captured. To evaluate the ordination of plots regarding amphibian species composition and abundance pooled for both rivers, we used a two-axis Non-Metric Multidimensional Scaling, NMDS, on a matrix of Bray-Curtis dissimilarity distances between plots (LEGENDRE & LEGENDRE 1998). The two NMDS axes were used as dependent variables in a factorial multivariate analysis of variance (MANOVA) using Wilks' λ to test if community composition represented by the two NMDS axes differs with respect to the rivers, riparian forest cover (lesser or greater), or an interaction of both (Báez et al. 2012). We conducted all analyzes in the R environment (version 2.13.0, R Development Core Team 2011).

RESULTS

We recorded a total of 133 individuals from 22 anuran species at both rivers. The most abundant species were *Cycloramphus brasiliensis* (Steindachner, 1864) (33.8% of all individuals), *Rhinella ornata* (Spix, 1824) (16.5%), and *Leptodactylus latrans* (Steffen, 1815) (12.8%) (Table I). In the Manoel Alexandre River, 70 individuals were recorded from 14 frog species, ten species (N = 39) in plots with higher forest cover, six (60%) only in these portions of the river. In plots with lower forest cover, eight species were recorded (N = 31), four (50%) only in these portions of the river. The most abundant species in the Manoel Alexandre River was *C. brasiliensis*, 48.7% in greater forest cover plots, 32.2% in lesser forest cover plots. In the Guapiaçu River, 63 individuals were recorded from 15 frog species, eight species (N = 31) in plots with greater forest cover, four (50%) exclusively in these portions of the river. In plots with lesser forest cover, 11 species were recorded (N = 32), seven of them (63.6%) exclusively in these portions of the river. The most abundant species in greater cover plots of the Guapiaçu River was also *C. brasiliensis* (48.4%), but in lesser cover plots *L. latrans* was the most abundant (37.5%).

The ordination of amphibian species composition and abundance (Stress = 0.124) showed that most plots with comparatively greater forest cover clustered together (Fig. 5). Conversely, the plots with lesser forest cover were more scattered

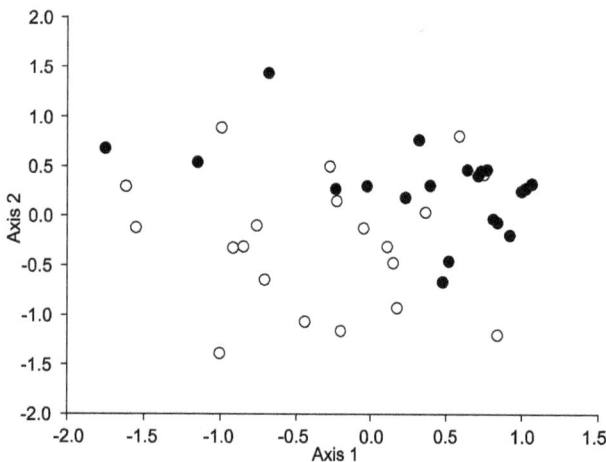

Figure 5. Ordination of sampling plots along the axes of Non-metric Multidimensional Scaling (NMDS) of anuran species composition and abundances in two large rivers in the municipality of Cachoeiras de Macacu, state of Rio de Janeiro, Brazil. Black circles are plots with greater riparian forest cover, and white circles represent plots with lesser riparian forest cover.

Table I. Number of individuals of amphibian species recorded during Visual Encounter Surveys in two rivers with high and low forest cover portions in the municipality of Cachoeiras de Macacu, state of Rio de Janeiro, Brazil.

	Manoel Alexandre River			Guapiaçu River		
	Greater	Lesser	Total	Greater	Lesser	Total
Brachycepahlidae						
Ischnocnema guentheri	0	0	0	1	0	1
Bufonidae						
Rhinella icterica	2	0	2	0	0	0
Rhinella ornata	4	5	9	3	10	13
Craugastoridae						
Haddadus binotatus	1	0	1	0	0	0
Cycloramphidae						
*Cycloramphus brasiliensis**	19	10	29	15	1	16
Hylidae						
Dendropsophus anceps	0	0	0	0	1	1
Dendropsophus berthalutzae	0	0	0	0	1	1
Hypsiboas albomarginatus	0	1	1	1	0	1
Hypsiboas faber	0	4	4	0	0	0
*Hypsiboas secedens**	2	0	2	0	0	0
Hypsiboas semilineatus	0	0	0	0	1	1
*Scinax albicans**	6	3	9	6	0	6
Scinax alter	0	3	3	0	0	0
*Scinax humilis**	1	2	3	0	1	1
*Scinax v-signatus**	1	0	1	0	0	0
Scinax aff. x-signatus	1	0	1	0	0	0
Trachycephalus mesophaeus	0	0	0	2	1	3
Leptodactylidae						
Adenomera cf. bokermanni	0	0	0	0	1	1
Leptodactylus latrans	0	3	3	2	12	14
Leptodactylus spixi	0	0	0	0	1	1
Physalaemus signifer	2	0	2	1	0	1
Microhylidae						
Stereocyclops parkeri	0	0	0	0	2	2

* Species endemic to state of Rio de Janeiro.

in the ordination biplot. The difference in species composition was nearly significant between rivers (MANOVA: Wilks' λ = 0.871, $F_{1,36}$ = 2.589, p = 0.089), but differences were significant between plots with different forest cover (Wilks' λ = 0.619, $F_{1,36}$ = 10.764, p < 0.001), and for the interaction between rivers and forest cover (Wilks' λ = 0.821, $F_{1,36}$ = 3.811; p = 0.032).

DISCUSSION

The results indicate that riparian habitats are extremely important for amphibian conservation since they maintain high species richness and abundance of frogs in spite of the relatively small area occupied by riparian forests. These specific environments may also provide more realistic assessments of extinction threats faced by stream-dwelling amphibians in the Atlantic Forest and other biomes (ALMEIDA-GOMES et al. 2014a). In a 10-year survey in the same region,(ALMEIDA-GOMES et al. 2014b), 73 amphibian species were registered, 10 of which are endemic to the state of Rio de Janeiro. The present study was comparatively shorter, yet it found one third of the species recorded in the region, including half of the endemic species recorded in that previous study.

Our results also demonstrate that a reduction in riparian forest cover has an effect on anuran species composition in both rivers. Riparian habitats can be used by amphibians for foraging, dispersal and reproduction, and the habitat size required to preserve some species is up to hundreds of meters (SEMLITSCH & BODIE 2003, FICETOLA et al. 2009). The removal of riparian vegetation can result in marked physical and structural changes in the river and on its margins (GOMI et al. 2006), and these changes generally cause a shift in amphibian species composition (OLSON et al. 2007, RODRÍGUEZ-MENDOZA & PINEDA 2010). The most abundant species in both rivers studied, *C. brasiliensis*, is a stream-dwelling amphibian typical of forested areas (MAIA-CARNEIRO et al. 2012), that is classified as Near Threatened (IUCN 2012). It was comparatively more abundant in the portions of the rivers that have greater forest cover. On the other hand, some species that are typical of open areas (natural or anthropogenic), for instance *Scinax alter* (Lutz, 1973), were found only in the plots with lesser riparian vegetation. The riparian vegetation in portions of the rivers outside the REGUA forest was mostly surrounded by pastures, which allows the occupation of these areas by frog species adapted to disturbed habitats, such as some hylids and leptodactylids (ALMEIDA-GOMES et al. 2008, 2010).

The value of riparian forests for the conservation of biodiversity and ecosystem services has already been recognized (RICKETTS 2004, LEES & PERES 2008, MENDENHALL et al. 2014), yet few studies have ascertained the effect riparian forest removal on local communities. This study demonstrates that riparian forests maintain a diverse community of amphibians that is drastically different from the community next to riparian pastures. Species that are more adapted to disturbed areas can benefit from forest reduction, which may cause the changes in species composition observed in our analysis. Our results exemplify the potential consequences of the proposed changes' the Brazilian forest code: population reductions and local extinctions of frog species that depend on these riparian environments.

ACKNOWLEDGMENTS

Mauricio Almeida-Gomes received Post-Doctoral scholarship from Programa Nacional de Pós-Doutorado (PNPD-CAPES). This study was supported by research grants from CNPq (processes 304791/2010-5 and 472287/2012-5) and FAPERJ through Cientistas do Nosso Estado Program (process E-26/102.404.2009 and E-26/102.765.2012) to C. F. D. Rocha and from CNPq (processes 307961/2011-7) and FAPERJ through Cientistas do Nosso Estado Program (process E-26/102.765/2012) to M.V. Vieira. We thank Nicholas J. Locke of the Reserva Ecológica de Guapiaçu (REGUA) for logistical support during our fieldwork in that area and all colleagues who helped us with data collection.

REFERENCES

ALMEIDA-GOMES M, ROCHA CFD (2014) Landscape connectivity may explain anuran species distribution in an Atlantic forest fragmented area. **Landscape Ecology 29**(1): 29-40. doi: 10.1007/s10980-013-9898-5

ALMEIDA-GOMES M, VRCIBRADIC D, SIQUEIRA CC, KIEFER MC, KLAION T, ALMEIDA-SANTOS P, NASCIMENTO D, ARIANI CV, BORGES-JÚNIOR VNT, FREITAS-FILHO RF, VAN SLUYS M, ROCHA CFD (2008) Herpetofauna of an Atlantic Rainforest area (Morro São João) in Rio de Janeiro State, Brazil. **Anais da Academia Brasileira de Ciências 80**(2): 291-300.

ALMEIDA-GOMES M, ALMEIDA-SANTOS M, GOYANNES-ARAÚJO P, BORGES-JÚNIOR VNT, VRCIBRADIC D, SIQUEIRA CC, ARIANI CV, DIAS AS, SOUZA VV, PINTO RR, VAN SLUYS M, ROCHA CFD (2010) Anurofauna of an Atlantic Rainforest fragment and its surroundings in Northern Rio de Janeiro State, Brazil. **Brazilian Journal of Biology 70**(3): 871-877.

ALMEIDA-GOMES M, LORINI ML, ROCHA CFD, VIEIRA MV (2014a) Underestimation of extinction threat to stream-dwelling amphibians due to lack of consideration of narrow area of occupancy. **Conservation Biology 28**(2): 616-619. doi: 10.1111/cobi.12196

ALMEIDA-GOMES M, SIQUEIRA CC, BORGES-JÚNIOR VNT, VRCIBRADIC D, FUSINATTO LA, ROCHA CFD (2014b) Herpetofauna of the Reserva Ecológica de Guapiaçu (REGUA) and its surrounding areas, in the state of Rio de Janeiro, Brazil. **Biota Neotropica 14**(3): 1-15. doi: 10.1590/1676-0603007813

BÁEZ S, COLLINS SL, POCKMAN WT, JOHNSON JE, SMALL EE (2012) Effects of experimental rainfall manipulations on Chihuahuan Desert grassland and shrubland plant communities. **Oecologia 172**(4): 1117-1127. doi: 10.1007/s00442-012-2552-0

BECKER CG, FONSECA CR, HADDAD CFB, BATISTA RF, PRADO PI (2007) Habitat split and the global decline of amphibians. **Science 318**(5857): 1775-1777. doi: 10.1126/science.1149374

BELL KE, DONNELLY MA (2006) Influence of forest fragmentation on community structure of frogs and lizards in northeastern Costa Rica. **Conservation Biology 20**(6): 1750-1760. doi: 10.1111/j.1523-1739.2006.00522.x

CRUMP ML, SCOTT JR NJ (1994) Visual encounter surveys, p. 84-92. In: HEYER WR, DONNELY MA, ROY WM, HAYEK LC, FOSTER MS (Eds) **Measuring and Monitoring Biological Diversity: Standard Methods for Amphibians.** Washington, DC, Smithsonian Institution Press, 384p.

Ficetola GF, Padoa-Schioppa E, Bernardi F. (2009) Influence of landscape elements in riparian buffers on the conservation of semiaquatic amphibians. **Conservation Biology 23**(1): 114-123. doi: 10.1111/j.1523-1739.2008.01081.x

Gomi T, Sidle RC, Noguchi S, Negishi JN, Nik AR, Sasaki S (2006) Sediment and wood accumulations in humid tropical headwater streams: Effects of logging and riparian buffers. **Forest Ecology and Management 224**(1-2): 166-175. doi: 10.1016/j.foreco.2005.12.016

Haddad CFB, Prado CPA (2005) Reproductive modes in frogs and their unexpected diversity in the Atlantic Forest of Brazil. **BioScience 55**(3): 207-217.

Hof C, Araújo MB, Jetz W, Rahbek C (2011) Additive threats from pathogens, climate and land-use change for global amphibian diversity. **Nature 480**: 516-519. doi: 10.1038/nature10650

IUCN (2012) **Red list of Threatened Species**. Version 2012.2. Available online at: www.iucnredlist.org [Accessed: 30 January 2013]

Keller A, Rödel M-O, Linsenmair E, Grafe TU (2009) The importance of environmental heterogeneity for species diversity and assemblage structure in Bornean stream frogs. **Journal of Animal Ecology 78**: 305-314. doi: 10.1111/j.1365-2656.2008.01457.x

Lees AC, Peres CA (2008) Conservation value of remnant riparian forest corridors of varying quality for amazonian birds and mammals. **Conservation Biology 22**: 439-449. doi: 10.1111/j.1523-1739.2007.00870.x

Legendre P, Legendre L (1998) **Numerical Ecology**. Amsterdam, Elsevier, 853p.

Maia-Carneiro T, Dorigo TA, Almeida-Gomes M, Van Sluys M, Rocha CFD (2012) Feedings habitats, microhabitat use, and daily activity of *Cycloramphus brasiliensis* (Anura: Cycloramphidae) from the Atlantic Rainforest, Brazil. **Zoologia 29**(3): 277-279. doi: 10.1590/S1984-46702012000300007

Mendenhall CD, Frishkoff LO, Santos-Barrera G, Pacheco JS, Mesfun E, Quijano FM, Ehrlich PR, Ceballos G, Daily GC, Pringle RM (2014) Countryside biogeography of Neotropical reptiles and amphibians. **Ecology 95**: 856-870. doi: 10.1890/12-2017.1

Metzger JP, Lewinsohn TM, Joly CA, Verdade LM, Martinelli LA, Rodrigues RR (2010) Brazilian Law: full speed in reverse? **Science 329**: 276-277.

Mittermeier RA, Gil RP, Hoffman M, Pilgrim J, Brooks T, Mittermeier CG, Lamoreux J, Fonseca GAB (2005) **Hotspots revisited: earth's biologically richest and most endangered terrestrial ecoregions**. Washington, DC, Conservation International, 392p.

Olson DH, Anderson PD, Frissel CA, Welsh Jr HH, Bradford DF (2007) Biodiversity management approaches for stream-riparian areas: Perspectives for Pacific Northwest headwater forests, microclimates, and amphibians. **Forest Ecology and Management 246**: 81-107. doi:10.1016/j.foreco.2007.03.053

Parris KM, McCarthy MA (1999) What influences the structure of frog assemblages at forest streams? **Australian Journal of Ecology 24**(5): 495-502. doi: 10.1046/j.1442-9993.1999.00989.x

Ribeiro MC, Metzger JP, Martensen AC, Ponzoni FJ, Hirota MM (2009) The Brazilian Atlantic Forest: how much is left, and how is the remaining forest distributed? Implications for conservation. **Biological Conservation 142**: 1144-1156. doi:10.1016/j.biocon.2009.02.021

Richards-Zawacki CL (2009) Effects of slope and riparian habitat connectivity on gene flow in an endangered Panamanian frog, *Atelopus varius*. **Diversity and Ditributions 15**: 796-806. doi: 10.1111/j.1472-4642.2009.00582.x

Ricketts TH (2004) Tropical forest fragments enhance pollinator activity in nearby coffee crops. **Conservation Biology 18**(5): 1262-1271. doi: 10.1111/j.1523-1739.2004.00227.x

Rodríguez-Mendoza C, Pineda E (2010) Importance of riparian remnants for frog species diversity in a highly fragmented rainforest. **Biology Letters 6**: 781-784. doi: 10.1098/rsbl.2010.0334

Semlitsch RD, Bodie JR (2003) Biological criteria for buffer zones around wetlands and riparian habitats for amphibians and reptiles. **Conservation Biology 17** (5): 1219-1228. doi: 10.1046/j.1523-1739.2003.02177.x

Todd BD, Luhring TM, Rothermel BB, Gibbons W (2009) Effects of forest removal on amphibian migrations: implications for habitat and landscape connectivity. **Journal of Applied Ecology 46**: 554-561. doi: 10.1111/j.1365-2664.2009.01645.x

Zimmerman BL, Bierregaard RO (1986) Relevance of the equilibrium theory of island biogeography and species-area relations to conservation with a case from Amazonia. **Journal of Biogeography 13** (2): 133-143. doi: 10.2307/2844988

5

Population biology of *Aegla platensis* (Decapoda: Anomura: Aeglidae) in a tributary of the Uruguay river, state of Rio Grande do Sul, Brazil

Marcelo M. Dalosto[1], Alexandre V. Palaoro[1], Davi de Oliveira[1], Évelin Samuelsson[2] & Sandro Santos[1,3]

[1] *Núcleo de Estudos em Biodiversidade Aquática, Programa de Pós-Graduação em Biodiversidade Animal, Centro de Ciências Naturais e Exatas, Universidade Federal de Santa Maria. Avenida Roraima 1000, 97105-900 Santa Maria, RS, Brazil.*
[2] *Programa de Pós-Graduação em Ecologia, Departamento de Ciências Biológicas, Universidade Regional Integrada do Alto Uruguai e das Missões. Avenida Sete de Setembro 1621, 99700-000 Erechim, RS, Brazil.*
[3] *Corresponding author. E-mail: sandro.santos30@gmail.com*

ABSTRACT. Aeglids are freshwater anomurans that are endemic from southern South America. While their population biology at the species-level is relatively well understood, intraspecific variation within populations has been poorly investigated. Our goal was to investigate the population biology of *Aegla platensis* Schmitt, 1942 from the Uruguay River Basin, and compare our data with data from other populations. We estimated biometric data, sex ratio, population density and size-class frequencies, and frequencies of ovigerous females and juveniles, from the austral spring of 2007 until autumn 2008. Sexual dimorphism was present in adults, with males being larger than females. Furthermore, males and females were significantly larger than previously recorded for the species. The overall sex ratio was 1.33:1 (male:female), and population density ranged from 1.8 (spring) to 3.83 ind.m^{-2} (winter). Data from this population differ from published information about *A. platensis* in almost all parameters quantified except for the reproductive period, which happens in the coldest months, and a population structure with two distinct cohorts. Difference among studies, however, may be in part due to methodological differences and should be further investigated in order to determine their cause. In addition to different methodologies, they may result from ecological plasticity or from the fact that the different populations actually correspond to more than one species.

KEY WORDS. Aeglids; ecological plasticity; intraspecific variation; population density; sampling methods; sex ratio.

Aeglids are freshwater anomuran crustaceans with benthonic habits, whose distribution is restricted to temperate and subtropical regions of South America (BUCKUP & BOND-BUCKUP 1999, BOND-BUCKUP 2003). These crustaceans occur in river basins in Southern Brazil, Uruguay, Argentina, Southern Bolivia, Paraguay and South-central Chile, with 70 species currently described (SANTOS *et al.* 2013). Albeit ubiquitous in well-oxygenated running waters in these regions (DALOSTO & SANTOS 2011), several species have a very restricted distribution (BOND-BUCKUP *et al.* 2008).

Understanding the basic traits of an organism's biology is important because it provides basic information for a wide array of studies. In the case of aeglids, these range from conservation efforts (PÉREZ-LOSADA *et al.* 2009) to the use of model species in laboratory studies (PALAORO *et al.* 2013, SIQUEIRA *et al.* 2013). There is a considerable amount of studies on the basic biology of aeglids, such as population structure and dynamics (e.g., BUENO & BOND-BUCKUP 2000, FRANSOZO *et al.* 2003, COHEN *et al.* 2011, GRABOWSKI *et al.* 2013). However, previous studies on aeglids focused solely on the population dynamics of one species in a

single location (e.g., BUENO & BOND-BUCKUP 2000, FRANSOZO *et al.* 2003, COHEN *et al.* 2011, GRABOWSKI *et al.* 2013). Ecological plasticity and variation in population biology parameters have been documented for other freshwater organisms, such as crayfish (HONAN & MITCHELL 1995, AUSTIN 1998, BEATTY *et al.* 2004, 2011), which share ecological similarities to aeglids (BOND-BUCKUP & BUCKUP 1994, NYSTRÖM 2002, AYRES-PERES *et al.* 2011, BURRESS *et al.* 2013, COGO & SANTOS 2013).

Unlike other aeglids, *Aegla platensis* Schmitt, 1942 has broad distribution and relatively large populations. This species is recorded for Paraguay, Uruguay, Argentina and Brazil, where it occurs in the states of Rio Grande do Sul and Santa Catarina (BOND-BUCKUP 2003). The population dynamics and growth of *A. platensis* have been studied for a population in the Guaíba Basin in the state of Rio Grande do Sul (BUENO & BOND-BUCKUP 2000 and BUENO *et al.* 2000, respectively). More recently, OLIVEIRA & SANTOS (2011) investigated the morphological sexual maturity of another population that inhabits the Uruguay River Basin, obtaining markedly different results from those reported by BUENO & BOND-BUCKUP (2000) and BUENO *et al.* (2000).

Our goal was to investigate several characteristics of the population biology of *A. platensis*, such as sex ratio, population structure, reproductive/recruitment seasons and population density. Also, we compare our results with data already available for this species, and with information available for other aeglids. Lastly, we discuss the variations in the population biology of this group of crustaceans and whether or not it is productive to compare among data obtained using different methods.

MATERIAL AND METHODS

The Lajeado Bonito stream (27°25'27"S; 53°24'39"W) is located in the municipality of Frederico Westphalen, state of Rio Grande do Sul. The dominant vegetation in the area is the Atlantic Forest and the climate is subtropical. The stream is a first order tributary of the Várzea River, in the Uruguay River Basin. The study site is located 470 m above sea level. Even though agricultural and livestock activities happen in the areas located upstream of the collection sites, the studied area harbors riparian vegetation on both margins of the stream. The streambed is composed of rocks of various sizes, sand, and bedrock.

Monthly collections of *A. platensis* were performed in a 160 m section of the stream from July 2007 to June 2008. This section was divided into 16 subunits. Aeglids were captured with traps (N = 16, one per subunit) placed before the dusk and revised in the morning of the following day. In order to sample the population more thoroughly, a 30 x 50 cm hand net with a 60 cm deep mouth and 1 mm mesh was also employed. The sampling effort, performed by two people, lasted approximately five minutes per subunit. Environmental variables (water temperature, dissolved oxygen, pH, flow speed, stream depth, stream width and conductivity) were measured monthly in three predetermined locations of the stream (Table I).

Table I. Enviromental parameters recorded for the Lajeado Bonito stream, Uruguay Basin, Rio Grande do Sul state, Brazil.

Parameters	Spring	Summer	Autumn	Winter
Temperature (°C)	19.830	19.830	16.270	15.200
Dissolved oxygen (mg/L)	6.250	6.280	8.040	7.990
Flow speed (m/s)	0.420	0.340	0.620	0.240
Conductivity (μS/cm)	68.680	88.510	71.990	76.960
pH	7.520	7.660	7.760	7.320
Stream depth (cm)	17.780	6.480	18.670	14.890
Stream width (m)	1.790	1.210	2.160	1.790
Discharge (m³/s)	0.277	0.024	0.075	0.096
Rainfall (mm/month)	246.000	97.000	198.000	147.330

After being separated from other animals captured accidentally, aeglids were identified, sexed, and females were checked for the presence of eggs. Sexing was based on morphological traits, such as the presence of pleopods in adult females, and their absence in adult males. Since younger individuals have inconspicuous pleopods, their sex was determined by observing the genital pores at the base of the third pereiopods (females) or their absence (males) (Bond-Buckup 2003). Biometric measurements were then taken with a digital caliper (0.01 mm accuracy), including carapace length (CL – from the tip of the rostrum to the posterior edge of the carapace), carapace width (CW – taken on the height of the upper suture of the gastric region), abdomen width (AW – measured on the second abdominal segment), length of the propodus of the left (LPL) and right (RPL) chelipeds (measured from the posterior proximal margin of the propodus to the tip of the fixed finger) and height of the chelar propodus (HCP – measured perpendicularly to the propodus length). Aeglids were classified according to their CL following Oliveira & Santos (2011): males larger than 19.15 mm were considered adults, and females larger than 16.5 mm were determined adults. Aeglids smaller than 8 mm CL were measured with the help of a stereomicroscope taken to the field site. In order to minimize impact on the studied population, most animals were released back at their capture sites after data recording, with the exception of a few large males from the first collections, which were preserved as vouchers in the scientific collection of the Núcleo de Estudos em Biodiversidade Aquática, Universidade Federal de Santa Maria (voucher number UFSM-C 298). To test for differences in the body measurements of males and females, a Mann-Whitney test was used due to heterocedasticity and non-normality of the data (Zar 2010). The test was performed in two different configurations: 1) using all captured individuals, and 2) using only adult individuals. The exception was the AW in the all-animals sample and the CL and CW in the adults-only, for which normality and homocedasticity could be attained through a log10 transformation, and for which a Welch two-sample t-test was used.

Sex ratio (male/female) was calculated for each season separately. A chi-square with Yates' correction for small samples was performed to test whether the sex ratio differed from the expected proportion of 1:1 within each season (Zar 2010). Additionally, data obtained from traps and from handnets were plotted separately to check for possible influences of the sampling method on the sex ratio. The chi-square test performs poorly with small sampling numbers, which can generate spurious results (Crawley 2012). Thus, the test was performed with pooled data from captures using traps and hand nets because of the low capture rates in certain seasons. The reproductive and recruitment seasons were estimated qualitatively through the frequency of ovigerous females and unsexed juveniles. Afterwards, we tested if the proportion of captures of ovigerous females and juveniles differed from the expected equal proportion of captures among the seasons using a binomial proportion test (Zar 2010).

To estimate population density for each season four field samplings (August and November 2007, February and May

2008) were performed differently. Traps were set on a given day and revisited the morning of the following day, as usual, and then all captured aeglids were marked with a plastic tag placed in their dorsal region. This tag indicated the initial capture site and month of capture. The aeglids were then released back in the stream. The writing on the tags was made with Nanking ink, and the tags were fixed on the aeglid's carapace with cianoacrilate glue. Differently from regular collections, traps were then put back on the stream and the sampling procedure was repeated the following day. The amount of recaptured tagged individuals was recorded. Peterson's estimate (BEGON 1979) was applied to estimate population size: $N = r*n/m$, where: (N) estimate of the population size, (r) number of animals marked in the first day, (n) number of animals collected in the second day, and (m) number of tagged animals recaptured in the second day.

All data were tested for normality and heterocedasticity with the tests of Shapiro-Wilk and Levene, respectively. All tests were performed in the BioEstat 5.0 software (ZAR 2010, AYRES *et al.* 2007), except for the Welch two-sample t-tests, which were performed in the R environment (R CORE TEAM 2013).

RESULTS

A total of 957 individuals were collected, of which 76 were non-sexed juveniles, 503 males (323 juveniles and 180 adults), 378 females (187 juveniles, 169 adults and 22 ovigerous) (Table II). The CL ranged from 6 to 31.75 mm for males (median ± SD: 15.09 ± 7.35 mm), and from 6.08 to 27.92 for females (median ± SD: 16.11 ± 5.95 mm). There were significant differences only for AW (t = 2.215, p = 0.027) between males and females when all aeglids were considered. However, when only adults were considered there was a significant difference between males and females in all dimensions compared (U = 3.793, 9.781, 11.150, 12.452, p < 0.001; for AW, LPL, RPL and HCP, respectively; and t = -12.226, -11.392, and p < 0.001 for CL and CW, respectively) (Table III). The frequency distribution of size-classes of males and females of *A. platensis* for each season presented a bimodal distribution in all seasons (Figs 1-4).

Table II. Number of individuals of *Aegla platensis* collected during the four seasons in the Lajeado Bonito stream, Uruguay Basin, Rio Grande do Sul state. (JM) Juvenile males, (AM) adult males, (JF) juvenile females, (AF) adult females, (OF) ovigerous females, (NS) non-sexed juveniles.

Seasons	JM	AM	JF	AF	OF	NS	Total
Spring	81	43	33	44	4	50	255
Summer	102	22	63	18	7	7	219
Autumn	89	51	61	68	1	13	283
Winter	51	64	30	39	10	6	200
Total	323	180	187	169	22	76	957

Table III. Medians of the biometric measurements (mm) of the adult individuals of *Aegla platensis* captured in the Lajeado Bonito stream, Uruguay Basin, Rio Grande do Sul state, Brazil. Different letters in the column indicate statistically significant differences (p < 0.05) in the Mann-Whitney (a,b) or t-tests (c,d).

	CL (mm)	CW (mm)	AW (mm)	RPL (mm)	LPL (mm)	ACP (mm)
Males	24.87[c]	14.67[c]	17.67[a]	14.12[a]	15.91[a]	9.55[a]
Females	21.29[d]	12.55[d]	16.48[b]	10.28[b]	10.74[b]	6.34[b]

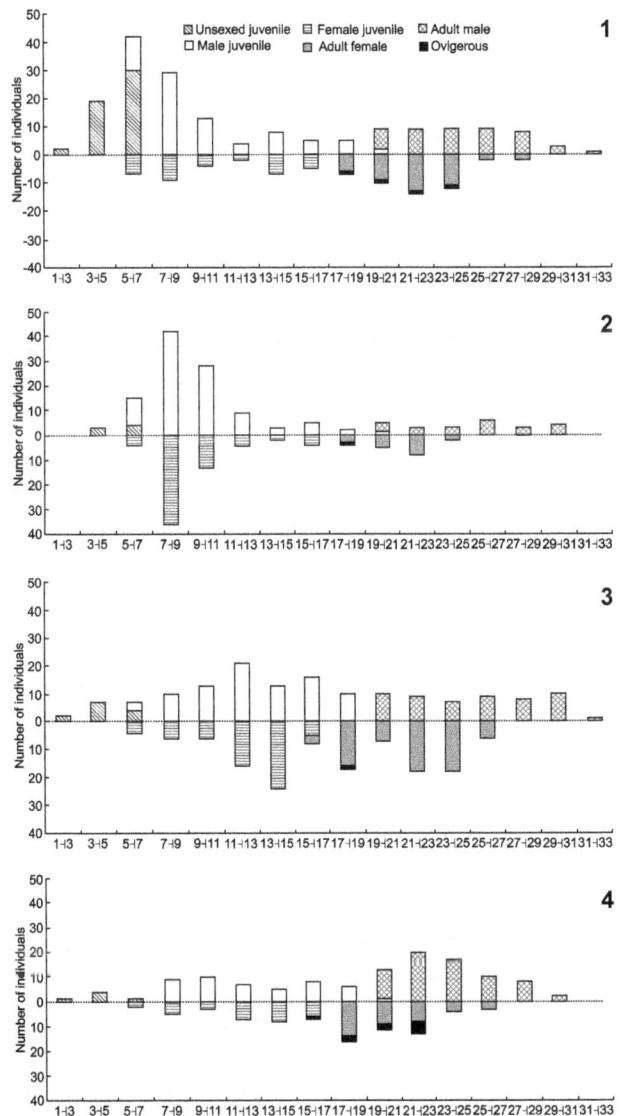

Figures 1-4. Absolute frequencies distribution of cephalothoracic length (CL) classes of individuals of *Aegla platensis* collected in the Lajeado Bonito stream, Uruguay Basin, Rio Grande do Sul state. Different letters indicate seasons: (1) spring; (2) summer; (3) autumn and (4) winter.

Males were more common in the spring and winter, with no difference for the other seasons, when all captures were considered (spring: $\chi^2 = 8.605$, df = 1, p = 0.003; summer: $\chi^2 = 2.151$, df = 1, p = 0.142; autumn: $\chi^2 = 0.034$, df = 1, p = 0.853; winter: $\chi^2 = 6.314$, df = 1, p = 0.012). The highest proportion of males was 60.48%, in the spring (Fig. 5). The number of ovigerous females caught (22) represented 5.82% of all the females. These were caught in all seasons, with a higher frequency in winter and summer, and the lowest frequency in the autumn. There was significant difference among the seasons, probably due to the small number of ovigerous females caught in the autumn (only 1; $\chi^2 = 24.735$, df = 3, p < 0.001; Fig. 6). Juveniles were caught throughout the sampling period, with a higher frequency during the spring ($\chi^2 = 63.104$, df = 3, p < 0.001; Fig. 6). Density ranged from 1.80 to 3.83 ind.m^{-2}, with the highest values in the winter (Table IV).

Table IV. Petersen's estimate of population size of *Aegla platensis* in the Lajeado Bonito stream, Uruguay Basin, Rio Grande do Sul state, Brazil.

Season	Marked in the 1st day	Captured 2nd day	Marked and recaptured	Population estimate	Individuals/m²
Winter	25	72	2	900	3.83
Spring	29	73	5	423	1.80
Summer	18	86	2	772	3.29
Autumn	23	95	4	547	2.33

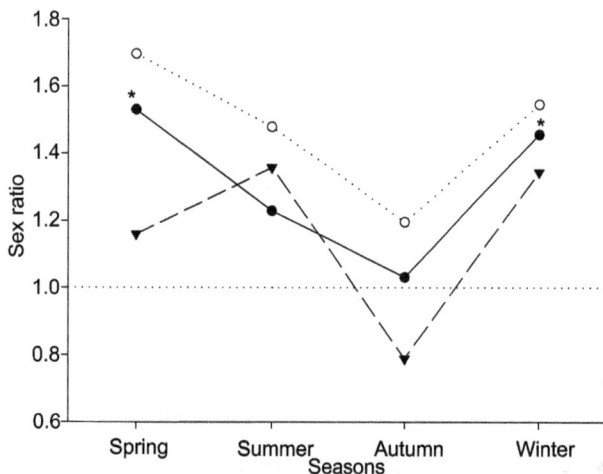

Figure 5. Relative frequency of males and females of *Aegla platensis* in the four seasons during the sampling period in the Lajeado Bonito stream, Uruguay Basin, Rio Grande do Sul state. Trap captures, handnet captures, and global (traps + handnet captures) are plotted separately. The asterisk (*) denotes statistical difference between the number of males and females by the Chi-square test (p < 0.05) for the global dataset. (●) Global, (○) hand net, (▼) trap.

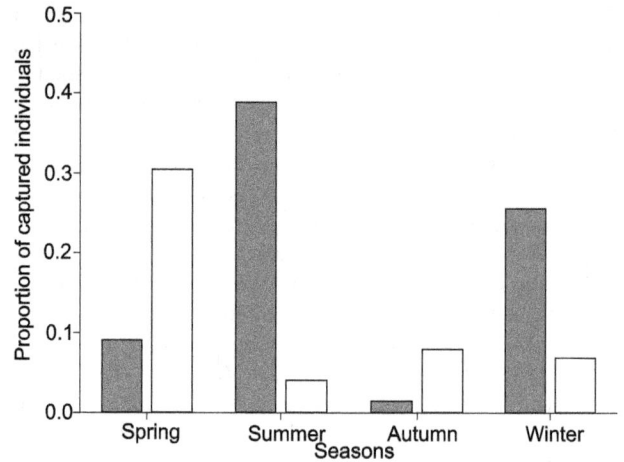

Figure 6. Number of ovigerous females and unsexed juveniles of *Aegla platensis* in the Lajeado Bonito stream, Uruguay Basin, Rio Grande do Sul state. Individuals were captured during four seasons. Frequencies of capture were distinct from the expected among the four seasons (Chi-square test, p < 0.05 for ovigerous females and juveniles). (■) Ovigerous, (□) unsexed juveniles.

DISCUSSION

In the population studied here (Uruguay River Basin, henceforth, UB), adult males had larger body dimensions than their female counterparts (except for AW, which is larger in females, even considering the juveniles). These data agree with the pattern described for most aeglids studied so far, where there is clear sexual dimorphism, with males prevailing in the larger size-classes (COLPO *et al.* 2005, TREVISAN & SANTOS 2012, TREVISAN *et al.* 2012), and females possessing broader abdomens for egg-incubation (LÓPEZ-GRECO *et al.* 2004). Interestingly enough, one of the few studies that does not fit this pattern is the other previously studied population of *A. platensis* in the Guaíba Basin (henceforth, GB), where females were larger than males (BUENO & BOND-BUCKUP 2000). Since sexual dimorphism with larger males is a characteristic of the group (BARRÍA *et al.* 2014, BOND-BUCKUP & BUCKUP 1994), a possible explanation for these opposite results might be the sampling methods adopted (see below for a more detailed discussion).

The maximum size of the animals also differed between UB and GB: maximum CL for males and females were of 17.39 and 19.12 mm for GB, and of 31.75 and 27.92 mm for UB, respectively (BUENO & BOND-BUCKUP 2000, Table V). Animals from UB were 50% larger than those registered for the GB. The different sampling and measuring methods adopted could account for this. In GB, the CL measurements did not include the rostrum, while our data include it. However, the rostrum would increase the CL size of the aeglids in approximately 10%, which is certainly not enough to compensate for a size difference of 66% in relation to animals from the UB. Additionally, BUENO &

BOND-BUCKUP (2000) only employed manual search, while we employed a combination of traps and handnet. The use of traps could affect the mean and maximum size of the animals captured. We found differences in the mean size of individuals captured by traps (21.83 ± 4.84 mm) and handnets (12.76 ± 6.34 mm), similar to what has been found for other decapod crustaceans. However, this was not the case for the maximum size, since the largest aeglid caught by traps had 31.75 mm CL, while the largest caught by hand-net had 30.65 mm CL. This shows that despite the fact that traps tend to capture larger individuals, handnet is still able to sample large specimens, and is thus considered an appropriate collecting gear to estimate size range. We think that it is safe to conclude that animals from the UB are considerably larger than GB.

Table V. Sampling methods and population parameters of *Aegla platensis* evaluated for the populations of the Uruguay River (this study) and Guaíba Basins (BUENO & BOND-BUCKUP 2000), respectively.

Parameters	Uruguay River Basin	Guaíba Basin
Sex ratio (M:F)	1.33:1	1.08:1
Population density (ind/m²)	1.8-3.83	8.7-19
Sexual dimorphism	Present, males larger	Present, females larger
Size of the largest male (mm)	31.75	17.39
Size of the largest female (mm)	27.92	19.12
Population structure	Bimodal	Bimodal
Reproductive period	Year-round, peak in coldest months	Year-round, peak in coldest months
Sampling technique	Handnet + traps	Handnet

The sex ratio also differed between GB and UB. In GB, it did not differ significantly from 1:1, while the opposite was found in UB, where it differed from the 1:1 expected proportion in the spring and winter, with an overall sex ratio of 1.33:1. Both results fit the pattern recorded for aeglids, in which the sex ratio ranges from 1:1 to values skewed towards males (Table VI). Once again, the effect of the sampling method makes it difficult to distinguish between actual differences between the species/populations, and the effects of the different sampling methods chosen by each author. The tendency to capture more large adult males using traps had already been demonstrated in crayfish surveys (BEATTY *et al.* 2004, 2011), and it is relatively safe to infer that the same is true for aeglids (BUENO *et al.* 2007, TEODÓSIO & MASUNARI 2009, GRABOWSKI *et al.* 2013). In this study, however, when analyzing the sex ratio of aeglids captured by hand-net and by traps (Fig. 5), we can see that the ratio was more skewed towards males in the hand-net captures than in the traps. Thus, we conclude that the difference regarding the sex ratio between GB and UB is not an effect of the methods chosen, but that it reflects an actual difference between these populations.

Table VI. Sex ratio and population density parameters of published studies on Brazilian species of *Aegla*.

Species	Sex ratio (M:F)	Density (ind/m²)	Authors
A. castro	1:1	–	SWIECH-AYOUB & MASUNARI (2001)
A. castro	1.08:1	–	FRANSOZO et al. (2003)
A. franca	–	2.2-2.7	BUENO et al. (2007)
A. franciscana	1:1	–	GONÇALVES et al. (2006)
A. leptodactyla	1.19:1	–	NORO & BUCKUP (2002)
A. longirostri	1:1	–	COLPO et al. (2005)
A. parana	2:1*	–	GRABOWSKI et al. (2013)
A. paulensis	1.66:1*	–	COHEN et al. (2011)
A. platensis	1.08:1	8.7-19	BUENO & BOND-BUCKUP (2000)
A. platensis	1.33:1*	1.8-3.83	Current study
A. schmitti	2:1*	–	TEODÓSIO & MASUNARI (2009)

* Sex ratio statistically different from an 1:1 expected proportion.

Ovigerous females were captured year-round, with a peak in the colder months (late winter and early spring). Thus, our data agree with the pattern known for other Brazilian species of *Aegla*, where the reproduction is either year-round with peaks in the colder months (BUENO & BOND-BUCKUP 2000, COLPO *et al.* 2005), or just concentrated in the colder months (TEODÓSIO & MASUNARI 2009, GRABOWSKI *et al.* 2013), including GB. The frequency of juveniles also follows a similar pattern: juveniles were captured year-round, being more abundant in the spring following the peak of the reproductive season (BUENO & BOND-BUCKUP 2000) (Fig. 6). The population structure was bimodal, with two age groups easily distinguishable in the size-class frequency distribution (Figs 1-4). This is in agreement with information for other aeglids, which also show two distinct cohorts in the population (e.g., BUENO & BOND-BUCKUP 2000, FRANSOZO *et al.* 2003).

The density also differed markedly between localities, being much lower in UB than in GB. In fact, the density of A. *platensis* in UB was much more similar to the density of another species, *A. franca* Schmitt, 1942, in the Barro Preto stream (Minas Gerais state, Brazil), than to the density of its conspecific in GB (Table V). Even though the capture methods used by BUENO & BOND-BUCKUP (2000) differ from ours, and may have affected our density results, the difference between both estimates is over 120%. Considering the markedly larger size of the aeglids in UB, one can expect that their populations will exhibit lower densities. This is even more likely if we consider the aggressive nature of aeglids (AYRES-PERES *et al.* 2011, PALAORO *et al.* 2013). In crayfish, spatial patterns investigated in natural environments show that dominant animals (i.e., the largest) are more spaced from other crayfish than smaller individuals (FERO & MOORE 2008). Although there are no such studies for aeglids, their ecological (BURRESS *et al.* 2013) and behavioral (MOORE 2007, AYRES-PERES *et al.* 2011) similarities with crayfish, along with our results, support the idea of a negative relationship between body size and density in *A. platensis*.

In general terms, the population structure of UB agrees with the known pattern for aeglids, presenting sexual dimorphism with larger males, a bimodal distribution of the size-class frequencies, reproduction concentrated in the coldest months of the year and release of juveniles in the following season (ROCHA *et al.* 2010, COHEN *et al.* 2011). When compared to GB, however, some differences can be highlighted: aeglids were much larger in UB than GB; the larger aeglids were males in UB and females in GB; the sex ratio was skewed towards males in UB, and similar to 1:1 in GB; and population density values were at least two times higher in GB than in UB. Differences in these population biology characteristics can also be a result of different environmental pressures. BÜCKER *et al.* (2008) have shown that the spatial micro distribution of *A. platensis* and *A. itacolomiensis* Bond-Buckup & Buckup, 1994 are correlated with the availability of coarse organic matter: the distribution of *A. platensis* was explained by the availability of twigs, followed by fragmented leaves, while the distribution of *A. itacolomiensis* was explained by fragmented leaves, followed by twigs. However, BUENO & BOND-BUCKUP (2000) do not present any environmental variable other than temperature. Furthermore, BÜCKER *et al.* (2008) make a much more detailed surveillance of environmental variables, but do not provide any data regarding the size, density, or other population biology parameters.

The only similarities between UB and GB were the bimodal population structure and the reproduction peak on the colder months (Table V). Nevertheless, these characteristics are shared by most *Aegla* species studied so far (e.g., FRANSOZO *et al.* 2003, GONÇALVES *et al.* 2006, TEODÓSIO & MASUNARI 2009), and thus, cannot be considered a species-specific characteristic. Conversely, the maximum size, sexual dimorphism, sex ratio and population density clearly differed between the two populations. Albeit variation in population parameters is expected, and sampling methods can bias the results, the differences between UB and GB are very marked, eventually presenting differences of over 100% in certain values. If we consider the geographical isolation between the two river basins (SCHWARZBOLD 2010), alongside the evidences for ecological differences, it becomes clear that molecular studies might be the best choice to elucidate if this is a case of ecological differences between populations, or if this a case of cryptic species (MARCHIORI *et al.* 2014).

The variety of methods employed by researchers is by far the greatest obstacle to reliable comparisons between population studies on *Aegla*. More specifically, the choice of the capture method (baited traps, manual search, handnet, Surber sampler, or any combination of these) seems to bias the results. A clear example can be seen in Table VI. The four studies of *Aegla* where the sex ratio differed significantly from 1:1 were the ones that employed traps, with the three with the more skewed sex ratios being those that relied solely on traps as the

sampling method. This issue has already been addressed by previous authors (e.g., BUENO *et al.* 2007, GRABOWSKI *et al.* 2013). Despite this, there is still no consensus among researchers on the best methods.

In conclusion, *A. platensis* presented marked differences from one population to another. These differences can be attributed partially to the different sampling methods used by different authors. These differences, along with isolation between the two river basins, suggest that molecular studies are needed to elucidate the taxonomic status of the populations of this species. The only similarities between the populations were common to many *Aegla* species, which highlights the need of a standardized technique to perform population studies in these anomurans, so that more reliable and less speculative comparisons can be made.

ACKNOWLEDGMENTS

We would like to thank CAPES for the scholarships for AVP and DO; CAPES/FAPERGS for the scholarship for MMD, and CNPq for the productivity grant for SS (308598/2011-3). We would also like to thank our colleagues at the Núcleo de Estudos em Biodiversidade Aquática for their help in the field work, T.M. Dias, two anonymous reviewers, and A.S. Melo for the helpful comments and suggestions that certainly improved the manuscript.

REFERENCES

AUSTIN, C.M. 1998. Intra-specific variation in clutch and brood size and rate of development in the yabby, *Cherax destructor* (Decapoda: Parastacidae). **Aquaculture 167**: 147-159. doi: 10.1016/S0044-8486(98)00306-8.

AYRES, M.; M. AYRES JR; D.L. AYRES & A.S. SANTOS. 2007. **Bioestat 5.0: aplicações estatísticas nas áreas das Ciências Biológicas e Médicas.** Belém, Sociedade Civil Mamirauá.

AYRES-PERES, L.; P.B. ARAUJO & S. SANTOS. 2011. Description of the agonistic behavior of *Aegla longirostri* (Decapoda: Aeglidae). **Journal of Crustacean Biology 31** (3): 379-388. doi: 10.1651/10-3422.1.

BARRÍA, E.M.; S. SANTOS; C.G. JARA & C.J. BUTLER. 2014. Sexual dimorphism in the cephalothorax of freshwater crabs of the genus *Aegla* Leach from Chile (Decapoda, Anomura, Aeglidae): an interspecific approach based on distance variables. **Zoomorphology.** doi: 10.1007/s00435-014-0231-x.

BEATTY S.J.; D.L. MORGAN & H.S. GILL. 2004. Biology of a translocated population of *Cherax cainii* Austin & Ryan, 2002 in a western Australian river. **Crustaceana 77** (11): 1329-1351. doi: 10.1163/1568540043166010.

BEATTY S.J.; M. DE GRAAF; B. MOLONY; V. NGUYEN & K. POLLOCK. 2011. Plasticity in population biology of *Cherax cainii* (Decapoda: Parastacidae) inhabiting lentic and lotic environments in south-western Australia: Implications for the sustainable management of the recreational ûshery. **Fisheries Research 110**: 312-324. doi: 10.1016/j.fishres.2011.04.021.

BEGON, M. 1979. **Investigating animal abundance: capture-recapture techniques for biologists.** London, Edward Arnold.

BOND-BUCKUP, G. 2003. A Família Aeglidae, p. 21-116. *In*: G.A.S. MELO (Ed.). **Manual de identificação dos Crustacea Decapoda de água doce do Brasil.** São Paulo, Editora Loyola.

BOND-BUCKUP, G. & L. BUCKUP. 1994. A Família Aeglidae (Crustacea, Decapoda, Anomura). **Arquivos de Zoologia 32** (4): 159-347.

BOND-BUCKUP, G.; C.G. JARA; M. PÉREZ-LOSADA; L. BUCKUP & K.A. CRANDALL. 2008. Global diversity of crabs (Aeglidae: Anomura: Decapoda) in freshwater. **Hydrobiologia 595**: 267-273. doi: 10.1007/s10750-007-9022-4.

BÜCKER, F; R. GONÇALVES; G. BOND-BUCKUP & A.S. MELO. 2008. Effect of the environmental variables on the distribution of two freshwater crabs (Anomura: Aeglidae). **Journal of Crustacean Biology 28** (2): 248-251. doi: 10.1651/0278-0372(2008) 028[0248:EOEVOT]2.0.CO;2.

BUCKUP, L. & G. BOND-BUCKUP. 1999. **Os crustáceos do Rio Grande do Sul.** Porto Alegre, Editora UFRGS.

BUENO, A.A.P. & G. BOND-BUCKUP. 2000. Dinâmica populacional de *Aegla platensis* Schmitt (Crustacea, Decapoda, Aeglidae). **Revista Brasileira de Zoologia 17** (1): 43-49. doi: 10.1590/S0101-81752000000100005.

BUENO, A.A.P.; G. BOND-BUCKUP & L. BUCKUP. 2000. Crescimento de *Aegla platensis* Schmitt em ambiente natural (Crustacea, Decapoda, Aeglidae). **Revista Brasileira de Zoologia 17** (1): 51-60. doi: 10.1590/S0101-81752000000100006.

BUENO, S.L.S.; R.M. SHIMIZU & S.S. DA ROCHA. 2007. Estimating the population size of *Aegla franca* (Decapoda: Anomura: Aeglidae) by mark-recapture technique from an isolated section of Barro Preto stream, county of Claraval, state of Minas Gerais, southeastern Brazil. **Journal of Crustacean Biology 27** (4): 553-559. doi: 10.1651/S-2762.1.

BURRESS, E.D.; M.M. GANGLOFF & L. SIEFFERMAN. 2013. Trophic analysis of two subtropical South American freshwater crabs using stable isotope ratios. **Hydrobiologia 702**: 5-13. doi: 10.1007/s10750-012-1290-y.

BYRON, C.J. & K.A. WILSON. 2001. Rusty crayfish (*Orconectes rusticus*) Movement within and between habitats in Trout Lake, Vilas County, Wisconsin. **Journal of the North American Benthological Society 20** (4): 606-614.

COHEN, F.P.A.; B.F. TAKANO; R.M. SHIMIZU & S.L.S. BUENO. 2011. Life cycle and population structure of *Aegla paulensis* (Decapoda: Anomura: Aeglidae). **Journal of Crustacean Biology 31** (3): 389-395. doi: 10.1651/10-3415.1.

COGO, G.B. & S. SANTOS. 2013. The role of aeglids in shredding organic matter in Neotropical streams. **Journal of Crustacean Biology 33** (4): 519-526. doi: 10.1163/1937240X-00002165.

COLPO, K.D.; L.R. OLIVEIRA & S. SANTOS. 2005. Population biology of the freshwater anomuran *Aegla longirostri* (Crustacea, Anomura, Aeglidae) from Ibicuí-Mirim River, Itaára, RS, Brazil. **Journal of Crustacean Biology 25** (3): 495-499. doi: 10.1651/C-2543.

CRAWLEY, M.J. 2012. **The R book.** Chichester, John Wiley, 2nd ed.

DALOSTO, M. & S. SANTOS. 2011. Differences in oxygen consumption and diel activity as adaptations related to microhabitat in Neotropical freshwater decapods (Crustacea). **Comparative Biochemistry and Physiology, Part A 160**: 461-466. doi: 10.1016/j.cbpa.2011.07.026.

FERO, K. & P.A. MOORE. 2008. Social spacing of crayfish in natural habitats: what role does dominance plays? **Behavioral Ecology Sociobiology 62**: 1119-1125. doi: 10.1007/s00265-007-0540-x.

FRANSOZO, A.; R.C. COSTA; A.L.D. REIGADA & J.M. NAKAGAKI. 2003. Population structure of *Aegla castro* Schmitt, 1942 (Crustacea: Anomura: Aeglidae) from Itatinga (SP), Brazil. **Acta Limnologica Brasiliensia 15** (2): 13-20.

GONÇALVES, R.S.; D.S. CASTIGLIONI & G. BOND-BUCKUP. 2006. Ecologia populacional de *Aegla franciscana* (Crustacea, Decapoda, Anomura) em São Francisco de Paula, RS, Brasil. **Iheringia, Série Zoologia, 96** (1): 109-114. doi: 10.1590/S0073-47212006000100019.

GRABOWSKI, R.C.; S. SANTOS & A.L. CASTILHO. 2013. Reproductive ecology and size of sexual maturity in the anomuran crab *Aegla parana* (Decapoda: Aeglidae). **Journal of Crustacean Biology 33** (3): 332-338. doi: 10.1163/1937240X-00002148.

HONAN, J.A. & B.D. MITCHELL. 1995. Reproduction of *Euastacus bispinosus* Clark (Decapoda: Parastacidae), and trends in the reproductive characteristics of freshwater crayûsh. **Marine and Freshwater Research 46**: 485-499. doi: 10.1071/MF9950485.

LÓPEZ-GRECO, L.; V. VIAU; M. LAVOLPE; G. BOND-BUCKUP & E.M. RODRIGUEZ. 2004. Juvenile hatching and maternal care in *Aegla uruguayana* (Anomura, Aeglidae). **Journal of Crustacean Biology 24** (2): 309-313. doi: 10.1651/C-2441.

MARCHIORI, A.B.; M.L. BARTHOLOMEI-SANTOS & S. SANTOS. 2014. Intraspecific variation in *Aegla longirostri* (Crustacea: Decapoda: Anomura) revealed by geometric morphometrics: evidence of an ongoing speciation process. **Biological Journal of the Linnean Society 112** (1): 31-39. doi: 10.1111/bij.12256.

MOORE, P.A. 2007. Agonistic behavior in freshwater crayûsh: the inûuence of intrinsic and extrinsic factors on aggressive encounters and dominance, p. 90-114. *In*: J.E. DUFFY & M. THIEL (Eds). **Evolutionary ecology of social and sexual systems – crustaceans as model organisms.** Oxford, Oxford University Press.

NORO, C.K. & L. BUCKUP. 2002. Biologia reprodutiva e ecologia de *Aegla leptodactyla* Buckup & Rossi (Crustacea, Anomura, Aeglidae). **Revista Brasileira de Zoologia 19** (4): 1063-1074. doi: 10.1590/S0101-81752002000400011.

Nyström, P. 2002. Ecology, p. 192-235. *In*: D.M. Holdich (Ed.). **Biology of Freshwater Crayfish.** Oxford, Blackwell Science.

Oliveira, D. & S. Santos. 2011. Maturidade sexual morfológica de *Aegla platensis* (Crustacea, Decapoda, Anomura) no Lajeado Bonito, norte do estado do Rio Grande do Sul, Brasil. **Iheringia, Série Zoologia, 101** (1-2): 127-130. doi: 10.1590/S0073-47212011000100018.

Palaoro, A.V.; L. Ayres-Peres & S. Santos. 2013. Modulation of male aggressiveness through different communication pathways. **Behavioral Ecology and Sociobiology 67** (2): 283-292. doi: 10.1007/s00265-012-1448-7.

Pérez-Losada, M.; G. Bond-Buckup; C.G. Jara & K.A. Crandall. 2009. Conservation assessment of southern South American freshwater ecoregions on the basis of the distribution and genetic diversity of crabs from the genus *Aegla*. **Conservation Biology 23** (3): 692-702. doi: 10.1111/j.1523-1739.2008.01161.x.

R Core Team. 2013. **R: A language and environment for statistical computing.** Vienna, R Foundation for Statistical Computing.

Rocha, S.S.; R.M. Shimizu & S.L.S. Bueno. 2010. Reproductive biology in females of *Aegla strinatii* (Decapoda: Anomura: Aeglidae). **Journal of Crustacean Biology 30** (4): 589-596. doi: 10.1651/10-3285.1.

Santos, S.; C.G.Jara; M.L. Bartholomei-Santos; M. Pérez-Losada & K.A. Crandall. 2013. New species and records of the genus *Aegla* Leach, 1820 (Crustacea, Anomura, Aeglidae) from the West-Central region of Rio Grande do Sul, Brazil. **Nauplius 21** (2): 211-223.

Schwarzbold, A. 2010. **Ciência & Ambiente n. 41 – Os Rios da América.** Santa Maria, Editora Universidade Federal de Santa Maria.

Siqueira, A.F.; A.V. Palaoro & S. Santos. 2013. Mate preference in the neotropical freshwater crab *Aegla longirostri* (Decapoda: Anomura): does the size matter? **Marine and Freshwater Behaviour and Physiology 46** (4): 219-227. doi: 10.1080/10236244.2013.808832.

Swiech-Ayoub, B.P. & S. Masunari. 2001. Flutuações temporal e espacial de abundância e composição de tamanho de *Aegla castro* Schmitt (Crustacea, Anomura, Aeglidae) no Buraco do Padre, Ponta Grossa, Paraná, Brasil. **Revista Brasileira de Zoologia 18** (3): 1003-1017. doi: 10.1590/S0101-81752001000300032.

Teodósio, E.A.O. & S. Masunari. 2009. Estrutura populacional de *Aegla schmitti* (Crustacea: Anomura: Aeglidae) nos reservatórios dos Mananciais da Serra, Piraquara, Paraná, Brasil. **Zoologia 26** (1): 19-24. doi: 10.1590/S1984-46702009000100004.

Trevisan, A. & S. Santos. 2012. Morphological sexual maturity, sexual dimorphism and heterochely in *Aegla manuinflata* (Anomura). **Journal of Crustacean Biology 32** (4): 519-527. doi:10.1163/193724012X635944.

Trevisan, A.; M.Z. Marochi; M. Costa; S. Santos & S. Masunari. 2012. Sexual dimorphism in *Aegla marginata* (Decapoda: Anomura). **Nauplius 20**: 75-86.

Zar, J. 2010. **Biostatistical analysis.** New Jersey, 5th ed., Prentice Hall.

6

Two new species of *Triplectides* (Trichoptera: Leptoceridae) from South America

Ana Lucia Henriques-Oliveira[1,2] & Leandro Lourenço Dumas[1]

[1] Universidade Federal do Rio de Janeiro, Laboratório de Entomologia, Departamento de Zoologia, Instituto de Biologia, Caixa Postal 68044, Cidade Universitária, 21941-971, Rio de Janeiro, RJ, Brazil.
[2] Corresponding author: anahenri@biologia.ufrj.br

ABSTRACT. *Triplectides*, with about 70 extant species, is the most diverse genus within the Triplectidinae. In the Neotropical Region there are 14 species distributed from southern Mexico to Patagonia. Two new species of *Triplectides* from the Neotropics are described and illustrated based on the male genitalia: *Triplectides cipo* **sp. nov.**, from state of Minas Gerais, southeastern Brazil, and *Triplectides qosqo* **sp. nov.**, from province of Cuzco, southern Peru. The news species can be distinguished by the male genitalia: *Triplectides cipo* **sp. nov.** can be recognized by having the inferior appendages with mesal lobes subacute and apical lobes short, and the tergum X robust, with a subtruncate apex and deep mesal notch; *Triplectides qosqo* **sp. nov.** can be recognized by the first article of inferior appendages long and narrow when compared to the others *Triplectides* species and by the tibial spur formula 2,2,4.

KEY WORDS. Brazil; Caddisflies; new species; Peru; Triplectidinae.

Leptoceridae is one of the most diverse families of caddisflies, with almost 2,000 described species (MORSE 2011). The family is divided into two subfamilies: Leptocerinae Leach, with cosmopolitan distribution, and Triplectidinae Ulmer, which is primarily distributed in the Southern Hemisphere (HOLZENTHAL et al. 2007). *Triplectides* Kolenati, 1859 is the most species-rich genus within Triplectidinae, with about 70 species (HOLZENTHAL 1988, MALM & JOHANSON 2008). The genus occurs in Central and South America, Southern-East Asia (India to Japan), and especially in Oceania, where it reaches its highest diversity, with 15 and 25 species recorded from New Caledonia and Australia, respectively (MALM & JOHANSON 2008).

MOSELY (1936) provided the first comprehensive revision of the genus. Later, the Neotropical species were reviewed by HOLZENTHAL (1988). Since then, only one species has been described from the Neotropics (DUMAS &NESSIMIAN 2010). Currently, there are 14 species described from the Neotropical Region, distributed from Southern Mexico to Southern Chile: *Triplectides chilensis* Holzenthal, 1988 (Argentina and Chile), *T. colombicus* Navás, 1916 (Colombia), *T. egleri* Sattler, 1963 (Brazil, Guyana, and Surinam), *T. flintorum* Holzenthal, 1988 (Colombia, Costa Rica, Ecuador, Guatemala, Honduras, Mexico, Nicaragua, Panama, Peru, and Surinam), *T. gracilis* (Burmeister, 1839) (Argentina, Brazil, Paraguay, and Surinam), *T. itatiaia* Dumas & Nessimian, 2010 (Brazil), *T. jaffuelli* Navás, 1918 (Argentina and Chile), *T. misionensis* Holzenthal, 1988 (Argentina and Brazil), *T. neblinus* Holzenthal, 1988 (Venezuela), *T. neotropicus* Holzenthal, 1988 (Brazil and Venezuela), *T. nevadus* Holzenthal, 1988 (Peru and Venezuela), *T. nigripennis* Mosely, 1936 (Argentina and Chile), *T. tepui* Holzenthal, 1988 (Venezuela), and *T. ultimus* Holzenthal, 1988 (Brazil).

In the present work, we describe and illustrate two new species from South America: *Triplectides cipo* **sp. nov.** from the state of Minas Gerais, southeastern Brazil, and *Triplectides qosqo* **sp. nov.** from province of Cuzco, southern Peru.

MATERIAL AND METHODS

Specimens were collected with malaise and light traps and were preserved in 80-96% ethanol. In order to observe the genital structures, the abdomen was removed and cleared using the lactic acid method (BLAHNIK et al. 2007). The abdomens were mounted on temporary slides with glycerin for viewing and drawing, and transferred back to ethanol and permanently stored in micro vials. Pencil illustrations were made under a stereomicroscope or under a compound microscope, both equipped with a camera lucida. Pencil drawings of genital structures were inked with a technical pen, and wing illustrations were made using vector lines in an Adobe Illustrator (v. 16, Adobe Inc.) document. The terminology used in this paper follows that presented by HOLZENTHAL (1988).

Type specimens are deposited in the following collections, as indicated in descriptions: Museo de Historia Natural "Javier Prado", Universidad Nacional Mayor de San Marcos, Lima (MUSM) and Coleção Entomológica Professor José Alfredo Pinheiro Dutra, Departamento de Zoologia, Universidade Federal do Rio de Janeiro, Rio de Janeiro (DZRJ).

TAXONOMY

Triplectides cipo sp. nov.
Figs. 1-7

Description. Adult male. General color brown (in alcohol). Antennae, palps and legs golden brown. Head and thorax mostly brown. Forewings with forks I and V present in males; discoidal cell apically large. Hind wings broad, with forks I, III, and V present; fork I with distinct petiole. Length of forewing 10.0-11.0 mm, length of hind wing 8.0-9.0 mm (n = 5) (Fig. 1). Tibial spur formula 2,2,4.

Genitalia. Segment IX, in lateral view, narrow with anterior margin almost straight and enlarged dorsally, posterior margin slightly concave medially (Fig. 3); tergum IX with posterior margin almost rounded, slightly protruding laterally, median process apparently absent (Figs. 2 and 3). Preanal appendages slender, digitate, slightly longer than half length of tergum X, bearing long setae (Fig. 2). Tergum X, in lateral view, wide at base, tapering apically, with apex rounded; in dorsal view, slightly widened apically, apex subtruncate, bearing small setae, with apicomesal excision extending anteriorly at half length of segment (Fig. 3). Inferior appendages, long, slightly surpassing tergum X, bearing long setae; 1st article, as viewed laterally, wide at base, constricted medially, with apical portion narrow; apicodorsal lobe club-like, with long setae; basoventral lobes digitate, bearing long setae; in ventral view, mesal lobes shorter than basoventral lobes, wide at base, tapering apically, with flattened aspect, apex subacute; 2nd article short, wide at base, tapering apically, gradually curved inward, with pointed apex (Fig. 5). Phallic apparatus simple, tubular, with phallotremal sclerite small, rod-like, apically positioned (Fig. 6).

Adult female. General color brown (in alcohol). Antennae, palps, and legs golden brown. Thorax and head brown. Length of forewing 13.5-14.5 mm, length of hind wing 10.5-11.5 mm (n = 3). Tibial spur formula 2,2,4.

Genitalia. Sternum VIII, in ventral view, with a sclerotized plate, dark brown; anterior margin deeply concave, posterior margin truncate with a small mesal cleft, with several short setae (Fig. 7). Sternum IX heavily sclerotized, with small transverse striae apically. Appendages of segment X, in lateral view, short, broad at base, subtriangular and setose (Fig. 6). Sensilla-bearing process absent. Valves ventrolateral, well developed, sclerotized, flap-like, slightly concave with fine setae (Fig. 6). Internal vaginal apparatus long, broad, and sclerotized (Fig. 7).

Holotype male: BRAZIL, *Minas Gerais*: Jaboticatubas (Parque Nacional da Serra do Cipó, Córrego das Pedras, 19°22′16.7″S, 48°36′2.8″W, 766 m), 9-13.xii.2011, Malaise trap, APM Santos, DM Takiya, RR Cavichioli & ML Monné *leg.* (DZRJ). Paratypes: same data as holotype, 1 male, 2 females (DZRJ); *Minas Gerais*: Jaboticatubas (Parque Nacional da Serra do Cipó, Córrego das Pedras, 19°22′16.7″S, 48°36′2.8″W, 766 m), 02-05.iii.2013, Mal-

aise trap, BHL Sampaio, BM Camisão, ALH Oliveira, APM Santos & DM Takiya *leg.* (15 males, 6 females) (DZRJ).

Distribution. Brazil (state of Minas Gerais).

Etymology. The specific epithet, *cipo*, refers to the type locality of the species, Parque Nacional da Serra do Cipó, located in Serra do Espinhaço mountain range.

Remarks. *Triplectides cipo* sp. nov. is closely related to *T. flintorum* Holzenthal, 1988 and *T. itatiaia* Dumas & Nessimian, 2010 as evidenced by the wing venation and mesal lobes of inferior appendages. In *T. cipo* the hind wing fork I is petiolate, as in the other two similar species cited above. However, the genital structure of the new species is quite distinct from those of *T. flintorum* and *T. itatiaia*. The mesal lobes of inferior appendages of *T. cipo* are less rounded apically, being subacute. Also, the apical lobes of the inferior appendages of the new species are comparatively shorter than those of *T. flintorum* and *T. itatiaia*. In addition, in *T. cipo* tergum X is robust, with a subtruncate apex and deep mesal notch, whereas in *T. itatiaia* it is rounded apically. *Triplectides flintorum* also has tergum X subtruncate apically but the mesal notch is less deep than in *T. cipo*.

Triplectides qosqo sp. nov.
Figs. 8-12

Description. Adult male. General color brown (in alcohol). Antennae and palps brown. Legs light brown. Forewings with forks I and V present in male; discoidal cell slightly narrower at apex. Hind wings broad, with forks I, III, and V present; fork I with distinct petiole. Length of forewing 12.0-13.0 mm, length of hind wing 9.5-10.0 mm (n = 6) (Fig. 8). Tibial spur formula 2,2,4.

Genitalia. Segment IX, in lateral view, annular, narrow with anterior margin almost straight and enlarged dorsally, posterior margin slightly protruded near dorsum (Fig. 10); tergum IX with posterior margin almost rounded, slightly protruding laterally (Figs. 9 and 10). Preanal appendages rounded and slightly flat, more than half length of tergum X, bearing long setae (Fig. 9). Tergum X, in lateral view, wide at base, anterodorsal area less sclerotized, slightly tapering apically, with apex rounded; in dorsal view, apex subtruncate, slightly narrower than base, bearing small setae, with apicomesal excision extending slightly beyond apical third of segment (Figs. 9 and 10). Inferior appendages long, surpassing tergum X, bearing long setae; 1st article, in lateral view, wide at base, constricted before half length of segment, with apical portion narrow; apicodorsal lobe digitate, with long setae; basoventral lobes digitate, tapering to apex, bearing long setae; in ventral view, mesal lobes shorter than basoventral lobes, wide at base, tapering apically, with pointed apex, bent outward, flattened in lateral view; 2nd article short, wide at base, tapering apically, gradually curved inward, with pointed apex (Figs. 10 and 12). Phallic apparatus simple, tubular, with phallotremal sclerite small, rod-like, mesally positioned (Fig. 11).

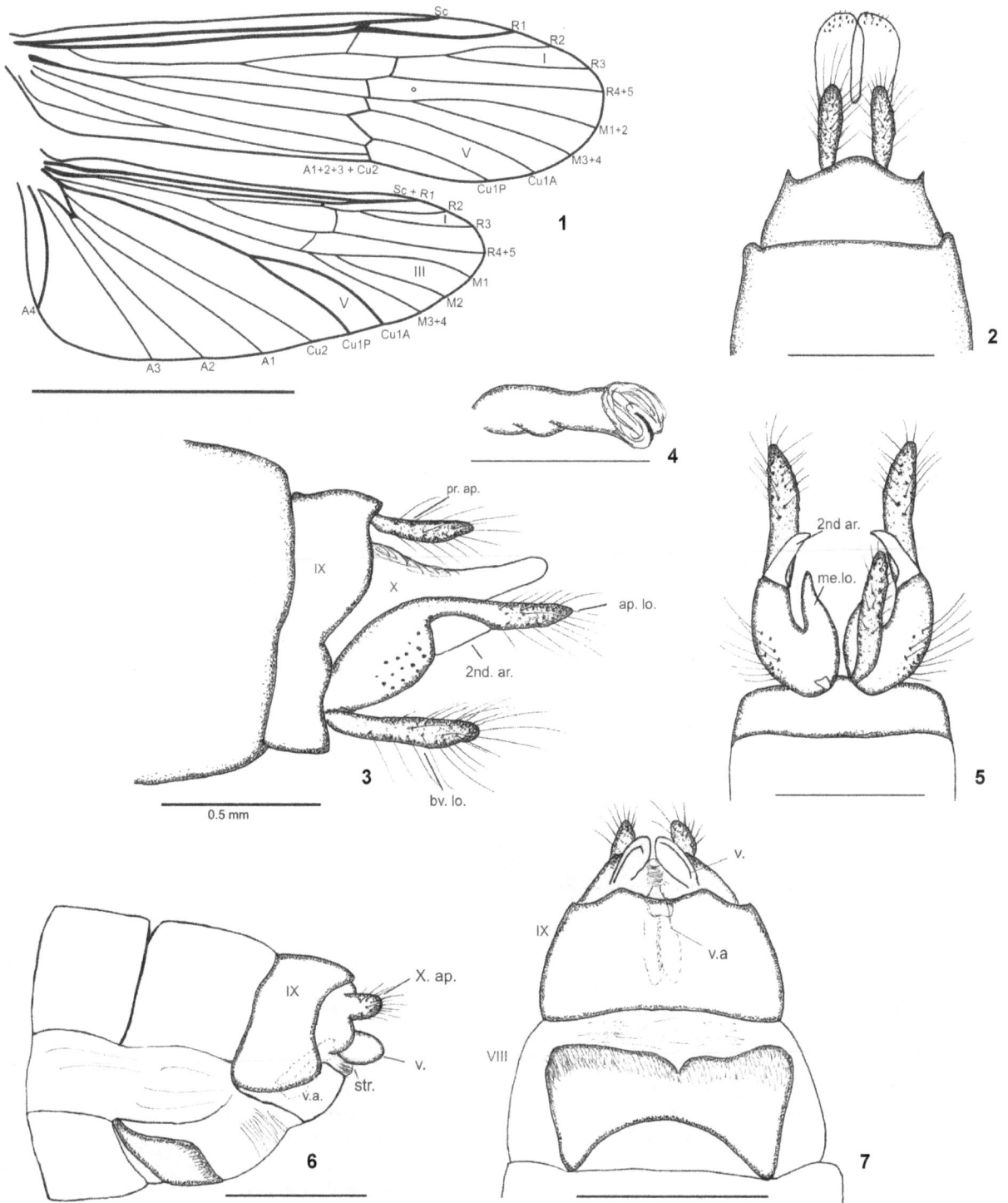

Figures 1-7. *Triplectides cipo* **sp. nov.** (1-5) Male: (1) fore and hind wings; (2) genitalia, dorsal view; (3) genitalia, lateral view; (4) phallic apparatus, lateral view; (5) genitalia, ventral view. (6-7) Female: (6) genitalia, lateral view; (7) genitalia, ventral view. (IX) Tergum IX, (X) Tergum X, (pr. ap.) preanal appendages, (ap. lo.) apicodorsal lobe, (bv. lo.) basoventral lobe, (me. lo.) mesal lobe, (2nd ar.) second article, (X. ap.) appendages of segment X, (v.) valves, (str.) striae, (v.a.) vaginal apparatus. Scale bars: 1 = 5.0 mm, 2-5 = 0.5 mm, 6-7 = 1.0 mm.

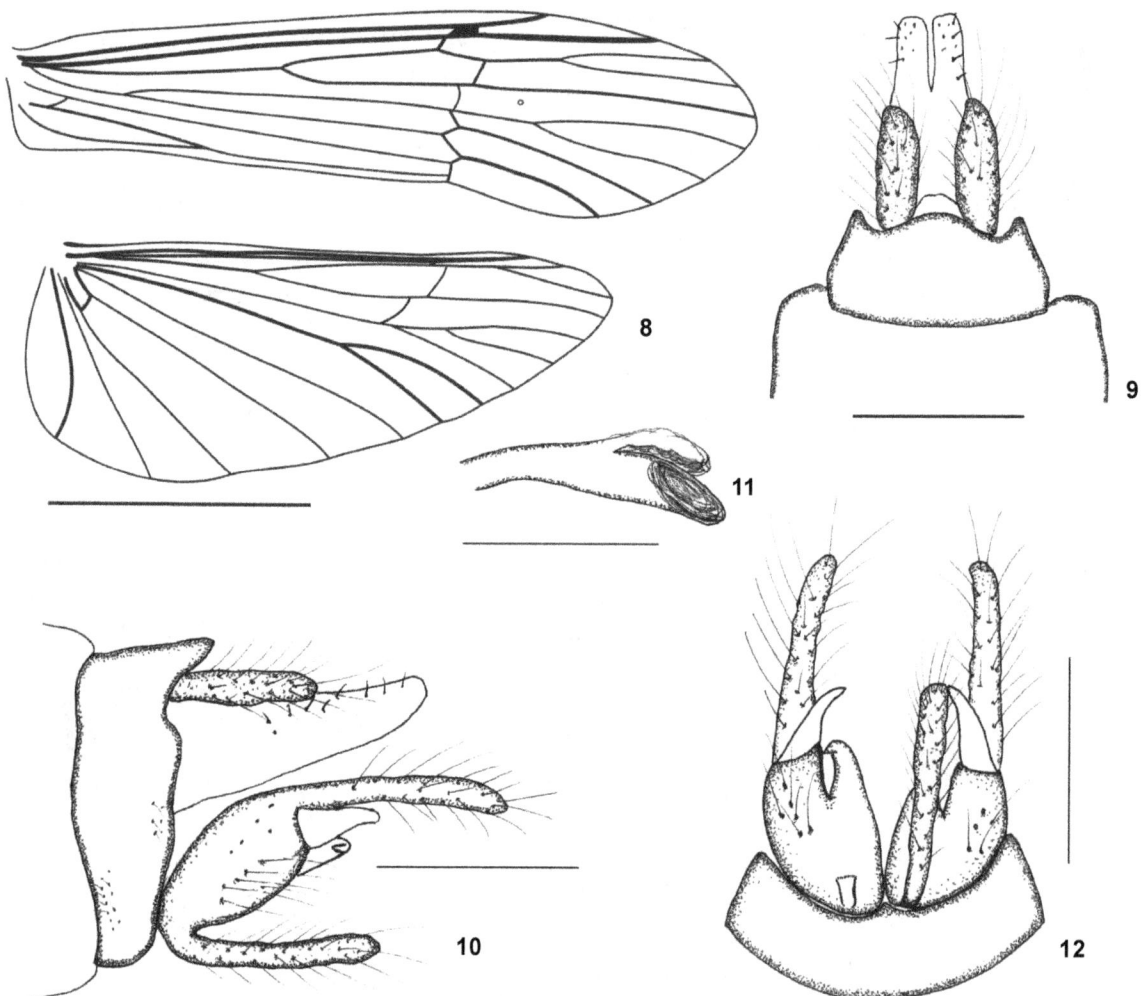

Figures 8-12. *Triplectides qosqo* **sp. nov.**, male: (8) Fore and hind wings; (9) genitalia, dorsal view; (10) genitalia, lateral view; (11) phallic apparatus, lateral view; (12) genitalia, ventral view.

Holotype male: Peru, *Cuzco*: (19rd km W Quincemil, Río Araza tributary, 874 m, 12°20'10"S, 70°50'5"W), Malaise trap, RR Cavichioli, JA Rafael, DM Takiya, APM Santos *leg.* (MUSM). Paratypes: same data as holotype, 1 male (MUSM), 4 males (DZRJ).

Distribution. Peru (province of Cuzco).

Etymology. The specific epithet, *qosqo*, refers to the type locality of the species, province of Cuzco. Cuzco, or Qosqo in Quechua language, is the ancient name for the capital of the great Incan Empire, meaning "navel of the world" in mystical terms.

Remarks. *Triplectides qosqo* **sp. nov.** is similar to *T. nevadus* Holzenthol, 1988, especially in the strongly hooked apices of the mesal lobes of inferior appendages. However, they can be easily separated by the structure of 1st article of inferior appendages, which are much longer and narrower in the new species. Additionally, the tibial spur formula in *T. nevadus* is 0,2,2 or 0,2,4, while in *T. qosqo* it is 2,2,4. The wing venation of the new species is virtually the same of *T. nevadus*.

ACKNOWLEDGEMENTS

We are grateful to Daniela M. Takiya and Allan P. Moreira dos Santos for collecting the specimens described here. The authors are also grateful to two anonymous reviewers for useful suggestions and improvements. Fundação Carlos Chagas Filho de Amparo à Pesquisa do Estado do Rio de Janeiro (FAPERJ) and Coordenação de Aperfeiçoamento de Pessoal de Nível Superior (CAPES) provided financial support. We also thank Instituto Chico Mendes de Conservação da Biodiversidade (ICMBio) for issuing collecting permits.

REFERENCES

BLAHNIK RJ, HOLZENTHAL RW, PRATHER A (2007) The lactic acid method for clearing Trichoptera genitalia, p. 9-14. In: BUENO-SORIA J, BARBA-ALVAREZ R, ARMITAGE B (Eds) **Proceedings of the 12th International Symposium on Trichoptera.** Columbus, The Caddis Press.

DUMAS LL, NESSIMIAN JL (2010) A new long-horned caddisfly in the genus *Triplectides* Kolenati (Trichoptera: Leptoceridae) from the Itatiaia massif, Southeastern Brazil. **Neotropical Entomology** 39(6): 949-951.

HOZENTHAL RW (1988) Sytematics of the Neotropical *Triplectides* (Trichoptera: Leptoceridae). **Annals of the Entomological Society of America** 81(2): 186-208.

HOLZENTHAL RW, BLAHNIK RJ, PRATHER AL, KJER KM (2007) Order Trichoptera Kirby, 1813 (Insecta), Caddisflies. In: ZHANG Z-Q, SHEAR WA (Eds) Linnaeus Tercentenary: Progress in Invertebrate Taxonomy. **Zootaxa 1668:** 639-698.

MALM T, JOHANSON KA (2008) Description of eleven new *Triplectides* species (Trichoptea: Leptoceridae) from New Caledonia. **Zootaxa 1816:** 1-34.

MORSE JC (2011) The Trichoptera world checklist. **Zoosymposia 5:** 372-380.

MOSELY ME (1936) A revision of the Triplectidinae, a subfamily of the Leptoceridae (Trichoptera). **Transactions of the Royal Entomological Society of London 85:** 91-129.

7

Occurrence and morphometrics of the brachioradialis muscle in wild carnivorans (Carnivora: Caniformia, Feliformia)

Paulo de Souza Junior[1,4], Lucas M.R.P. dos Santos[1], Daniele M.P. Nogueira[1], Marcelo Abidu-Figueiredo[2] & André L.Q. Santos[3]

[1]*Laboratório de Anatomia Animal, Universidade Federal do Pampa. Rodovia BR-472, km 585, Caixa postal 118, 97500-970 Uruguaiana, RS, Brazil.*
[2]*Departamento de Biologia Animal, Universidade Federal Rural do Rio de Janeiro. Rodovia BR-465, km 07, 23890-000 Seropédica, RJ, Brazil.*
[3]*Laboratório de Ensino e Pesquisa em Animais Silvestres, Universidade Federal de Uberlândia. Avenida Amazonas 2245, 38405-302 Uberlândia, MG, Brazil.*
[4]*Corresponding author. E-mail:* paulosouza@unipampa.edu.br

ABSTRACT. The brachioradialis is an important muscle that acts in the external rotation of the forearm (supination). However, its occurrence is controversial and little studied in the order Carnivora. Thus, this study investigates the occurrence and anatomo-functional arrangement of this muscle in wild carnivorans species. Fifty-eight thoracic limbs of specimens from species of Canidae, Procyonidae, Mustelidae and Felidae were dissected. Measurements of the length of the muscle (ML), the length of the forearm (FL), latero-medial width of the muscle (MW) and the lateral-medial diameter of the forearm (FD) were obtained to establish the ratios MW/FD and ML/FL in order to investigate the relative proportion of the muscle in relation to the forearm of each species. The brachioradialis muscle was identified in all species, although it was unilaterally or bilaterally absent in some canid individuals. The ratios demonstrated significant differences in the anatomical proportions among the families, with greater functional importance in the mustelids, procyonids, and felids because of a set of elaborate movements in the thoracic limb of representatives of these families when compared to canids.

KEY WORDS. Comparative anatomy; forelimb; myology.

Since thoracic limbs are not only used in locomotion, but also in prey capture and grooming and mating behavior, their morphology can be a good predictor of numerous ecological variables, such as the size and kind of prey, the variety of movements, the role in supporting body mass (ANDERSSON 2004a, FABRE et al. 2013a) and the habitat (DAVIS 1964, EWER et al. 1973, TAYLOR 1989, POLLY 2007, MEACHEN-SAMUELS & VAN-VALKENBURGH 2009, FABRE et al. 2013a, et al. 2013b, MELORO et al. 2013, MARTÍN-SERRA et al. 2014). Together with cranio-dental data, data on thoracic limbs are also used to extrapolate the predatory behavior of extinct species (IWANIUK et al. 1999, ANDERSSON & WERDELIN 2003). Most ecomorphology studies have prioritized the osteological characteristics of the humerus, whereas the shape and other features of the radio-ulnar joint remain largely unstudied (FABRE et al. 2013a, b, 2014). Moreover, the muscular arrangement is rarely taken into account. Knowledge about muscular disposition associated with some biomechanical findings can better elucidate the function of some important bone structures that would otherwise be neglected (JULIK et al. 2012). Several studies in this field have biases because of misunderstanding of the muscular topography. Thus, the addition of quantitative and qualitative data on muscular anatomy would contribute to improve the capacity to characterize forelimb morphology in the context of locomotion, grasping ability and dexterity of the species (IWANIUK et al. 2001, FABRE et al. 2013b). By homology, soft tissue information from extant species can help making well-founded or even speculative inferences about extinct species (WITMER 1995).

The musculoskeletal system forms an arrangement based on levers in which the joints act as fulcra (HERMANSON 2013). The mechanical benefits of its configuration depend on the positions of the muscle attachments (relative to the fulcrum) and the usage of the load. A muscle attached close to a fulcrum is less powerful than a comparable muscle inserted at a greater distance, although the former produces its effects faster. This reflects a conflict between the requirements of speed and power (DYCE et al. 2010).

The brachioradialis muscle (formerly called the supinator longus) usually consists of a narrow muscular band situated at the flexor angle of the humerus-radius-ulna joint (Fig. 1) (BUDRAS et al. 2012, HERMANSON 2013). It is positioned between the superficial and deep layers of antebrachial fascia and adheres to the surface of the deep fascia's leaflet (MILLS 2003, DYCE et al. 2010, HERMANSON 2013), together with the cephalic vein

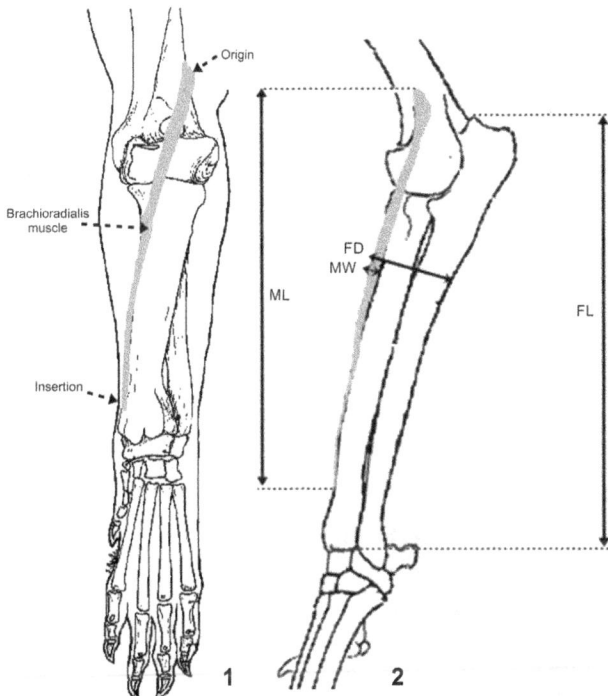

Figures 1-2. (1) Schematic representation of the basic arrangement of brachioradialis muscle in the domestic dog. (2) Schematic representations of the measurement points. ML: muscle length; FL: forearm length; MW: muscle width; FD: forearm diameter.

and the superficial branch of the radial nerve (Saint Clair 1986, Sebastiani & Fishbeck 2005), therefore being the most cranial and superficial muscle of the craniolateral group of the forearm (Bohensky 2002, Sebastiani & Fishbeck 2005).

In domestic carnivorans, this muscle has its origin at the proximal extremity of the humeral lateral supracondylar crest, immediately proximal and superficial to the extensor carpi radialis muscle (Schwarze 1984, Mills 2003, Liebich et al. 2011, Hermanson 2013). The muscle extends cranially over the proximal part of the extensor carpi radialis muscle, crosses the forearm medially, and extends distally in the groove between the extensor carpi radialis muscle and the radius (Schwarze 1984, Liebich et al. 2011, Hermanson 2013). It ends in the periosteum of the radius at the level of the third or fourth distal parts, by a thin aponeurosis (Bohensky 2002, Mills 2003, Budras et al. 2012, Hermanson 2013). Some authors describe its insertion into the styloid process of the radius (Leach 1976, Sebastian & Fishbeck 2005, Liebich et al. 2011, Moore et al. 2013, Ercoli et al. 2014).

The function of the brachioradialis muscle is to perform the craniolateral rotation of the radius (supination) (Bohensky 2002, Sebastiani & Fishbeck 2005, Hermanson 2013).

Supination is a movement of flipping the distal radius over the distal ulna, rotating the radius craniolaterally around its long axis (Andersson 2004b). The movement starts with the

contraction of forearm muscles (supinator and brachioradialis), transmitting external rotation also to the manus. Cursorial mammals often have restricted pronation-supination, whereas scansorial mammals can usually completely supinate the manus (Polly 2007). Thus, the brachioradialis functionally belongs to the group of muscles that act on the radio-ulnar joint, and is expected to be well developed only in carnivorans. In domestic ungulates, this muscle is vestigial or absent due to the reduced or lost capacity of movement between these two bones (Polly 2007, Liebich et al. 2011).

In domestic dogs, the occurrence and antimeric distribution of the brachioradialis muscle has been well documented by Wakuri & Kano (1966), Santos Junior et al. (2002), and Pestana et al. (2009). There are no reports in the literature, as far as we know, about the occurrence of the brachioradialis muscle in wild carnivorans, although there are studies with variable contexts containing references to the presence of the muscle in Carnivora (Davis 1964, Arlamowska-Palider 1970, Julik et al. 2012, Sánchez et al. 2013, Ercoli et al. 2014).

Carnivorans form a successful and functionally diverse clade, with close to 300 living species (Ewer 1973, Wilson & Mittermeier 2009, Hunter 2011). Despite this diversity, the accumulated knowledge of myological variation within the order is still incomplete (Macalister 1873a, Mackintosh 1875, Windle & Parsons 1897, Hall 1926, 1927, Howard 1973, Leach 1976, Fischer et al. 2009, Moore et al. 2013, Ercoli et al. 2014). Hence, these animals provide a good model for this study, as they represent one of the most successful cases of repeated and independent evolution of similar morphologies in a great range of ecologies (Andersson 2004a, b, 2005, Nowak 2005, Sato et al. 2009, 2012, Slater et al. 2012, Fabre et al. 2013a, 2014, Samuels et al. 2013, Martín-Serra et al. 2014). The locomotor range of movements of carnivorans includes, to varying extents, climbing, digging, running and swimming (Andersson & Werdelin 2003). Furthermore, carnivorans species show different degrees of supination and some species cannot even use the forelimbs for grappling with or handling prey (Ewer 1973, Andersson & Werdelin 2003). We hypothesized that the life style of carnivorans, including cursoriality and food procurement strategies, should be reflected in changes in brachioradialis muscle arrangement. This muscle is expected to be frequently found and to be relatively larger in species that need to rotate the forelimbs. Thereby, the aim of this study is to verify the occurrence, anatomo-functional arrangement and sexual dimorphism of the brachioradialis muscle in wild carnivoran species, thus contributing to studies in ecomorphology.

MATERIAL AND METHODS

This study was carried out with 29 carnivorans cadavers of Canidae: *Cerdocyon thous* (Linnaeus, 1766) (three males and six females) and *Lycalopex gymnocercus* (G. Fischer, 1814) (four males and one female); Mustelidae: *Galictis cuja* (Molina, 1782) (three males and two females) and *Lontra longicaudis* (Olfers, 1818) (one female); Procyonidae: *Procyon cancrivorus* (G.[Baron] Cuvier, 1798) (one male and two females) and *Nasua nasua* (Linnaeus, 1766) (one female); and Felidae: *Leopardus geoffroyi* (d'Orbigny & Gervais, 1844) (four females) and *Leopardus colocolo* (Molina, 1782) (one female). These specimens were collected dead from highways in the southwest region of the state of Rio Grande do Sul (Pampa biome) between July 2012 and November 2013 (IBAMA/SISBIO authorization number 33667-1). Only adult individuals were included, based on inspection of permanent dentition.

After collection, the specimens were fixed in a formaldehyde solution (50%) and conserved in opaque polyethylene tanks with the same solution at 10% for at least 14 days, until they were dissected. The skin and fascia of the thoracic limbs were carefully removed and the superficial intrinsic muscles identified. The brachioradialis muscle, whenever present, was dissected until its origin and insertion were exposed. Then measurements were taken (Fig. 2) of the length of the brachioradialis muscle from its origin until its insertion (ML) and the length of the forearm from the olecranon tuberosity until the radiocarpian joint (FL). Also, lateral-medial width of the brachioradialis muscle (MW) and the lateral-medial diameter of the forearm (FD) were obtained at the level of their middle thirds. The measurements were performed by a single examiner using a digital pachymeter (resolution 0.01mm, accuracy ±0.02 mm, ZAAS Precision Amatools®). Thereafter, two ratios were calculated: MW/FD and ML/FL. The MW/FD ratio was calculated to reflect the relative proportion of the functional participation of the brachioradialis muscle in the forearm region of the specimens. The ML/FL ratio represents the proportion of muscle length in relation to the forearm length and can be associated with added speed during contraction. The 29 specimens are deposited in the Laboratory of Animal Anatomy of the Universidade Federal do Pampa, Uruguaiana, RS, Brazil. The deposit numbers of the specimens analyzed are available in the appendix.

Descriptive statistical data (mean, standard deviation, variance and coefficient of variation) were calculated. The ratios were compared among species and families using analysis of variance (one-way ANOVA) and significant differences between the means were determined by using the Tukey test at 99% probability. In species with enough samples for comparison of the ratios between genders (*C. thous*, *G. cuja* and *P. cancrivorus*), the t-test was performed at 99% probability. These tests were executed by the BioEstat 5.3® program. Photomacrographs were taken with a Sony Cybershot DSC-TF1® camera with 16.1 MP and the images were treated with the Photoscape® v.3.5 software.

RESULTS

Except for two specimens of *L. gymnocercus*, in which one male only had the muscle on the right antimere and a female only on the left, along with a female of *C. thous* that did not

have the muscle in any antimere, all the other specimens (26) had the brachioradialis muscle in both antimeres. Therefore, among all 58 thoracic limbs analyzed, 54 presented the muscle (Table I).

The results of the MW/FD and ML/FL ratios grouped by family (Canidae, Mustelidae, Procyonidae and Felidae) are presented in Table II and in Figs. 3 and 4.

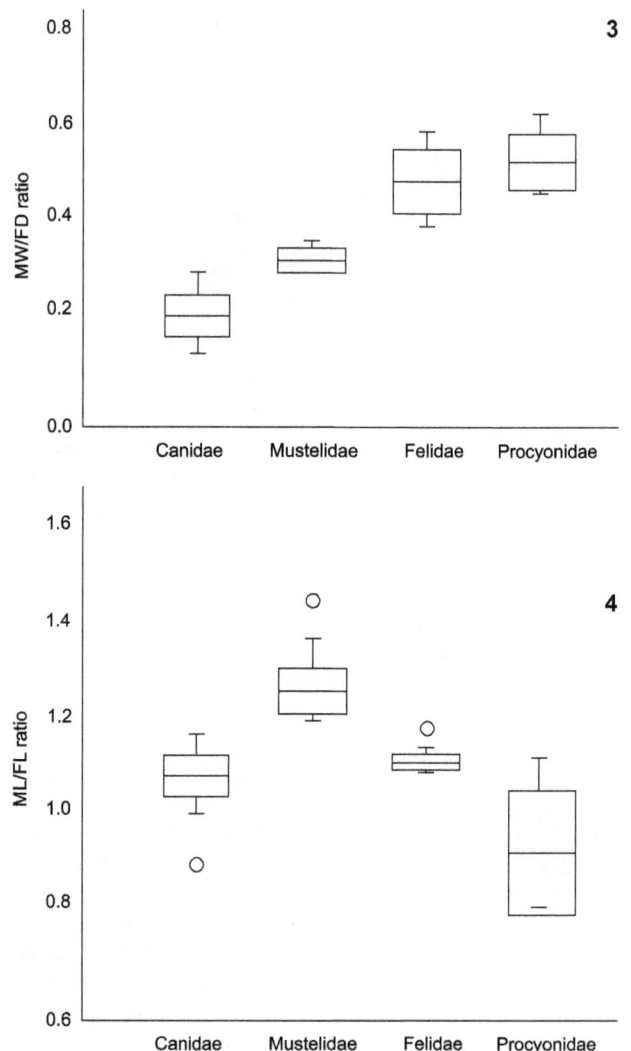

Figures 3-4. Box-plots showing the mean ± SD of the MW/FD ratio (3) and ML/FL ratio (4) grouped by families of carnivorans analyzed.

In canids, felids and procyonids, the brachioradialis muscle originated on the lateral supracondylar crest of the humerus and its insertion was on the medial surface of the radial distal extremity (medial styloid process) in every specimen analyzed (Figs. 5-12), except one male specimen of *L. gymnocercus*, in which

Occurrence and morphometrics of the brachioradialis muscle in wild carnivorans...

43

Table I. Means and standard deviations (mm) of the measurements and MW/FD and ML/FL ratios obtained from the forelimbs (n = 54) of carnivorans specimens that presented the brachioradialis muscle.

Family	Species	n	Male	Female	ML	MW	FL	FD	MW/FD Ratio	ML/FL Ratio
Canidae	L. gymnocercus	8	7	1	125.65 ± 8.05	3.46 ± 0.98	123.72 ± 3.59	22.54 ± 1.77	0.15 ± 0.04 [d]	1.01 ± 0.06 [c]
Canidae	C. thous	16	6	10	127.43 ± 11.7	4.93 ± 0.77	117.42 ± 11.94	24.60 ± 2.49	0.20 ± 0.03 [cd]	1.08 ± 0.04 [bc]
Mustelidae	L. longicaudis	2	0	2	92.69 ± 0.75	11.66 ± 0.24	76.87 ± 0.71	40.43 ± 0.60	0.29 ± 0.00 [bc]	1.21 ± 0.21 [ab]
Mustelidae	G.cuja	10	6	4	63.81 ± 2.65	6.83 ± 0.53	50.13 ± 3.59	22.28 ± 1.99	0.31 ± 0.02 [b]	1.27 ± 0.07 [a]
Felidae	L. geoffroyi	8	0	8	112.32 ± 3.60	11.63 ± 1.65	102.39 ± 3.38	25.63 ± 1.73	0.45 ± 0.06 [a]	1.09 ± 0.01 [bc]
Procyonidae	N. nasua	2	0	2	110.94 ± 0.49	15.73 ± 0.46	100.52 ± 0.82	32.89 ± 0.42	0.48 ± 0.02 [a]	1.11 ± 0.01 [bc]
Procyonidae	P. cancrivorus	6	2	4	112.78 ± 4.64	13.72 ± 1.47	134.54 ± 7.88	25.97 ± 0.73	0.53 ± 0.06 [a]	0.84 ± 0.05 [d]
Felidae	L. colocolo	2	0	2	123.59 ± 1.31	11.99 ± 0.07	107.61 ± 1.43	21.79 ± 0.30	0.55 ± 0.04 [a]	1.15 ± 0.03 [abc]

Values followed by different letters in the same column show statistically significant differences according to the Tukey test (p < 0.01). Brachioradialis muscle length (ML); brachioradialis muscle width (MW); length of the forearm (FL); diameter of the forearm (FD).

Table II. Descriptive statistics regarding the MW/FD and ML/FL ratios obtained for the thoracic limbs (n = 54) of carnivorans specimens grouped by families (SD) Standard deviation, (CV) Coefficient of variation.

Family	MW/FD ratio						ML/FL ratio					
	n	Mean	Variance	SD	CV (%)	Minimum	Maximum	n	Mean	Variance	SD	CV (%)	Minimum	Maximum
Procyonidae	8	0.51[a]	0.0036	0.06	11.62	0.45	0.62	12	1.27 [a]	0.0050	0.07	5.62	1.19	1.44
Felidae	10	0.47[a]	0.0048	0.07	14.58	0.38	0.58	10	1.11 [b]	0.0008	0.02	2.48	1.08	1.17
Mustelidae	12	0.30[b]	0.0005	0.02	7.44	0.28	0.35	24	1.06 [b]	0.0031	0.05	5.28	0.88	1.16
Canidae	24	0.18[c]	0.0020	0.45	24.22	0.11	0.28	8	0.91 [c]	0.0171	0.13	14.38	0.79	1.11

Values followed by different letters in the same column show statistically significant differences according the Tukey test (p < 0.01).

the right brachioradialis muscle was shortened and joined to the middle third of the extensor carpi radialis muscle. In mustelids, the muscle originated from the caudal surface of the humeral neck and caudomedially to the brachial muscle point of origin, and also was inserted in the styloid process of the radius, being very fleshy in its course. In procyonids, especially in P. cancrivorus, the muscular part was restricted until the middle third of the forearm, where it narrowed into a thin insertion tendon (Fig. 7). In every specimen it was the most superficial muscle in the forearm, transiting near the cephalic vein, and was innervated by branches of the radial nerve.

The t-test (p < 0.01) for comparison of means of the MW/FD ratios revealed similarity between genders in C. thous (p = 0.2219), G. cuja (p = 0.7273) and P. cancrivorus (p = 0.0986). The same test for comparison of means of the ML/FL ratios revealed similarity between genders in C. thous (p = 0.4026) and G. cuja (p = 0.1743) and difference in P. cancrivorus (p = 0.0024).

DISCUSSION

The presence in nearly all the specimens assessed in this study reflects the functional relevance of the brachioradialis muscle to the order Carnivora. Its occurrence is expected in species that require significant mobility in the radius-ulnar joint, especially in external rotation (supination) of the hand

(paw) and forearm, which does not happen, for instance, in ungulates (NICKEL et al. 1986, LIEBICH et al. 2011). According to SALADIN (2010), the brachioradialis muscle also acts as a synergist in the flexion of the humerus-radio-ulnar joint, but by itself it is not able to generate enough strength because its insertion is far from the fulcrum.

Only among canids were individuals identified that did not present the muscle unilaterally or bilaterally (one C. thous and two L. gymnocercus). Besides this, in one male specimen of L. gymnocercus the brachioradialis muscle was not long enough to be inserted in the radius, joining the extensor carpi radialis muscle in a clearly accessory position. The absence or presence, unilaterally or bilaterally, of the brachioradialis muscle has also been reported in C. familiaris by WAKURI & KANO (1966), SANTOS JUNIOR et al. (2002) and PESTANA et al. (2009). Considered together, these three studies reveal that the brachioradialis muscle is absent in 46 to 62% of dogs, appearing unilaterally in 15 to 20% of cases. In running dogs of the greyhound breed, the muscle was not found in ten individuals dissected by WILLIAMS et al. (2008). These findings corroborate the well-recognized observation that canids have lost some of the ability to supinate their manus (EWER 1973, ANDERSSON & WERDELIN 2003). Therefore, in a superficial analysis considering the canids, the brachioradialis muscle seems to be more frequent in wild ones than in domestic ones, perhaps because

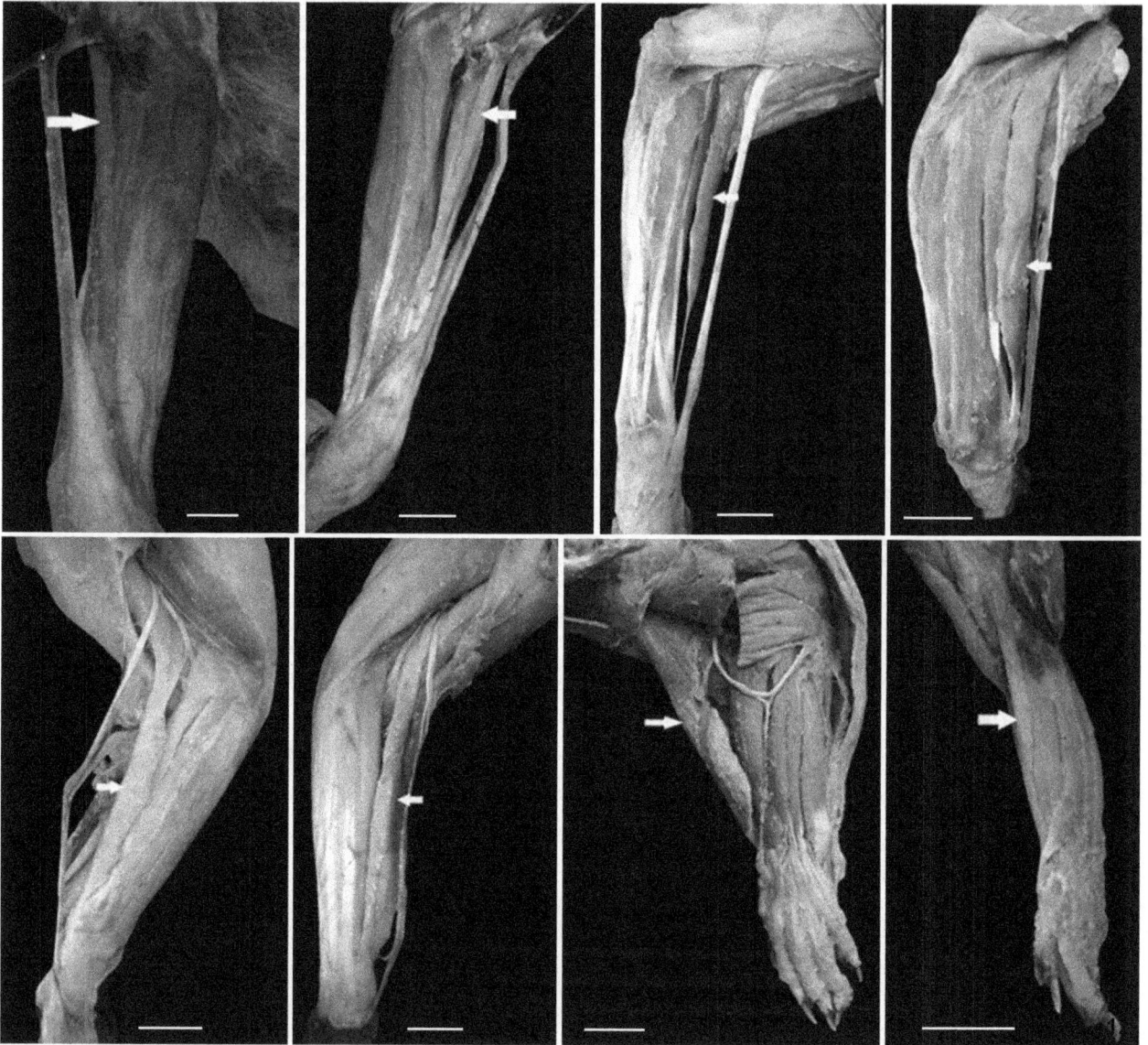

Figures 5-12. Photomacrographs showing the comparative anatomic arrangement of the brachioradialis muscle (white arrow) in the forearm region of: (5) *Cerdocyon thous*, right forearm; (6) *Lycalopex gymnocercus*, left forearm; (7) *Procyon cancrivorus*, right forearm; (8) *Nasua nasua*, right of forearm; (9) *Leopardus geoffroyi*, left forearm; (10) *Leopardus colocolo*, right forearm; (11) *Lontra longicaudis*, left forearm; (12) *Galictis cuja*, left forearm. Scale bars: 20 mm.

the former animals retain a greater need to perform supination movements to capture prey. Nevertheless, the unilateral and even bilateral absence of this muscle in some individuals suggests that it may no longer perform a useful function and instead is a rudimentary muscle that is on its way out. The apparent absence of this muscle in greyhound dogs could reflect the increased intensity of artificial selection to eliminate a useless muscle, since it can interfere with a desirable running ability in wild canids.

After examining some specimens, W.J. Gonyea (unpubl. data) noted that the brachioradialis muscle was present in arboreal and fossorial taxa and absent in cursorial ones. In support of this hypothesis, the muscle was not identified in the cursorial canids *Canis latrans* (Say, 1823) (n = 1), *Urocyon cinereoargenteus* (Schreber, 1775) (n = 4) and *Vulpes vulpes* (Linnaeus, 1758) (n = 5) (S.A. Feeney unpubl. data). The appearance of the brachioradialis muscle in almost all cursorial canids *C. thous* (n = 8/9) and *L. gymnocercus* (n = 5/5) analyzed in this study contrasts with previous observation (W.J. Gonyea, unpubl. data). Until there are more investigations, we can propose that the presence of this muscle in canids may be a characteristic shared by Neotropical species with close phylogenetic relationship rather than just a morphofunctional issue. Supporting this hypothesis, the presence of the brachioradialis muscle was also mentioned by Vaz et al. (2011) in an adult female *Atelocynus microtis* (Sclater, 1883) and in an adult male *C. thous*. Indeed, *C. thous*, *L. gymnocercus* and *A. microtis* inhabit the Neotropics and have the same number of chromosomes (74) (Pessutti et al. 2001), unlike *C. latrans*, *U. cinereoargenteus* and *V. vulpes* (S.A. Feeney, unpubl. data).

Despite disagreement over the cursorial canids, observations of the presence of the muscle in arboreal and fossorial carnivorans are compatible with the findings of this study. In fact, the mustelids, procyonids and felids analyzed here have scansorial, arboreal and fossorial habits (Reis et al. 2010, Hunter 2011). Mustelids and procyonids in particular can display great ability to perform different kinds of movements such as grasping, swimming and food manipulation (McClearn 1992, Iwaniuk et al. 1999, Fabre et al. 2013b), and their lack of cursorial adaptation means that none of them have lost the ability to supinate their paws, in contrast to other carnivorans such as canids and some hyaenids (Iwaniuk et al. 1999, Polly 2007, Fabre et al. 2014). Although with different purposes and small samples, some other studies have mentioned the occurrence of this muscle in species of these families, for instance in the mustelids *Aonyx cinerea* (Illiger, 1815) (Macalister 1873b), *Lutra lutra* (Linnaeus, 1758) (Windle & Parsons 1897), *Enhydra lutris* (Linnaeus, 1758) (Howard 1973), *Martes pennanti* (Erxleben, 1777) (n = 1) (S.A. Feeney, unpubl. data) and *G. cuja* (n = 3) (Ercoli et al. 2014); in the procyonid *Procyon lotor* (Linnaeus, 1758) (n = 2) (S.A. Feeney, unpubl. data); and in the felids *Puma concolor* (Linnaeus, 1771) (n = 2) (Concha et al. 2004), *Acinonyx jubatus* (Schreber, 1775) (n = 8) (Hudson et al. 2011), *Leopardus pardalis* (Linnaeus, 1758) (n = 1) (Julik et al. 2012) and *Panthera onca* (Linnaeus, 1758) (n = 2) (Sánchez et al. 2013).

The superficial location in the forearm, the proximity of the cephalic vein and innervation by branches of the radial nerve were common to all specimens analyzed, resembling the description of domestic carnivores (Sebastiani & Fishbeck 2005, Budras et al. 2012).

The MW/FD ratio was calculated to reflect the relative proportion of its participation in the forearm region of the specimens. This way, when comparing species and/or families, those with the highest ratios should be the ones in which the brachioradialis muscle has greater relative contribution through the group of antebrachium muscles. Although calculation of physiological cross-section area (PCSA) has been used to estimate the maximum isometric force of muscles (Williams et al. 2008), in this study we considered this determination to be less important since it generates an absolute value which is highly influenced by the body size of the individuals. For ex-

ample, the brachioradialis of a crab-eating fox (*C. thous*) would exhibit both higher absolute PCSA and maximum isometric force values than the muscle in a lesser grison (*G. cuja*). However, the muscle is proportionally weaker in the former. Since the brachioradialis has parallel fibers, determination of the ratio between cross section widths (MW) to the forearm diameter would give a more proportional estimate of functional relevance, at least in its topographic region (forearm). In fact, the results of MW/FD ratio reflected observations from the usage of the forelimbs in each species.

The significantly lower MW/FD ratio in canids compared to the other three families used in this study can be explained by the fact that canids are essentially terrestrial and have developed a highly specialized lifestyle among carnivorans. In more cursorial taxa, the functional adaptations prioritize, among other aspects, the movements of the limbs in the sagittal plane, disfavoring supination or pronation (Ewer 1973). In contrast to felids, canids have a limited ability to subdue and grapple with other animals by using their forelimbs. Instead they engage in sustained pursuit predation, an activity whose success depends on the number of animals participating in the hunt (Andersson 2005). In addition, canids rarely climb or manipulate prey to a higher extent. These habits are correlated to a less functional, or even absent, brachioradialis muscle.

In the mustelids *G. cuja* and *L. longicaudis* the MW/FD ratio, significantly higher only than in the canids, comparatively reflects greater recruitment of the muscle to help in specific swimming movements and fossorial habits. At times, they also use their thoracic limbs to drag prey out of the water (Reis et al. 2010).

The felids and procyonids of this study showed the highest MW/FD ratios, with no differences among them. The repertoire of manual movements of these two families is more complex because, besides being fast runners, they are also able to swim, climb trees and remain balanced at tall heights (Reis et al. 2010, Hunter 2011). Undeniably, scansorial habits and grasping requires accurate three-dimensional movements, which demand morphofunctional adaptations and higher recruitment and precision in muscular contraction (Ewer 1973, Fabre et al. 2013b). Procyonids even use their hands to precisely bring food into their mouths, requiring a greater capacity for supination (Paranaíba et al. 2012). Species showing well-developed grasping ability potentially have a wide range of pronation-supination movements, which can confer greater mobility to the forearm and the hand (Fabre et al. 2013b). This was the case of *P. cancrivorus*, which exhibited a high MW/FD ratio. In felids, essentially carnivores, rotation movements are even more necessary to capture, overwhelm and manipulate prey extensively (Hudson et al. 2011).

These functional correlations are coherent with those previously proposed for domestic carnivorans. According to Saint Clair (1986), the distal muscles of the thoracic limbs are more developed in cats than in dogs to assure the greater rotation of the distal portion of the limb. The brachioradialis muscle presents a correlation not only with the forelimb usage in carnivorans. ANDERSSON (2004b) stated that manipulation and locomotion are conflicting functions, since elbow-joint morphology supports a division between grapplers (i.e., ambushers) and nongrapplers (i.e., pursuers). Joints of the former are relatively wide, while in the latter they are relatively narrow and box-like with pronounced stabilizing features. Concerning forepaw dexterity, IWANIUK et al. (2001) considered that manus and carpal shape and myology may play a more critical role than manus proportions. According to them, behavioral observations also suggest that manus proportions correlated more closely with locomotion than non-locomotory forepaw usage.

The ML/FL ratio represents the proportion of muscle length in relation to the forearm length. This ratio was significantly higher in mustelids. Longer muscles, especially with parallel fibers (which is the case of the brachioradialis), have more sarcomeres in series, which means added speed during contraction (KARDONG 2011). Therefore, in the mustelids one can assume that the brachioradialis muscle has a higher shortening speed than in the other families, an aspect that may be functionally important during swimming. This relatively longer length is a reflection of a more proximal level of origin in the humerus than in the other families, in other words, on the caudal surface of the humeral neck instead of the lateral supracondylar crest. Strengthening this hypothesis, in procyonids the ratio was significantly lower, because these animals have the slowest movements among the analyzed families. This shorter length was determined by the fact that its muscular part extends only until the forearm's middle third. This trait was also verified in two specimens of *P. lotor* (S.A. Feeney, unpubl. data).

From a mechanical point of view, it can be supposed that the arrangement of the mustelid brachioradialis muscle raises the power leverage, creating a low power ratio, which increases the strength and becomes important in fossorial habits. In felids, in contrast, the origin at a more distal level makes the power leverage lower, creating a higher power ratio, which increases speed, an important aspect for chasing prey. This confirms the findings of HUDSON et al. (2011) for *A. jubatus*, that the internal architecture with long fibers of the brachioradialis muscle is an adaptation that allows the muscle to contract at high speeds and extensively rotate the radio-ulnar joint.

In the mustelids of the subfamily Lutrinae *A. cinerea*, *L. lutra* and *E. lutris* (MACALISTER 1873b, WINDLE & PARSONS 1897, HOWARD 1973), the brachioradialis muscle has its origin proximal to the humeral diaphysis, which also happened with *L. longicaudis*, a member of the same subfamily analyzed in this study. The origin at a proximal level could be a synapomorphy of Lutrinae. In the *Lontra canadensis* (Schreber, 1777), a representative species of the most basal lineage of otters, FISHER (1942) described the brachioradialis muscle as originating from the proximal region of the humerus and running to the lateral su-

pracondylar crest. From an evolutionary perspective, this condition may reflect an intermediate position between the mustelids of the ferret type (Mustelinae) and lineages like otters (Lutrinae). Based on Bayesian inference methods, KOEPFLI et al. (2008) stated these two subfamilies diverged in the late Miocene (10 MYA) during the first burst of diversification among Mustelidae.

Among the mustelids of Mustelinae, in the five specimens of ferrets (*G. cuja*) dissected in this study, the muscle extended from the proximal part of the humerus to the lateral supracondylar ridge. However, in six ferrets of the same species dissected by ERCOLI et al. (2014), the muscle had proximal origin in one-half (similar to the findings of this study) and in the other half the origin was restricted to the supracondylar crest. In four individuals of the species *M. pennanti*, the origin of the muscle occurred just proximally to the lateral humeral supracondylar ridge (S.A. Feeney, unpubl. data).

In fact, a great number of morphologic characteristics shared between the subfamilies Lutrinae and Mustelinae can be understood as favorable for both aquatic habits and locomotion in tunnels (ERCOLI et al. 2014). This peculiar arrangement, elongated and wide, of the brachioradialis muscle in this family reinforces this observation. Furthermore, a more proximal origin means the levers are more equilibrated. Thus it can be assumed that the muscle also acts as an important flexor of the humerus radio-ulnar joint.

SCHWARZE (1984) and DYCE et al. (2010) reported that the muscle is small and almost never identified in dogs, especially in small ones. In *L. gymnocercus*, a canid slightly smaller than *C. thous*, the averages of the MW/FD and ML/FL ratios were also lower (significantly at 95% probability). However, the body size should not be a condition to predict the occurrence or functional relevance of the brachioradialis muscle in carnivorans, since small specimens such as *G. cuja* and felines exhibited a bilaterally well developed muscle.

In species with enough samples for comparison of the ratios between genders (*C. thous*, *G. cuja* and *P. cancrivorus*), only ML/FL ratio was significantly lower ($p = 0.024$) in female (0.81) than male (0.92) *P. cancrivorus*. Though it would be desirable to study a larger number of specimens, this may reflect the need for more developed skills for pray chasing in males.

Finally, the study allowed establishing that the brachioradialis muscle occurs in individuals of the eight carnivorous species analyzed. This muscle originated on the lateral supracondylar crest of the humerus and was inserted in the medial surface of the radial distal extremity in canids, felids and procyonids, while its origin was on the caudal humeral neck in mustelids. Signs of sexual dimorphism were only detected in the relative length of the muscle in male *P. cancrivorus*. Overall, the mustelids, procyonids and felids have a proportionally more developed muscle than canids. As expected, these findings are consistent with the complexity and diversity of movements executed by the forelimbs of these species in the wild.

ACKNOWLEGMENTS

We acknowledge grants received from the Scholarships for Academic Development Program (PBDA) of Universidade Federal do Pampa and from the Young Talents for Science Program awarded to the second and third authors, respectively.

REFERENCES

Andersson K (2004a) Predicting carnivoran body mass from a weight-bearing joint. **Journal of Zoology** 262(2): 161-172. doi:10.1017/S0952836903004564

Andersson K (2004b) Elbow-joint morphology as a guide to forearm function and foraging behaviour in mammalian carnivores. **Zoological Journal of the Linnean Society** 142: 91-104. doi: 10.1111/j.1096-3642.2004.00129.x

Andersson K (2005) Were there pack-hunting canids in the Tertiary, and how can we know? **Paleobiology** 31(1): 56-72. doi:10.1666/0094-8373(2005)031<0056:WTPCIT>2.0.CO;2

Andersson K, Werdelin L (2003) The evolution of cursorial carnivores in the Tertiary: implications of elbow joint morphology. **Proceeding of the Biological Society** 270 (Suppl.): S163-S165. doi:10.1098/rsbl.2003.0070

Arlamowska-Palider A (1970) Morphological Studies on the Main Branches of the Radial Nerve in Mammals. **Acta Theriologica** 15(2): 185-197.

Bohensky F (2002) **Photo manual and dissection Guide of the Cat.** New York, Square One Publishers, 167p.

Budras KD, McCarthy PH, Fricke W, Richter R, Horowitz A, Berg R (2012) **Anatomia do Cão, Texto e Atlas.** Barueri, Manole, 219p.

Concha I, Adaro L, Borroni C, Altamirano C (2004) Consideraciones anatómicas sobre la musculatura intrínseca del miembro torácico del puma (*Puma concolor*). **International Journal of Morphology** 22(2): 121-125. doi: 10.4067/S0717-95022004000200004

Davis DD (1964) The giant panda: a morphological study of evolutionary mechanisms. **Fieldiana: Zoology Memors 3:** 1-339.

Dyce KM, Sack WO, Wensing CJG (2010) **Tratado de Anatomia Veterinária.** São Paulo, Elsevier, 856p.

Ercoli MD, Álvarez A, Stefanini MI, Busker F, Morales MM (2014) Muscular Anatomy of the Forelimbs of the Lesser Grison (*Galictis cuja*), and a Functional and Phylogenetic Overview of Mustelidae and Other Caniformia. **Journal of Mammalian Evolution** 22(1): 57-91. doi: 10.1007/s10914-014-9257-6

Ewer RF (1973) **The Carnivores.** New York, Cornell University Press, 494p.

Fabre AC, Cornette R, Peigné S, Goswami A (2013a) Influence of body mass on the shape of forelimb in musteloid carnivorans. **Biological Journal of the Linnean Society** 110: 91-103. doi: 10.1111/bij.12103

Fabre AC, Cornette R, Slater G, Argot C, Peigné S, Goswami A, Pouydebat E (2013b) Getting a grip on the evolution of grasping in musteloid carnivorans: a three-dimensional analysis of forelimb shape. **Journal of Evolutionary Biology** 26: 1521-1535. doi: 10.1111/jeb.12161

Fabre AC, Goswami A, Peigné S, Cornette R (2014) Morphological integration in the forelimb of musteloid carnivorans. **Journal of Anatomy** 225: 19-30. doi: 10.1111/joa.12194

Fisher EM (1942) **The Osteology and Myology of the California River Otter.** Stanford, Stanford University Press, 66p.

Fisher RE, Adrian B, Barton M, Holmgren J, Tang SY (2009) The phylogeny of the red panda (Ailurus fulgens): evidence from the forelimb. **Journal of Anatomy 215** (6): 611-635. doi:10.1111/j.1469-7580.2009.01156.x

Hall ER (1926). The muscular anatomy of three mustelid mammals, Mephitis, Spilogale, and Martes. **University of California Publications in Zoology** 30(2): 7-39.

Hall ER (1927) The muscular anatomy of the American badger (Taxidea taxus). **University of California Publications in Zoology** 30(8): 205-219.

Hermanson JW (2013) The Muscular System, p. 185-280. In: Evans HE, DeLahunta A (Eds) **Miller's Anatomy of the Dog.** Missouri, Elsevier, 872p.

Howard LD (1973) Muscular anatomy of the fore-limb of the sea otter (*Enhydra lutris*). **Proceedings of the California Academy of Science** 39(4): 411-500.

Hudson PE, Corr SA, Payne-Davis RC, Clancy SN, Lane E, Wilson AM (2011) Functional anatomy of the cheetah (*Acinonyx jubatus*) forelimb. **Journal of Anatomy** 218(4): 375-385. doi: 10.1111/j.1469-7580.2011.01344.x

Hunter L (2011) **Carnivores of the world.** Princeton, Princeton University Press, 240p.

Iwaniuk AN, Pellis SM, Whishaw IQ (1999) The relationship between forelimb morphology and behaviour in North American carnivores (Carnivora). **Canadian Journal of Zoology** 77(7): 1064-1074. doi: 10.1139/z99-082

Iwaniuk AN, Pellis SM, Whishaw IQ (2001) Are long digits correlated with high forepaw dexterity? A comparative test in terrestrial carnivores (Carnivora). **Canadian Journal of Zoology** 79(5): 900-906. doi: 10.1139/z01-058

Julik E, Zack S, Adrian B, Maredia S, Parsa A, Poole M, Starbuck A, Fisher RE (2012) Functional Anatomy of the Forelimb Muscles of the Ocelot (*Leopardus pardalis*). **Journal of Mammalian Evolution** 19(4): 277-304. doi:10.1007/s10914-012-9191-4

Kardong KV (2011) **Vertebrados: anatomia comparada, função e evolução.** São Paulo, Roca, 913p.

Koepfli KP, Deere KA, Slater GJ, Begg C, Begg K, Grassman L, Lucherini M, Veron G, Wayne RK (2008) Multigene phylogeny of the Mustelidae: resolving relationships, tempo and biogeographic history of a mammalian adaptive radiation. **BMC Biology** 6(10): 122. doi: 10.1186/1741-7007-6-10

Leach D (1976) The forelimb musculature of marten (Martes americana Turton) and fisher (Martes pennanti Erxleben). **Canadian Journal of Zoology** 55(1): 31-41. doi:10.1139/z77-00

Liebich HG, Maierl J, König HE (2011) Membros Torácicos ou Anteriores (Membra Thoracica), p. 165-234. In: König HE,

LIEBICH HG (Eds) **Anatomia dos Animais Domésticos: Texto e Atlas Colorido.** Porto Alegre, Artmed, 787p.

MACALISTER A (1873a) The muscular anatomy of the civet and tayra. **Proceedings of the Royal Irish Academy Academy Series 2:** 506-513.

MACALISTER A (1873b) On the anatomy of *Aonyx.* **Proceedings of the Royal Irish Academy Series 2:** 539-547.

MACKINTOSH BA (1875) Notes on the myology of the coati mondth (Nasua narica and N. fusca) and common marten (Martes foina). **Proceedings of the Royal Irish Academy Series 2:** 48-55.

MARTÍN-SERRA A, FIGUEIRIDO B, PALMQVIST P (2014) A Three-Dimensional Analysis of Morphological Evolution and Locomotor Performance of the Carnivoran Forelimb. **PLoS ONE 9:** e85574. doi:10.1371/journal.pone.0085574

MCCLEARN D (1992) Locomotion, Posture, and feeding behavior of kinkajous, coatis, and raccoons. **Journal of Mammalogy 73** (2): 245-261.

MEACHEN-SAMUELS JA, VAN-VALKENBURGH B (2009) Forelimb indicators of prey-size preference in the Felidae. **Journal of Morphology 270** (6): 729-744. doi: 10.1002/jmor.10712

MELORO C, ELTON C, LOUYS J, BISHOP LC, DITCHFIELD P (2013) Cats in the forest: predicting habitat adaptations from humerus morphometry in extant and fossil Felidae (Carnivora). **Paleobiology 39**(3): 323-244. doi:10.1666/12001

MILLS P (2003) **Comparative Animal Anatomy.** Queensland, Gatton Desktop Publishing, 316p.

MOORE AL, BUDNY JE, RUSSEL AP, BUTCHER MT (2013) Architectural specialization of the intrinsic thoracic limb musculature of the American badger (Taxidea taxus). **Journal of Morphology 274**(1): 35-48. doi: 10.1002/jmor.20074

NICKEL R, SCHUMMER A, SEIFERLE E, FREWEIN J, WILKENS H, WILLE K (1986) **The Locomotor System of Domestic Mammals.** Berlin, Verlag Paul Parey, 515p.

NOWAK RM (2005) **Walker's Carnivores of the World.** Baltimore, The Johns Hopkins University Press, 328p.

PARANAIBA JF, HELRIGLE C, ARAÚJO EG, PEREIRA KF (2012) Aspectos morfológicos da mão e pé de *Procyon cancrivorus.* **Natureza On line 10** (4): 165-169. Available online at: http://www.naturezaonline.com.br/natureza/conteudo/pdf/03_ParanaibaJFetal_165_169.pdf. [Accessed: 31 July 2014]

PESSUTTI C, SANTIAGO MEB, OLIVEIRA LTF (2001) Order Carnivora, Family Canidae (Dogs, foxes and maned wolves), p. 279-290. In: FOWLER ME, CUBAS ZS (Eds) **Biology, Medicine and Surgery of South American Wild Animals.** Ames, Iowa State University Press, 550p.

PESTANA FM, SILVA BX, CHAGAS MA, BABINSKI MA, ABIDU-FIGUEIREDO M (2009) Distribuição antimérica do músculo braquiorradial em cães sem raça definida. **Revista de Ciências da Vida 29**(1): 55-59.

POLLY PD (2007) Limbs in mammalian evolution, p. 245-268. In: Hall BK (Ed.) **Fins into Limbs: Evolution, Development, and Transformation.** Chicago, University of Chicago Press, 344p.

REIS NR, PERACCHI AL, FREGONEZI MN, ROSSANEIS BK (2010) **Mamíferos do Brasil: guia de identificação.** Rio de Janeiro, Technical Books, 560p.

SAINT CLAIR LE (1986) Músculos do carnívoro, p. 1416-1444. In: GETTY R (Ed.) **Anatomia dos Animais Domésticos.** Rio de Janeiro, Guanabara Koogan, 2048p.

SALADIN KS (2010) **Anatomy and Physiology: The Unity of Form and Function.** New York, McGraw Hill, 1248p.

SAMUELS JX, MEACHEN JA, SAKAI SA (2013) Postcranial morphology and the locomotor habits of living and extinct carnivorans. **Journal of Morphology 274**(2): 121-146. doi:10.1002/jmor.20077

SÁNCHEZ HL, SILVA LB, RAFASQUINO ME, MATEO AG, ZUCCOLILLI GO, PORTIANSKY EL, ALONSO CR (2013) Anatomical study of the forearm and hand nerves of the domestic cat (*Felis catus*), puma (*Puma concolor*) and jaguar (*Panthera onca*). **Anatomia Histologia Embryologia 42** (2): 99-104. doi: 10.1111/j.1439-0264.2012.01170.x

SANTOS JUNIOR I, RODRIGUES CA, CAMPOS A, SANTOS D (2002) Presença do músculo braquiorradial em cães. **Bioscience Journal 18**(1): 79-83.

SATO JJ, WOLSAN M, MINAMI S, HOSODA T, SINAGA MH, HIYAMA K, YAMAGUCHI Y, SUZUKI H (2009) Deciphering and dating the red panda's ancestry and early adaptive radiation of Musteloidea. **Molecular Phylogenetics and Evolution 53**(3): 907-922. doi:10.1016/j.ympev.2009.08.019

SATO JJ, WOLSAN M, PREVOSTI FJF, D'ELÍA G, BEGG C, BEGG K, HOSODA T, CAMPBELL KL, SUZUKI H (2012) Evolutionary and biogeographic history of Weasellike carnivorans (Musteloidea). **Molecular Phylogenetics and Evolution 63**(3): 745-757. doi:10.1016/j.ympev.2012.02.025

SCHWARZE E (1984) **Compendio de Anatomia Veterinaria.** Zaragoza, Editora Acribia, 318p.

SLATER GJ, HARMON LJ, ALFARO ME (2012) Integrating fossils with molecular phylogenies improves inference of trait evolution. **Evolution 66**(12): 3931-3944. doi:10.1111/j.1558-5646.2012.01723.x

SEBASTIANI AM, FISHBECK DW (2005) **Mammalian anatomy: the cat.** Colorado, Morton Publishing Company, 184p.

TAYLOR ME (1989) Locomotor adaptations by carnivores, p. 382-409. In: GITTLEMAN JL (Ed.) **Carnivore Behavior, Ecology, and Evolution.** Cornell, Comstosck Publishing Associates, 620p.

VAZ MGR, LIMA AR, SOUZA ACB, PEREIRA LC, BRANCO E (2011) Estudo morfológico dos músculos do antebraço de cachorro-do-mato-de-orelhas-curtas (*Atelocynus microtis*) e cachorro-do-mato (*Cerdocyon thous*). **Biotemas 24**(4): 121-127. doi:10.5007/2175-7925.2011v24n4p121

WAKURI H, KANO Y (1966) Anatomical studies on the brachioradial muscle in dogs. **Acta Anatomica Nipponica 41:** 222-231.

WILLIAMS SB, WILSON AM, RHODES L, ANDREWS J, PAYNE RC (2008) Functional anatomy and muscle moment arms of the thoracic limb of an elite sprinting athlete: the racing greyhound (*Canis familiaris*). **Journal of Anatomy 213**(4): 361-372. doi: 10.1111/j.1469-7580.2008.00962.x

Wilson DE, Mittermeier RA (2009) **Handbook of the Mammals of the World.** Barcelona, Lynx Edicions, vol. 1, 727p.

Windle BCA, Parsons FG (1897) On the myology of the terrestrial Carnivora. Part I: muscles of the head, neck, and fore-limb. **Proceedings of Zoological Society of London 65**: 370-409.

Witmer LM (1995) The extant phylogenetic bracket and the importance of reconstructing soft tissues in fossils, p. 19-33. In: Thomason JJ (Ed.) **Functional morphology in vertebrate paleontology.** Cambridge, Cambridge University Press, 277p.

A new genus and new species of spittlebug (Hemiptera: Cercopidae: Ischnorhininae) from Southern Brazil

Andressa Paladini[1] & Rodney Ramiro Cavichioli[1,2]

[1]Departamento de Zoologia, Universidade Federal do Paraná. Caixa Postal 19020, 81531-980 Curitiba, PR, Brazil.
E-mail: andri_bio@yahoo.com.br
[2]Corresponding author. E-mail: cavich@ufpr.br

ABSTRACT. A new genus of spittlebug is described to include *Gervasiella oakenshieldi* **sp. nov.** (holotype male from Brazil, state of Paraná, municipality of Piraquara, Mananciais da Serra at 25°29'46"S, 48°58'54"W, 1000 m a.s.l., 15.XI.2008, P.C. Grossi *leg.*, deposited in DZUP). In addition, *Aeneolamia bucca* Paladini & Cavichioli, 2013 is transferred to *Gervasiella* **gen. nov.** based on the results of a cladistic analysis. *Gervasiella* **gen. nov.** can be distinguished from the other cercopid genera by the following: postclypeus inflated with upper portion black and basal one yellowish; color of tylus distinct from color of head and rostrum, barely reaching mesocoxae. *Gervasiella oakenshieldi* **sp. nov.** is diagnosed by having the head black with tylus white, postclypeus in profile inflated and convex with a prominent longitudinal carina; tegmina black with two elongate white maculae near costal margin, one on anterior third and the other on posterior third.

KEY WORDS. Auchenorrhyncha; Neotropical Region; phylogeny; taxonomy.

Insects belonging to Cercopidae are known as spittlebugs due to the bubble nest produced by the nymphs. This family forms a large group of xylem feeding insects with approximately 1500 worldwide species included in 150 genera. Most species are distributed in the tropical and subtropical regions. Adults feed on leaves or stems of a wide variety of plants, nymphs can feed on roots and in some cases they complete their development above the ground (CARVALHO & WEBB 2005).

The Neotropical genera of Cercopidae have been usually defined by characters of the head and pronotum, and by the number of spines on the hind leg. The same set of characters also form the basis of the tribal classification proposed by FENNAH (1968).

An ongoing study on Neotropical cercopids has revealed a new genus of Ischnorhininae. The new genus and the new species are described and illustrated. Also, we propose a new combination: *Aeneolamia bucca* Paladini & Cavichioli, 2013 is transferred to *Gervasiella* **gen. nov.** The species of *Gervasiella* **gen. nov.** are distinguished based on a comparative diagnosis.

MATERIAL AND METHODS

The specimens studied are deposited in the Coleção Entomológica Padre Jesus Santiago Moure, Departamento de Zoologia, Universidade Federal do Paraná, Curitiba, Paraná, Brazil (DZUP). Morphological terminology follows FENNAH (1968) and PALADINI & CRYAN (2012). Techniques for preparation of genital structures follow OMAN (1949). The dissected parts were stored in micro vials with glycerin. Photographs were obtained with a Leica DFC-550 digital camera attached to the stereomicroscope (Leica MZ16) and captured with the software IM50 (Image Manager; Leica Microsystems Imaging Solutions Ltd, Cambridge, UK), after montage using Auto-Montage Syncroscopy of Taxonline (Rede Paranaense de Coleções). Illustrations were made with the aid of a camera lucida and the final art were finalized using vectors with the software Corel-Draw version X5.

Terminal taxa. Besides the Neotropical cercopids present in the matrix analyzed by PALADINI et al. (2015) two species of *Gervasiella* **gen. nov.** were also included, to test the validity of the new genus proposed here.

Characters were coded to include most of the morphological variation of the external morphology of the adult, and male and female genitalia. We included and reanalyzed 108 characters from the cladistics analysis of PALADINI et al. (2015), and added two characters totaling 110 characters. Each character was considered a hypothesis of grouping. Primary homologies were proposed by similarity or topological correspondence (DE PINNA 1991). The contingent coding was used when novel features appeared and evolved, and this feature shows variation (SERENO 2007). Multistate characters were treated as unordered (nonadditive) (FITCH 1971). Character state polarity was determined by outgroup rooting (NIXON & CARPENTER 1993). Missing data were coded as '?' and nonapplicable characters were coded as '–'. The data matrix was built using Winclada v1.00.08 (NIXON 2002).

Analyses were performed using two character weighting schemes: equal weight and implied weight. Analyses were conducted using TNT version 1.1 (Goloboff et al. 2008) using Traditional Search basing the heuristic search strategies on RAS + TBR (random addition sequences plus swap by tree bisection and reconnection), with 1,000 replications with 100 trees saved per replication. The choice for the best constant of concavity (K) values range for the data followed the methodology of Paladini et al. (2015). The best K range for the data matrix presented here was 8-13. Branch support was calculated using the relative Bremer support (Goloboff & Farris 2001). Nonparametric Bootstrap support values were computed running 1000 bootstrap pseudoreplicates (Felsenstein 1985).

TAXONOMY

Gervasiella gen. nov.

Type species. *Gervasiella oakenshieldi* **sp. nov.** by original designation.

Diagnosis. *Gervasiella* **gen. nov.** can be distinguished from all other cercopids genera by the following combination of characters: 1) postclypeus inflated with upper portion black and basal one yellowish; 2) tylus with a distinct color from the head; 3) rostrum barely reaching mesocoxae; 4) subgenital plates shorter than pygofer, in ventral view quadrangular with apex truncate; 5) paramere long and slender, apex rounded, a unique subapical spine quadrangular; 6) paramere's spines located upon a lateral concavity similar to a hole; 7) aedeagus slender with quandrangular and wide base, one pair of dorsal processes long and slender turned upward.

Description. Head triangular with two deep impressions on the vertex near the median line; tylus quadrangular; ocelli near to each other than the compound eyes. Antennae with pedicel visible in dorsal view, flagellum normal in length, with ovoid basal body and an arista almost as long as pedicel; postclypeus inflated, convex in profile with a well-marked longitudinal carina; rostrum barely reaching mesocoxae. Pronotum hexagonal, surface smooth; anterior and lateral anterior margins straight; posterior and lateroposterior margins slightly sinuated; tegmina long and slender. Hindwings with Cu1 not thickened at base. Hind tibiae with two lateral spines and a row of apical spines; hind basitarsus with apical spines distributed in two irregular rows. Pygofer with one process between anal tube and subgenital plate in lateral view; subgenital plate quandrangular in ventral view. Aedeagus long, slender with one pair of dorsal processes; parameres slender, dorsal margin with two processes, one subapical spine quadrangular and sclerotized. First valvulae of ovipositor with two processes near the base; second valvulae with dorsal margin smooth.

Etymology. The genus is named in honor of Prof. Dr. Gervásio Silva Carvalho, a specialist of Neotropical cercopids, in recognition of his expertise and several contributions to the taxonomy of the group.

Remarks. Based on the tree resulting from the cladistics analysis (Figs. 12 and 13), *Gervasiella* **gen. nov.** is sister group of the clade including *Prosapia* Fennah, 1949, *Aeneolamia* Fennah, 1949 and *Isozulia* Fennah, 1953. In these genera the aedeagus presents a long and slender dorsal process inserted medially. *Gervasiella* **gen. nov.** is supported by two synapomorphies: paramere with a concavity under the spine (Figs. 14 and 15) and aedeagus base quadrangular (Figs. 16 and 17).

Gervasiella oakenshieldi **sp. nov.**
Figs. 1-11, 19

Measurements. Length, male 6.8 mm; females 6.7-7.8 mm.

Diagnosis. Head black with tylus white, postclypeus in profile, inflated and convex with a prominent longitudinal carina; tegmina black with two elongate white maculae near the costal margin, one on the anterior third and the other on the posterior third.

Description. Head triangular black; rostrum yellowish with the third segment black; compound eyes black, rounded, arranged transversely; vertex smooth, rectangular, with a prominent median carina; ocelli reddish near to each other than to compound eyes; tylus white, smooth, quadrangular, lacking a median carina; antennae black, pedicel scarcely setose, basal body of flagellum ovoid with an arista almost as long as the pedicel; postclypeus inflated, convex in profile with a wide longitudinal carina, lateral grooves slightly marked, apical portion black and two basal thirds yellowish. Thorax black; pronotum black, flattened, hexagonal, lacking median carina, anterior margin straight, lateral-anterior margins straight, lateral-posterior margin slightly sinuous, posterior margin with a light groove; scutellum black, with a slight central concavity and transversal grooves. Tegmina black with two white maculae: the first one elongated, located near the costal margin, extending from its base until the median third of the tegmina; the second one rounded located between the median and apical third; apical plexus of vein poorly developed; hindwings hyaline with brown venation; vein Cu1 not thickened at base; legs brownish; metathoracic tibia with two lateral spines (basal spine equal in size to spines in apical crown; apical spine larger than spines in apical crown); apical crown of spines on tibia consisting of two rows; basitarsus with one row of spines covered by sparse setae; subungueal process absent.

Male genitalia. Pygofer with one quadrangular process between the anal tube and subgenital plates; subgenital plates short, quadrangular with a rounded apex, dorsal margin produced in a rectangular process (Fig. 3); parameres long and slender with a quadrangular sclerotized spine turned backwards located over a concavity on the external side, dorsal margin with a finger like process turned to the inner side (Figs. 7 and 8); aedeagus cylindrical with a pair of dorsal processes long and slender turned upward, aedeagus base quadrangular and wide, apex quadrangular (Figs. 5 and 6).

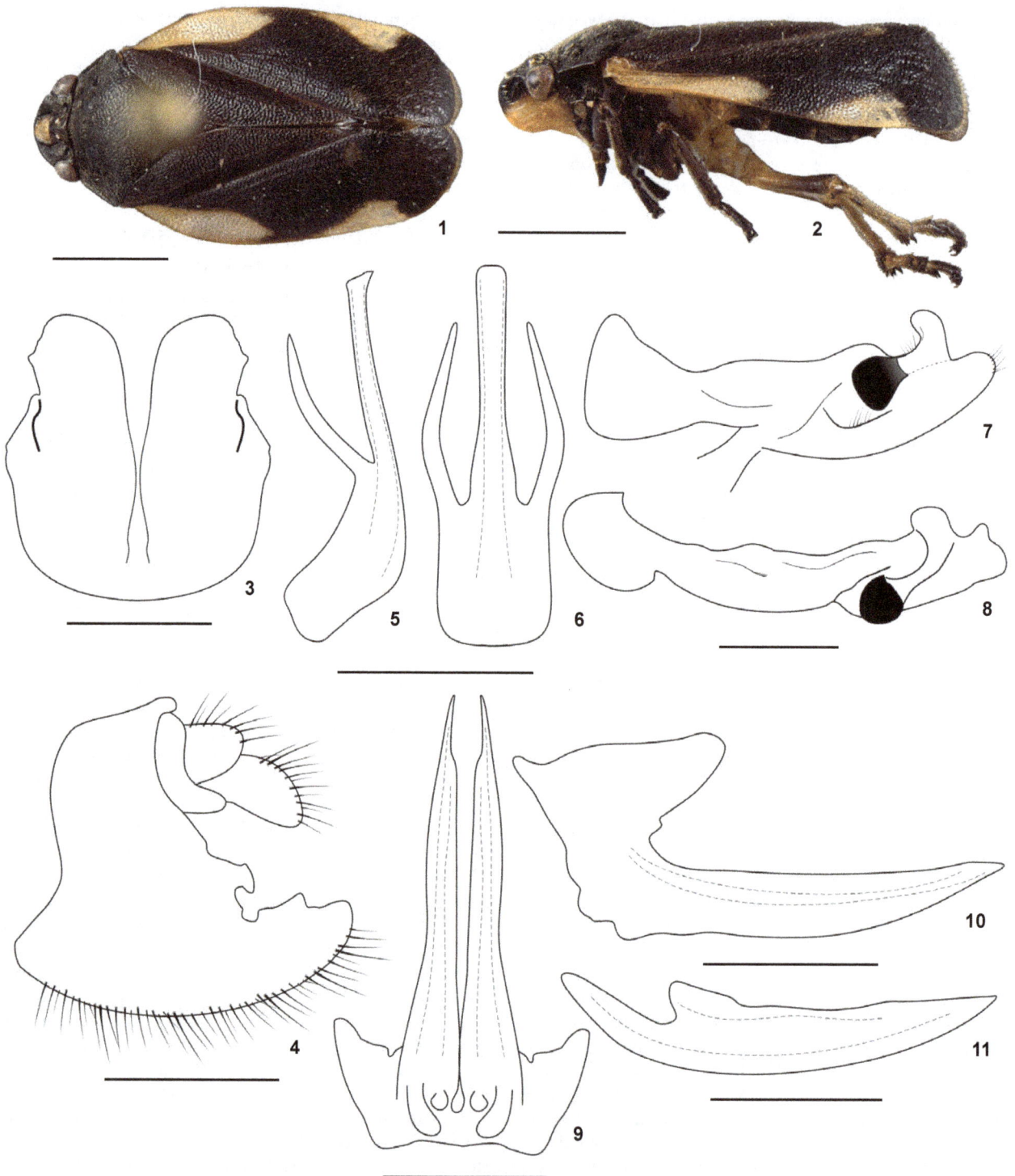

Figures 1-11. *Gervasiella oakenshieldi* **sp. nov.**: (1-2, 9-11) female paratype, (3-8) male holotype: (1) habitus, dorsal view; (2) habitus, lateral view; (3) subgenital plates and pygofer, ventral view; (4) pygofer and subgenital plates, lateral view; (5) aedeagus, lateral view; (6) aedeagus, dorsal view; (7) paramere, lateral view; (8) paramere, dorsal view; (9) first valvulae of ovipositor, ventral view; (10) first valvulae of ovipositor, lateral view; (11) second valvulae of ovipositor, lateral view. Scale bars: 1-2 = 2 mm, 3, 4, 9-11 = 0.5 mm, 5-8 = 0.25 mm.

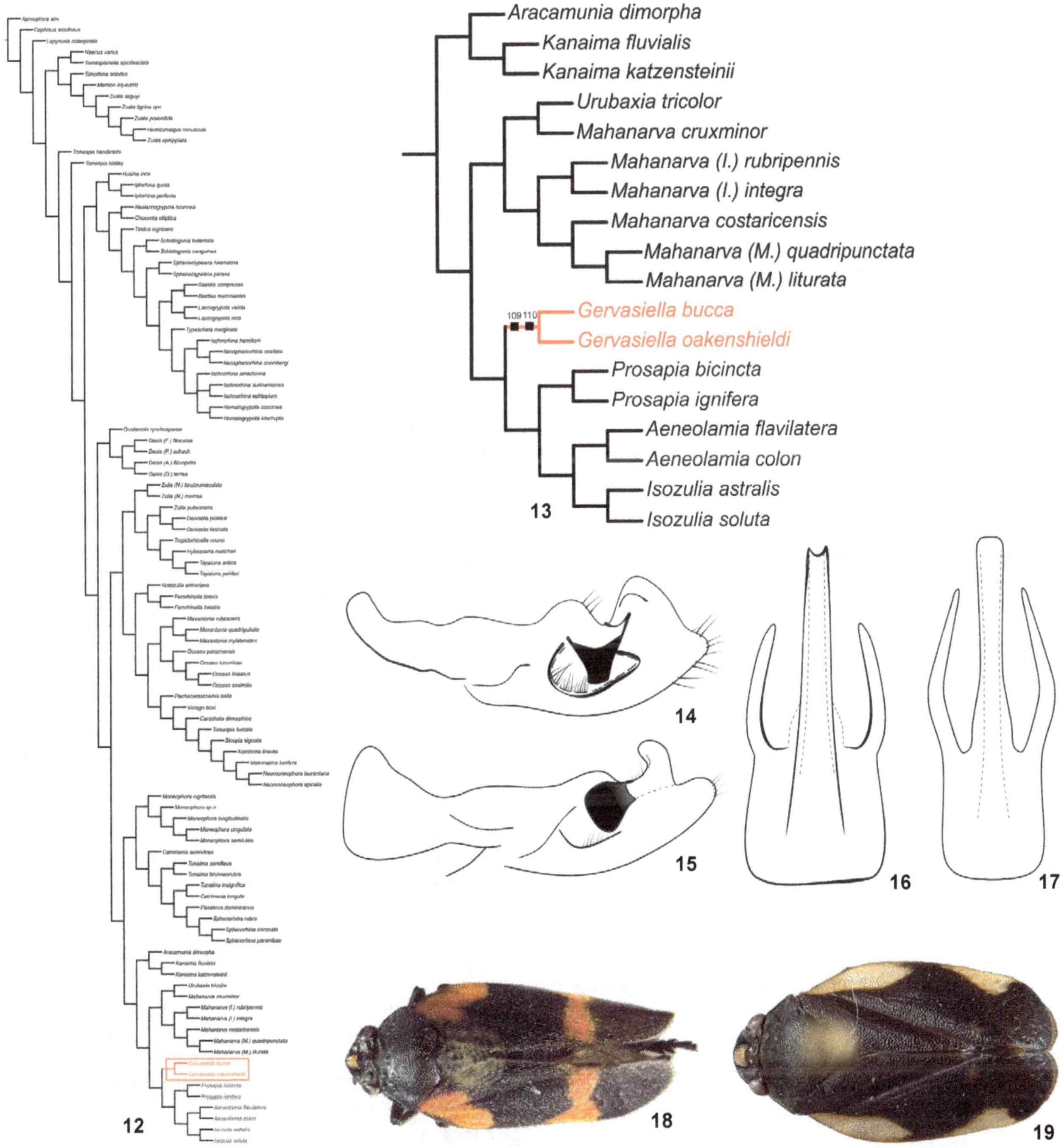

Figures 12-19. Phylogenetic relationships of Ischnorhininae and diagnostic characters of *Gervasiella* **gen. nov.**: (12) unique most parsimonious tree resulting from the analysis of morphological data with the implied weighting scheme using the optimal constant of concavity interval of K8-12, highlighting the position of *Gervasiella* **gen. nov.**; (13) clade including *Gervasiella* **gen. nov.**; (14) *Gervasiella bucca* **comb. nov.** paramere in lateral view; (15) *Gervasiella oakenshieldi* **sp. nov.** parameres in lateral view; (16) *Gervasiella bucca* **comb nov.** aedeagus in dorsal view; (17) *Gervasiella oakenshieldi* **sp. nov.** aedeagus in dorsal view; (18) *Gervasiella bucca* dorsal habitus; (19) *Gervasiella oakenshieldi* **sp. nov.** dorsal habitus.

Female. First valvulae of ovipositor long and slender with acute apex and two basal process poorly developed, rounded, directed ventrally (Figs. 9 and10); second valvulae long and slender, dorsal margin smooth (Fig. 11), third valvulae short and wide, with long setae ventrally.

Etymology. Noun in genitive singular after to a fictional character surname of the novel The Hobbit, Thorin Oakenshield; in honor to J.R.R Tolkien an English writer known as the author of classic fantasy books.

Remarks. *Gervasiella oakenshieldi* **sp. nov.** (Figs. 1-2 and 19) superficially resembles *Gervasiella bucca* **comb. nov.** (Fig. 18) in having the same color pattern but the paramere is slender, with the concavity under the spine less pronounced (Fig. 15).

Examined material. Holotype male from BRAZIL, *Paraná*: Piraquara (Mananciais da Serra 25°29'46"S, 48°58'54"W, 1000 m a.s.l., 15.XI.2008), P.C. Grossi *leg*. Paratypes: 1 female same data as holotype; 1 female, same locality as holotype but 25.III.2012, light trap; 1 female same locality as holotype but flight interception [trap], XI.2007, P.C. Grossi & D. Parizotto *leg*. All deposited in DZUP.

Gervasiella bucca
(Paladini & Cavichioli, 2013) **comb. nov.**
Figs. 14, 16, 18

Aeneolamia bucca Paladini & Cavichioli, 2013: 353.

Diagnosis. General coloration black; tegmina with basal red macula and one apical red stripe; Pygofer short with finger-like process between anal tube and subgenital plates; aedeagus with one pair of dorsal, slender processes directed upward.

Remarks. *Gervasiella bucca* was originally described in *Aeneolamia* due to a superficial resemblance in the morphology of the male genitalia, although other features indicated that those species were not congeneric. Subsequently, *Gervasiella* **gen. nov.** was erected to accommodate *G. bucca* and the newly described *G. oakenshieldi* **sp. nov.**, based on the examination of additional specimens from Southern Brazil (from the municipality of Piraquara, Paraná State). Diagnostic traits of this genus were previously mentioned in the generic diagnosis. These features were later recovered as synapomorphies validating the monophyly of *Gervasiella* **gen. nov.**, as inferred in our morphology-based phylogenetic analysis of Ischnorhininae that included both *G. bucca* and *G. oakenshieldi* **sp. nov.**

CLADISTIC ANALYSIS

List of new characters. The complete list of characters can be found in PALADINI et al. (2015) and the full data matrix is available in Appendix S1[1]. The two new characters included in the analysis are: Male genitalia: (109) Paramere, lateral view, concavity under the main spine: (0) absent; (1) present; and (110) Aedeagus, shape of base in dorsal view: (0) rectangular; (1) quadrangular. The data matrix included 102 taxa and 110 characters (Appendix S1[1]). The equal weights analysis produced 30 equally parsimonious trees (length = 1061 steps, CI = 14, RI = 61). The strict consensus cladogram had 59 collapsed nodes. The relationships among genera were poorly resolved. In the analysis with implied weighting scheme the best K range for the data matrix presented here was 8-12; this range was chosen based on PALADINI et al. (2015). The five trees obtained with the best K range were had the same topology,(Fig. 12) which will be used as the hypothesis to infer the phylogenetic relationship and monophyly of *Gervasiella* **gen. nov.** The topology obtained in the present analysis is similar to that of PALADINI et al. (2015) except for the inclusion of the new genus. Only unambiguous characters were optimized in the resultant cladogram.

The main goal of this cladistics analysis was to evaluate and to support the description of a new genus. The clade *Gervasiella* **gen. nov.** (Fig. 13) has a relative Bremer support of 71 and a Bootstrap support of 99. The genus is supported by two synapomorphies: paramere with a concavity located under the main spine (109_1) and aedeagus with a quadrangular base (110_1). and 10 a homoplasious character-state transformations: vertex shape narrow (3_0); antennae with basal body of flagellum ovoid (6_1); posterior margin of pronotum slightly grooved (31_0); tegmina venation almost indistinct (37_2); tegmina with apical plexus of veins reduced (40_1); basitarsus of the posterior leg with two rows of spines (47_1); subgenital plates short compared to pygofer (53_0); apex of subgenital plates truncated (50_2); spine of paramere oriented vertically (69_1); ovipositor with two basal processes (103_1).

Gervasiella **gen. nov.** is included in the clade Tomaspidini and is sister group to *Prosapia, Aeneolamia,* and *Isozulia.*

ACKNOWLEDGEMENTS

We thank Paschoal C. Grossi (UFRPE) and Daniele Parizzoto for collecting and generously providing specimens; Olivia Envangelista (MZUSP) for her valuable suggestions; the anonymous reviewers and associate editor for their constructive comments and significant improvement on an earlier version of this manuscript. This work was supported by a CNPq postdoctoral grant (process 150163/2013-4) to the senior author. This research is also partially funded by the advisor's grant (RRC) from PROTAX/CNPq (processes 561298/2010-6 and 303127/2010-4). This paper is the contribution number 1917 of the Departamento de Zoologia, Universidade Federal do Paraná.

REFERENCES

CARVALHO GS, WEBB MD (2005) **Cercopid Spittlebugs of the New World (Hemiptera, Auchenorrhyncha, Cercopidae).** Sofia, Pensoft, 271p.

DE PINNA MCC (1991) Concepts and tests of homology in the cladistics paradigm. **Cladistics 7**(4): 367-394.

FENNAH RG (1968) Revisionary notes on the new world genera of cercopid froghoppers (Homoptera, Cercopoidea). **Bulletin of Entomological Research 58**: 165-190.

FELSENSTEIN J (1985) Confidence limits on phylogenies: an approach using the bootstrap. **Evolution 39**: 783-791.

FITCH WN (1971) Toward defining the course of evolution, minimum change for a specified tree topology. **Systematic Zoology 20**: 406-416.

GOLOBOFF PA, JS FARRIS (2001) Methods for quick consensus estimation. **Cladistic 17**(1): S26-S34. doi: 10.1111/j.1096-0031.2001.tb00102.x

GOLOBOFF PA, FARRIS JS, NIXON KC (2008) TNT, a free program for phylogenetic analysis. **Cladistics 24**(5): 774-786. doi: 10.1111/j.1096-0031.2008.00217.x

NIXON KC (2002) **Winclada.** New York, Published by the Author, v. 1.00.08.

NIXON KC, CARPENTER JM (1993) On outgroups. **Cladistics 9**(4): 413-426. doi: 10.1111/j.1096-0031.1993.tb00234.x

OMAN PW (1949) The Nearctic leafhoppers (Homoptera: Cicadellidae). A generic classification and check list. **Memoirs of the Entomological Society of Washington 3:** 1-253.

PALADINI A, CAVICHIOLI RR (2013) A new species of *Aeneolamia* (Hemiptera: Cercopidae: Tomaspidinae) from the Neotropical Region. **Zoologia 30**(3): 353-355. doi: 10.1590/S1984-46702013000300016

PALADINI A, CRYAN JR (2012) Nine new species of Neotropical spittlebugs. **Zootaxa 3519**: 53-68.

PALADINI A, TAKIYA DM, CAVICHIOLI RR, CARVALHO GS (2015) Phylogeny and biogeography of Neotropical spittlebugs (Hemiptera: Cercopidae: Ischnorhininae): revised tribal classification based on morphological data. **Systematic Entomology 40**(1): 82-108. doi: 10.1111/syen.12091

SERENO PC (2007) Logical basis for morphological characters in phylogenetics. **Cladistics 23**(6): 565-587.

Description of the first species of *Metharpinia* (Crustacea: Amphipoda: Phoxocephalidae) from Brazil

Luiz F. Andrade[1], Rodrigo Johnsson[2] & André R. Senna[2]

[1]*Programa de Pós-graduação em Biologia Animal, Universidade Federal Rural do Rio de Janeiro. Rodovia BR 465, km 7, 23890-000 Seropédica, RJ, Brazil. E-mail: lzflp.andrade@hotmail.com*
[2]*Laboratório de Invertebrados Marinhos: Crustacea, Cnidaria & Fauna Associada, Instituto de Biologia, Universidade Federal da Bahia. Rua Barão de Jeremoabo 147, Ondina, 40170-290 Salvador, BA, Brazil. E-mail: r.johnsson@gmail.com; senna.carcinologia@gmail.com*

ABSTRACT. A new amphipod species of *Metharpinia* Schellenberg, 1931 is described from Campos Basin, southeastern Brazilian coast. The material was collected with van Veen grab from unconsolidated substratum, off the mouth of the Paraíba do Sul River. The new species can be distinguished from its congeners by presenting a strongly constricted rostrum and a slender palp of maxilla 1. There are four species in *Metharpinia* from the South Atlantic: *M. dentiurosoma* Alonso de Pina, 2003, *M. grandirama* Alonso de Pina, 2003 and *M. iado* Alonso de Pina, 2003, and *Metharpinia taylorae* **sp. nov.** This is the first record of a species of the genus from Brazilian waters.

KEY WORDS. Amphipod; Campos Basin; Habitats Project; *Metharpinia taylorae* **sp. nov.**; taxonomy.

Phoxocephalidae Sars, 1895, one of the most diverse amphipod taxa in terms of taxonomic characters, is characterized by the following: antennae 1 and 2 haustorioid in shape and with multiarticulate accessory flagellum; gnathopods 1 and 2 subchelated or chelated; pereopod 7 distinct from pereopod 5-6, shortened, article 2 expanded posteriorly; uropod 3 biramous; telson deeply cleft (BARNARD & DRUMMOND 1978, 1982). The Phoxocephalidae are benthic-burrowing amphipods (HURLEY 1954). They are widely distributed from shallow to deep waters. According to BARNARD & DRUMMOND (1978), Australia is the evolutionary center of the Phoxocephalidae, and there are two areas of dispersion of shallow water phoxocephalids, the Magellanic region plus the Falkland islands, and the Antarctica (including South Geogia Islands). However, this hypothesis needs to be tested by modern phylogenetic methods.

Phoxocephalidae currently includes more than 460 species around the world, grouped in 11 subfamilies and 74 genera (HORTON & DE BROYER 2014).Most species are found exclusively in the deep sea of the southern hemisphere (BARNARD & DRUMMOND 1978). According to SENNA & SOUZA-FILHO (2011) there are 13 Phoxocephalidae species recorded from Brazilian waters: *Bathybirubius margaretae* Senna, 2010, *Coxophoxus alonso* Senna, 2010, *Harpiniopsis galera* Barnard J.L., 1960, *Hererophoxus videns* Barnard KH, 1930, *Leptophoxoides marina* Senna, 2010, *Microphoxus breviramus* Bustamante, 2002, *M. cornutus* (Schellenberg, 1931), *M. moraesi* Bustamante, 2002, *M. uroserratus* Bustamante, 2002, *Phoxocephalus homilis* Barnard JL, 1960, *Pseudharpinia berardo* Senna, 2010, *P. ovata* Senna, 2010, and *P. tupinamba* Senna & Souza-Filho, 2011.

Metharpinia Schellenberg, 1931 and its sister-group, *Microphoxus* Barnard, 1960, are among the most primitive genera in the birubiin-parharpiniin group of the Americas. *Metharpinia* has nine species, all distributed along the west and east coasts of North and Central America, and Argentina (BARNARD & DRUMMOND 1978, BARNARD & KARAMAN 1991, ALONSO DE PINA 2001, 2003a, b). According to ALONSO DE PINA (2003a), species of *Metharpinia* are characterized by the following characters: antenna 1, article 2 with ventral setae placed proximally; maxilliped with dactylar nail partially fused and immersed; gnathopods 1 and 2, palms acute and propodus poorly setose anteriorly; and pereopods 3 and 4, propodus with facial setal formula composed of stout setae and dactyli with inner acclivity sharp, produced as tooth.

We describe a new species of *Metharpinia*. This is the first record of the genus from Brazilian waters, increasing the Phoxocephalidae diversity in Brazil to 14 species in nine genera.

MATERIAL AND METHODS

The material examined was collected during the Habitats Project (Environmental Heterogeneity of Campos Basin), coordinated by the Brazilian Oil Company (CENPES/PETROBRAS). Collecting trips were conducted at the Campos Basin, off the mouth of the Paraíba do Sul River, between the states of Rio de Janeiro and Espírito Santo, southeastern Brazil, in the Summer and Winter of 2009. Collections were made aboard of the R/V Gyre, from unconsolidated substratum, using a van Veen grab.

The specimens were dissected under a stereoscopic microscope Motic K-401L and mounted in glycerine gel slides. The illustrations were produced under an optic microscope with a camera lucida Motic BA-310. The type material is deposited at the Crustacea Collection of the Museu Nacional, Universidade Federal do Rio de Janeiro (MNRJ). It is preserved in 70% ethanol or glycerine gel slides. The setal classification adopted in this paper follows WATLING (1989). Nomenclature of the gnathopod palm is based on POORE & LOWRY (1997).

TAXONOMY

Metharpinia Schellenberg, 1931

Diagnosis. See BARNARD & KARAMAN (1991).

Composition of the genus: *M. coronadoi* Barnard, 1980; *M. dentiurosoma* Alonso de Pina, 2003; *M. floridana* (Shoemaker, 1933); *M. grandirama* Alonso de Pina, 2003; *M. iado* Alonso de Pina, 2003; *M. jonesi* (Barnard, 1963); *M. longirostris* Schellenberg, 1931; *M. oripacifica* Barnard, 1980; *M. protuberantis* Alonso de Pina, 2001; *M. taylorae* sp. nov.

Metharpinia taylorae sp. nov.
Figs. 1-25

Diagnosis. Rostrum strongly constricted. Right mandible, incisor with two spines, one large, apically bifid, and one small and subrounded. Left mandible, incisor with three teeth, one large and two small, one of them apically bifid, lacinia mobilis well developed, apically smooth and subrounded. Maxilla 1, palp very slender and setose. Maxilliped, inner plate with three apical plumose setae. Gnathopods 1-2 poorly setose. Pereopod 7, basis with posteroventral lobe rounded. Epimeral plate 3, ventral margin with five submarginal pectinate setae, posterior margin slightly serrate, with 13 long slender setae, posteroventral corner broadly rounded. Urosomite 3 without dorsal hook. Uropod 3, inner ramus bearing few plumose setae, outer ramus with one plumose seta. Telson, deeply cleft, about 90% of its length, apical margin sinuous, each lobe with one lateral small plumose seta, subapical margin bearing five slender setae on each lobe, inner teeth of apex naked.

Description. Based on the holotype (MNRJ 477) and allotype (MNRJ 479). Head (Figs. 1 and 2), eyes present, rostrum strongly constricted, narrow, spatulate, elongate, about 1.3X longer than antenna 1 peduncular article 1. Antenna 1 (Fig. 3), peduncle article 1 about 1.4X longer than wide, without setae; article 2 anterodorsal corner with one small slender seta, ventral margin with seven slender setae, 1.3X longer than wide; article 3 shortened with two ventral slender setae, 1.2X wider than long; flagellum 16-articulate, poorly setose; accessory flagellum 14-articulate, elongate, poorly setose. Antenna 2 (Fig. 4), peduncle, article 4 about 1.5X longer than wide ventral margin with a row of 10 slender setae, 11 stout facial setae arranged in three rows, anterodorsal corner with one stout seta and one

simple seta; article 5, ventral margin setose, facial row of setae with nine medium to small stout setae, apical margin with two medium slender setae; flagellum 20-articulate, poorly setose. Right mandible (Fig. 5), incisor with two teeth, one large, apically bifid, and one small and blunt; accessory setal row with seven stout multi-cuspidate setae; molar not triturative with four slender setae; palpar hump small, lacinia mobilis absent, palp 3-articulate, article 3 apically setose. Left mandible (Fig. 6), incisor with three teeth, one large and two small, one of them apically bifid; accessory setal row with eight stout multi-cuspidate setae; molar not triturative, with nine apical slender setae; lacinia mobilis well developed, apically smooth and subrounded. Maxilla 1 (Fig. 7), inner plate 1.1X wider than long with five apical plumose setae; outer plate 1.3X longer than wide with 10 multi-cuspidate robust apical setae; palp 2-articulate, very slender, article 2, outer margin with five slender setae, three proximal and two distal, inner margin with four slender setae, and apical margin with three slender setae. Maxilla 2 (Fig. 8), inner plate about 1.6X longer than wide and slightly shorter than outer plate, apical margin with nine slender setae; outer plate about 2.6X longer than wide, apical margin with eight slender setae. Maxilliped (Fig. 9), inner plate subrectangular, with three apical plumose setae; outer plate lanceolate with seven medial setae, two apical setae, and one small lateral setae; palp, 4-articulate, article 2, suboval, about 2.2X longer than wide medially setose; article 3 suboval, medially setose, with one lateral slender setae in notch, about 1.9X longer than wide; article 4 simple, curved and slender.

Gnathopod 1 (Fig. 10) poorly setose, coxa weakly expanded anteriorly, posteroventral corner with six slender setae; basis about 3.5X longer than wide, subrectangular, posterior margin with 13 short slender setae, ventral margin without setae; ischium, small, posteroventral corner with three slender setae; merus, small, subtriangular, posterior margin with three slender setae; carpus, about 1.4X longer than wide posterior margin medially setose; propodus, about 1.4X longer than wide, anterior margin without setae, anterodistal corner with four slender setae, posterior margin straight with 10 slender setae, palm almost transverse, palmar corner defined by a small and slightly upwards curved spine with one lateral stout seta with accessory seta; dactylus, curved, simple, subequal in length to palm. Gnathopod 2 (Fig. 11) poorly setose, coxa weakly expanded anteriorly, posteroventral corner with 12 slender setae, posterior margin with two pairs of small slender setae, anterior margin with one pair of small slender setae; basis about 4.1X longer than wide, subrectangular, anterior margin with two slender setae, anteroventral corner with six slender setae, posterior margin with two slender setae; ischium, small, without setae; merus, small, subtriangular, posterior margin with six slender setae; carpus, elongate, about 3X longer than wide posterior margin setose; propodus, about 1.9X longer than wide, anterodistal corner setose, posterodistal margin setose, palm almost transverse, palmar corner defined by a small spine

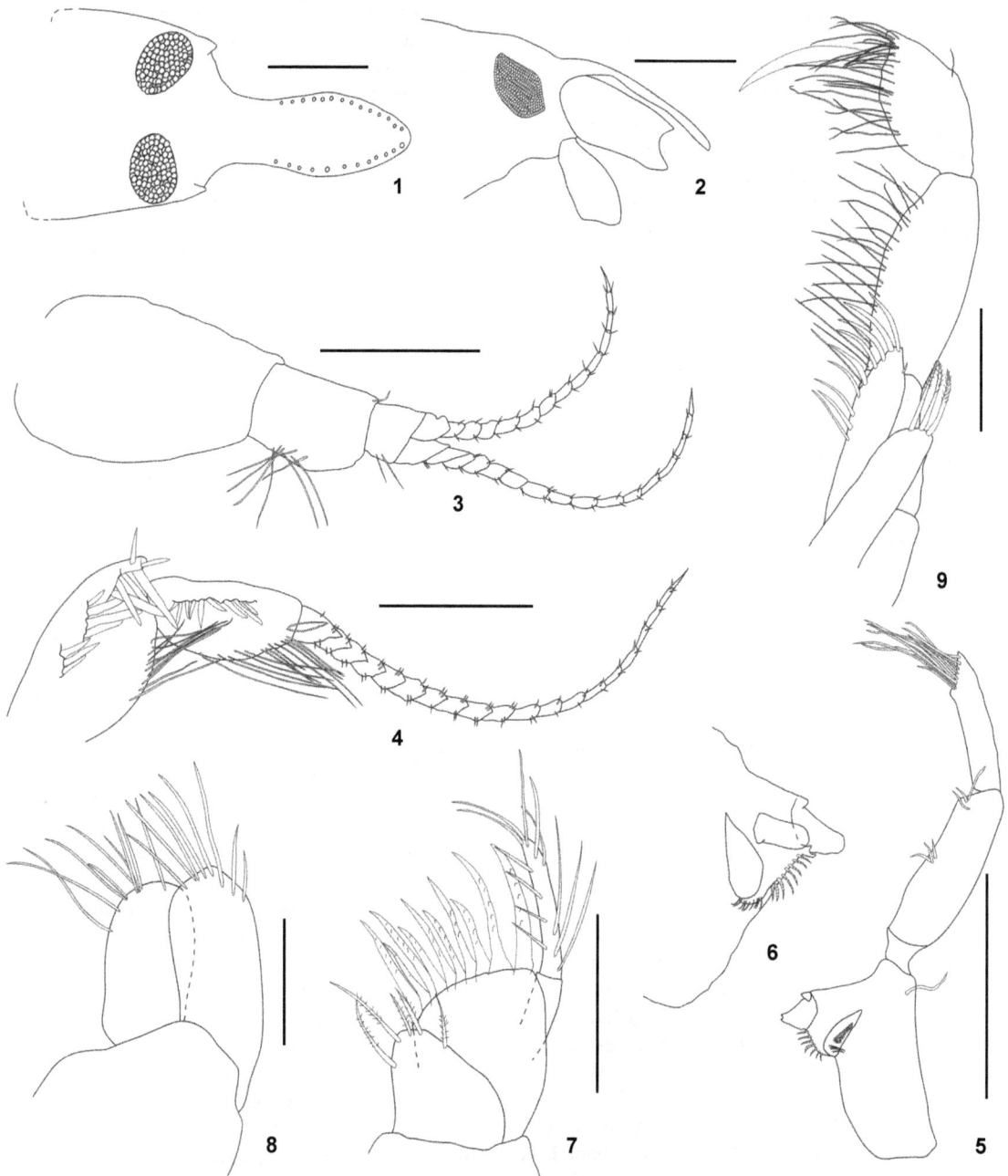

Figures 1-9. *Metharpinia taylorae* **sp. nov.**, holotype, female: (1) head, dorsal view; (2) head, lateral view; (3) antenna 1; (4) antenna 2; (5) right mandible; (6) left mandible; (7) maxilla 1; (8) maxilla 2; (9) maxilliped. Scale bars: 0.2 mm for maxilla 1-2; 0.5 mm for the remainder.

slightly curved upwards; dactylus curved, simple, slightly longer than palm. Pereopod 3 (Fig. 12), coxa subrectangular, about 1.7X longer than wide, ventral margin subrounded, without setae; basis, about 3X longer than wide, subrectangular, posterior margin with five long slender setae, anterior margin with seven small setae; ischium small, posteroventral corner with one slender setae; merus elongate, about twice longer than wide, subrectangular, posterior margin with three sets of long setae (2-1-2), posterodistal corner with a row of seven long setae, anterodistal corner with two small setae; carpus broad, about 1.2X longer than wide ventral margin setose; propodus elongate, about 3.8X longer than wide, posterior margin with eight sets of setae (6-2-3-2-5-2-2-1); dactylus simple, about 0.4X as long as propodus. Pereopod 4 (Fig. 13), coxa suboval, ventral margin rounded, posteroventral corner with two small slender setae; basis, 3.2X longer than wide, posterior margin with five long slender setae, posteroventral corner with three slender setae; ischium, small, posterior margin with five slender setae; merus, elongate, about 2.1X longer than wide, subrectangular, anterodistal corner with two slender setae, posterior margin with eight pairs of long slender setae, posterodistal corner with four slender setae; carpus, broad, about 1.4X longer than wide, posterior margin setose; propodus elongate, about 3.9X longer than wide, posterior margin with four sets of slender setae (4-2-4-2); dactylus robust, simple, about half length of propodus. Pereopod 5 (Fig. 14), coxa wider than long, with two lobes, posterior lobe with sinuous ventral margin, deeply produced, bearing four slender setae; basis, about twice longer than wide, subrectangular, posteriorly slightly expanded, anterior margin setose, anterodistal corner with five slender setae, posterior margin naked and slightly concave; ischium, small, naked; merus, about 1.7X longer than wide posteriorly expanded, anterior margin with one slender seta, posterior margin setose; carpus, about 1.4X longer than wide, anterior margin setose, posterior margin with three sets of slender setae (5-1-11), posteroventral corner with two long plumose setae; propodus, about 4.1X longer than wide, anterior margin with two sets of slender setae (4-5), anteroventral corner with five slender setae, posteroventral corner with six setae; dactylus slender, simple, about half length of propodus. Pereopod 6 (Fig. 15), coxa with a subacute posterior lobe, posterior margin setose; basis, about 1.2X longer than wide expanded posteriorly, anterior margin setose, posterior margin naked; ischium, small, naked, about twice wider than long; merus, wide, about 1.6X longer than wide anterior margin with four sets of setae (2-4-2-4), posterior margin with four sets of setae (3-4-5-5); carpus, about 2.2X longer than wide, anterior margin with three sets of setae (2-2-4), posterior margin with four sets of setae (1-2-2-6); propodus, elongate, about 8.2X longer than wide, anterior margin with two sets of setae (3-2), anterodistal corner with two slender setae and one stout setae with accessory seta, posterior margin with six sets of setae (3-2-2-2-3-1), posteroventral corner with five slender setae; dactylus, about 0.3X as long as propodus. Pereopod 7 (Fig. 16), coxa, posterior margin minutely setose; basis strongly expanded posteriorly, posterior margin serrate, posteroventral lobe rounded, smooth, and naked; ischium small, anterodistal corner with one slender seta; merus, anterior margin with two sets of setae (2-2), posterior margin with four sets of setae (1-2-2-2); carpus, about 1.5X longer than wide, anterior margin with four slender setae, anterodistal corner with a set of four slender setae, posterior margin with two sets of setae (3-2), posteroventral corner with a set of four slender setae; propodus, about 4X longer than wide, anterior margin with one slender seta, anterodistal corner with one slender seta, posterior mar-

gin with two sets of setae (2-4), posteroventral corner with three slender setae; dactylus slightly robust, about 0.6X as long as propodus.

Epimeral plate 1 (Fig. 17), anterior margin with one slender seta, ventral margin with two slender and 13 plumose setae, posterior margin slightly serrate, with 11 long slender setae, posteroventral corner rounded. Epimeral plate 2 (Fig. 18), anterior margin naked, ventral margin with seven medium and four long plumose submarginal setae, posterior margin with slightly serrate, with 10 long slender setae, posteroventral corner subrounded. Epimeral plate 3 (Fig. 19), anterior margin with seven small slender setae, ventral margin with five submarginal pectinate setae, posterior margin with slightly serrate, with 13 long slender setae, posteroventral corner broadly rounded. Uropod 1 (Fig. 20), peduncle elongated, about 2.1X longer than wide, dorsal margin with four stout setae; outer ramus slightly longer than inner ramus, about 7.9X longer than wide, dorsal margin with 11 stout setae, plus one subapical long stout seta; inner ramus subequal in length to peduncle, about 6X longer than wide, dorsal margin with three stout setae, plus one subapical long stout seta. Uropod 2 (Fig. 21), peduncle about 1.7X longer than wide, dorsal margin with three stout setae, dorsoapical corner with one stout setae, apicolateral corner with one stout setae; outer ramus, slightly longer than inner ramus, about 6.7X longer than wide, dorsal margin with nine stout setae, plus one subapical long stout seta; inner ramus, about 1.2X longer than peduncle, about 5.7X longer than wide, dorsal margin naked, with one subapical long stout seta. Urosomite 3 (Fig. 22) without dorsal hook. Uropod 3 (Fig. 23), peduncle short, about 1.3X longer than wide, apicolateral corner with five stout setae; outer ramus 2-articulate, about 1.6X longer than inner ramus, about 2.7X longer than peduncle, article 1 elongated, about 4.7X longer than wide, about 3.4X longer than article 2, dorsal margin bearing two stout setae and one small distal seta, ventral margin with one long distal plumose seta; article 2, about 4.7X longer than wide, bearing two apical slender setae; inner ramus, about 1.7X longer than peduncle, about 4.7X longer than wide, bearing seven apical and one subapical stout plumose setae. Male uropod 3 (Fig. 24), peduncle short, about 2.2X longer than wide, apicolateral corner with three stout setae with accessory setae, lateral margin with one stout setae with accessory setae and one short setae, facial margin with one slender setae; outer ramus 2-articulate, about 1.1X longer than inner ramus, about 2.2X longer than peduncle, article 1 elongated, about 4.6X longer than wide, about 4.4X longer than article 2, dorsal margin bearing two sets of stout setae with accessory setae (2-2) and three stout setae with accessory setae distally, ventral margin with five long plumose setae; article 2, about 3.4X longer than wide, bearing two apical slender setae; inner ramus, about 1.9X longer than peduncle, about 5.6X longer than wide, dorsal margin bearing two long plumose setae, ventral margin bearing five long plumose setae and one short simple setae, apical margin bearing two long plumose setae. Telson (Fig.

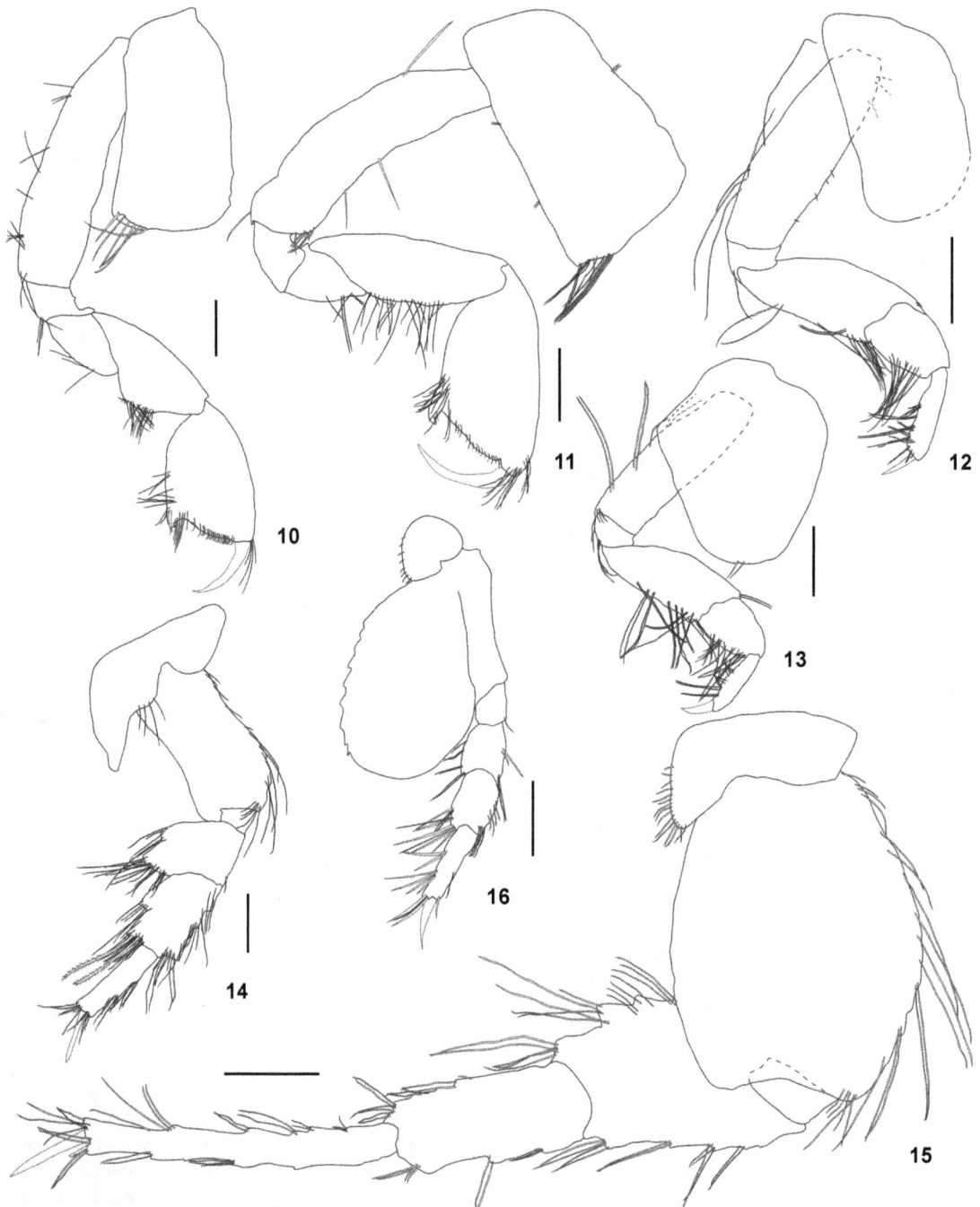

Figures 10-16. *Metharpinia taylorae* **sp. nov.**, holotype, female: (10) gnathopod 1; (11) gnathopod 2; (12) pereopod 3; (13) pereopod 4; (14) pereopod 5; (15) pereopod 6; (16) pereopod 7. Scale bars: 1.0 mm for gnathopods 1-2; 0.5 mm for the remainder.

25), about 1.3X longer than wide, deeply cleft, about 90% of its length, apical margin truncate with blunt cusp, each lobe with one lateral small plumose seta, subapical margin bearing five slender setae on each lobe, apex naked.

Material examined. Holotype female, 8.5 mm, BRAZIL, *Rio de Janeiro*: Campos Basin (21°33′52.574″S, 40°42′53.900″W, 22 m depth), 10 March 2009, R/V Gyre *leg.*, MNRJ 477. Allotype male, BRAZIL, *Espírito Santo*: Campos Basin (21°11′0,850″S,

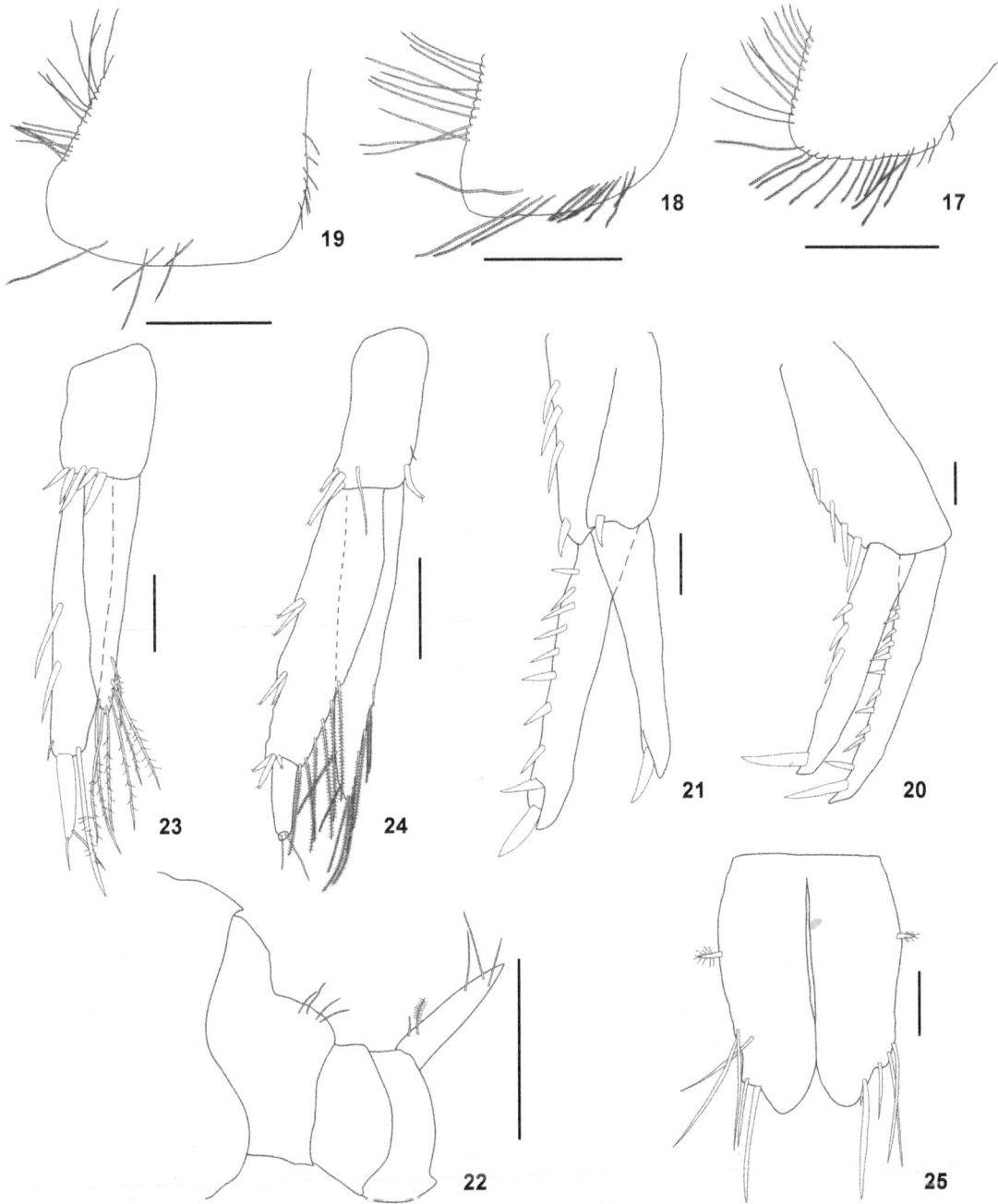

Figures 17-25. *Metharpinia taylorae* **sp. nov.**, holotype, female: (17) epimeral plate 1; (18) epimeral plate 2; (19) epimeral plate 3; (20) uropod 1; (21) uropod 2; (22) urosomite 3; (23) uropod 3; (25) telson; paratype, male: (24) uropod 3. Scale bars: 0.2 for male and female uropod 3; 0.5 mm for the remainder.

40°28′27.125″W, 26 m depth); 5 March 2009, R/V Gyre *leg.*, MNRJ 479. Paratypes: 1 male, BRAZIL, *Espírito Santo*: Campos Basin (21°17′51.743″S, 40°30′59.011″W, 29 m depth), 07 March 2009, R/V Gyre *leg.*, MNRJ 478; 1 male and 12 juveniles, *Rio de Janeiro*: Campos Basin (21°33′53.096″S, 40°42′55.466″W, 21 m depth, van Veen), 10 March 2009, R/V Gyre *leg.*, MNRJ 480; 2 females and 13 juveniles, (21°39′11.066″S, 40°48′49.898″W, 21 m depth, van Veen), 11 March 2009, R/V Gyre *leg.*, MNRJ 481; 1 ovigerous female, 1 male and 8 juveniles, (21°39′9.790″S, 40°48′50.234″W, 22 m depth, van Veen), 19 July 2009, R/V Gyre *leg.*, MNRJ 482.

Geographic distribution. Brazil, north coast of Rio de Janeiro State and south coast Espírito Santo State, Campos Basin, off the mouth of the Paraíba do Sul River (Fig. 26). Type locality: 21°33'52.574"S, 40°42'53.900"W.

Bathymetric range. Collected from 21 to 29 m depth.

Etymology. The species epithet, *taylorae*, is dedicated to Dr. Joanne Taylor, from the Museum Victoria, Australia, to honor her important contributions to the knowledge on the amphipod family Phoxocephalidae.

DISCUSSION

Metharpinia taylorae **sp. nov.** shares the diagnostic characters of the genus, such as the constricted, narrow, spatulated and elongated rostrum, and uropod 3 with one of rami longer than peduncle, bearing article 2 on outer ramus, with two apical setae (BARNARD & KARAMAN 1991). Although *M. taylorae* **sp. nov.** shares some characters with species of *Microphoxus* (see comparison between the two genera in ALONSO DE PINA 2003a), we placed the new species in *Metharpinia* because, for the most part, it fits the diagnosis of this genus.

The new species is easily distinguished from *M. dentiurosoma* and from *M. grandirama* by in lacking the dorsal hook on urosomite 3, a unique feature of *M. dentiurosoma* and *M. grandirama* (ALONSO DE PINA 2003a).

Metharpinia taylorae **sp. nov.** differs from *M. protuberantis* by the following combination of characters (*M. protuberantis* characters in parenthesis): rostrum strongly constricted and highly developed (weakly constricted, poorly developed); gnathopod 2, basis, slightly elongate (strongly elongate); pereopod 5, coxa posterior lobe deeply produced (slightly produced); epimeral plate 3, posteroventral corner broadly rounded (strongly produced into a large tooth) (ALONSO DE PINA 2001).

Metharpinia taylorae **sp. nov.** differs from *M. coronadoi* by the following characters (*M. coronadoi* characters in parenthesis): coxa 1 weakly expanded anteriorly (anterior margin straight); right lacinia mobilis absent (present); left lacinia mobilis well developed, apically smooth and subrounded (with 2-3 teeth plus 1-2 accessory teeth); epimeral plate 3, posteroventral margin rounded (rounded-quadrate) (BARNARD 1980).

Metharpinia taylorae **sp. nov.** differs from *M. jonesi* by the following characters (*M. jonesi* characters in parenthesis): pereopods 3 and 4 very similar in shape (pereopod 4 stouter and longer than pereopod 3); epimeral plate 3, ventral margin with five submarginal pectinate setae (with large tooth, ventral margin with 4 setae) (BARNARD 1963).

Metharpinia taylorae **sp. nov.** differs from *M. floridana* by the following characters (*M. floridana* characters in parenthesis): epimeral plate 2 rounded (rounded-subquadrate); telson, both male and female, each lobe with 1 lateral small plumose seta (each lobe with 1 lateral and 1 subapical plumose setae) (SHOEMAKER 1933).

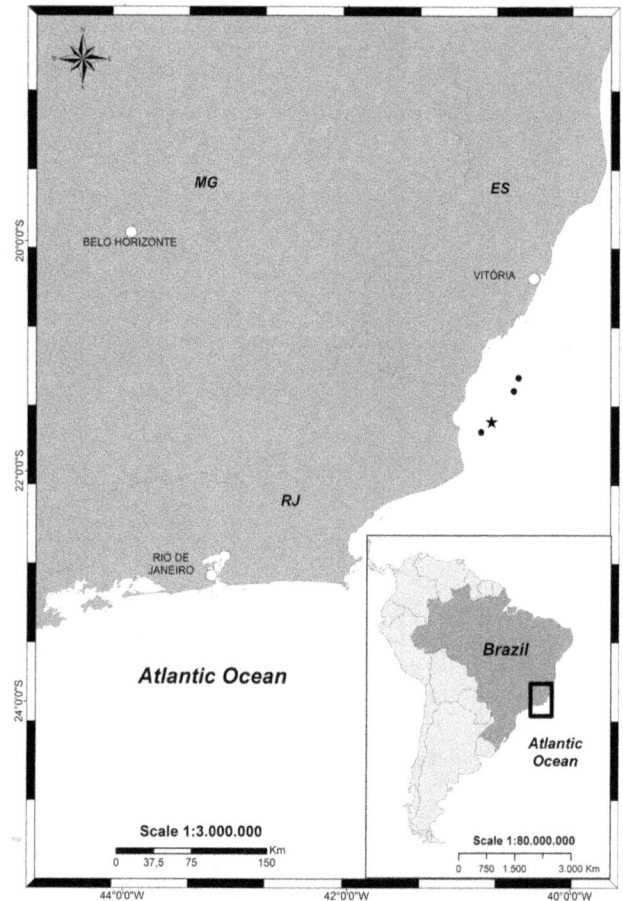

Figure 26. Distribution of *Metharpinia taylorae* **sp. nov.** Star: type locality, 21°33'52.574"S, 40°42'53.900"W; Circle: ocurrence of paratypes. RJ: Rio de Janeiro State; ES: Espírito Santo State; MG: Minas Gerais State (Distribution map by Danielle P. Cintra).

Metharpinia taylorae **sp. nov.** differs from *M. oripacifica* by (*M. oripacifica* characters in parenthesis): right lacinia mobilis absent (present); left lacinia mobilis well developed, apically smooth and subrounded (with five teeth, middle teeth scarcely shortened); telson, each lobe bearing 5 slender setae (dorsolateral brush of 7 setae) (BARNARD 1980).

Metharpinia iado, recorded from Argentina, is probably the most different from *M. taylorae* **sp. nov.**, due to the following characters (*M. iado* characters in parenthesis): left mandible, lacinia mobilis apically smooth and blunt (multi-cuspidate); maxilla 2, inner plate without plumose setae (with apical and medial plumose setae), outer plate, outer margin naked (setose); maxilliped, inner plate, with three apical plumose setae (eight apical plumose setae, plus one stout seta); gnathopods 1-2 weakly setose (strongly setose), palm of gnathopods 1-2 sinuous (almost straight); pereopod 5, coxa posterior lobe produced and subacute (not produced, round), pereopods 5-6, merus and carpus without facial setae (with facial sets of stout setae); telson, deeply cleft, about 90% (three-quarters cleft) (ALONSO DE PINA 2003b).

The description of the type-species of *Metharpinia*, *M. longirostris*, is insufficient and the species is poorly illustrated. However, we can distinguish the new species from it by the following characters (characters of *M. longirostris* within parenthesis): pereopods 5-6 without facial setae (with facial setae); posteroventral lobe of basis of pereopod 7 round (truncated); coxa 4, posteroventral corner with two small slender setae (ventral and posterior margins setose); each telson lobe with one lateral small plumose seta, apex naked (without lateral plumose setae, apex setose) (SCHELLENBERG 1931, BARNARD 1980).

Two characteristics are exclusive of the new species among the representatives of *Metharpinia*: palp of maxilla 1 slender, article 2 bearing slender setae on both sides and at apical margin; and the lacinia mobilis absent in right mandible and smooth and subrounded in the left mandible.

ACKNOWLEDGMENTS

We thank Coordenação de Aperfeiçoamento de Pessoal de Nível Superior (CAPES) and Fundação Carlos Chagas Filho de Amparo à Pesquisa do Estado do Rio de Janeiro (FAPERJ) for the financial support and fellowships. The material examined was provided by Centro de Pesquisas e Desenvolvimento Leopoldo Américo Miguez de Mello (CENPES-PETROBRAS). The distribution map was made by Danielle P. Cintra from Instituto de Geociências, Universidade Federal do Rio de Janeiro (IGEO-UFRJ).

REFERENCES

ALONSO DE PINA, GM (2001) Two new phoxocephalids (Crustacea: Amphipoda: Phoxocephalidae) from the south-west Atlantic. **Journal of Natural History 35**: 515-537.

ALONSO DE PINA GM (2003a) Two new species of *Metharpinia* Schellenberg (Amphipoda: Phoxocephalidae) from the southwest Atlantic. **Journal of Natural History 37**: 2521-2545.

ALONSO DE PINA GM (2003b) A new species of Phoxocephalidae and some other records of sand-borrowing Amphipoda (Crustacea) from Argentina. **Journal of Natural History 37**: 1029-1057.

BARNARD JL (1963) Relationship of benthic Amphipoda to invertebrate communities of inshore sublittoral sands of southern California. **Pacific Naturalist 3**(15): 437-467.

BARNARD JL (1980) Revision of *Metharpinia* and *Microphoxus* (marine phoxocephalid Amphipoda from the Americas). **Proceedings of the Biological Society of Washington 93**(1): 104-135.

BARNARD JL, DRUMMOND MM (1978) Gammaridean Amphipoda of Australia, part III: The Phoxocephalidae. **Smithsonian Contributions to Zoology 245**: 1-551.

BARNARD JL, DRUMMOND MM (1982) Gammaridean Amphipoda of Australia, Part V: Superfamily Haustorioidea. **Smithsonian Contributions to Zoology 360**: 1-148.

BARNARD JL, KARAMAN GS (1991) The Families and Genera of Marine Gammaridean Amphipoda (Except Marine Gammaroidea). **Records of the Australian Museum 13**: 1-866.

HORTON T, DE BROYER C (2014) Phoxocephalidae Sars, 1891. In: HORTON T, LOWRY J, BROYER C DE (Ed.). **World Amphipoda Database**. Available online at: http://www.marinespecies.org/aphia.php?p=taxdetails&id=101403 [Accessed: 22 May 2014]

HURLEY DE (1954) Studies on the New Zealand Amphipodan Fauna 3. The Family Phoxocephalidae. **Transactions of the Royal Society of New Zealand 81**: 579-599.

POORE AGB, LOWRY JK (1997) New ampithoid amphipods from Port Jackson, New South Wales, Australia (Crustacea: Amphipoda: Ampithoidae). **Invertebrate Taxonomy 11**: 897-941.

SENNA AR, SOUZA-FILHO JF (2011) A new species of *Pseudharpinia* (Amphipoda: Haustorioidea: Phoxocephalidae) from Southeastern Brazilian continental shelf. **Nauplius 19**(1): 7-16.

SCHELLENBERG A (1931) Gammariden und Caprelliden des Magellangebietes, Südgeorgiens und der Westantarktis. **Further Zoological Results of the Swedish Antarctic Expedition 1901-1903 2**(6): 1-290.

SHOEMAKER CR (1933) Amphipoda from Florida and the West Indies. **American Museum Novitates 598**: 1-24.

WATLING L (1989) A classification of crustacean setae based on the homology concept, p. 15-26. In: FELGENHAUER BE, THISTLE AB, WATLING L (Ed.) **Functional Morphology of Feeding and Grooming in Crustacea**. New York, CRC Press, Crustacean Issues, vol. 6.

10

Missing for the last twenty years: the case of the southernmost populations of the Tropical Mockingbird *Mimus gilvus* (Passeriformes: Mimidae)

Mariana S. Zanon[1,2], Mariana M. Vale[3] & Maria Alice S. Alves[2,4]

[1]*Programa de Pós-graduação em Ecologia e Evolução, Instituto de Biologia Roberto Alcantara Gomes, Universidade do Estado do Rio de Janeiro. Rua São Francisco Xavier 524, 20550-011 Rio de Janeiro, RJ, Brazil.*
[2]*Departamento de Ecologia, Instituto de Biologia Roberto Alcantara Gomes, Universidade do Estado do Rio de Janeiro. Rua São Francisco Xavier 524, 20550-011 Rio de Janeiro, RJ, Brazil.*
[3]*Departamento de Ecologia, Instituto de Biologia, Centro de Ciências da Saúde, Universidade Federal do Rio de Janeiro. Avenida Carlos Chagas 373, Ilha do Fundão, Cidade Universitária, 21941-902 Rio de Janeiro, RJ, Brazil.*
[4]*Corresponding author. E-mail: masaal@globo.com*

ABSTRACT. The Tropical Mockingbird *Mimus gilvus* (Vieillot, 1808) is a widespread species in the Neotropics, but its southernmost populations in Brazil are ecologically (and possibly taxonomically) distinct, occurring only along the coast in restinga vegetation. Once considered the most common bird in restinga, it is becoming increasingly rare, likely due to habitat loss and illegal capture of nestlings. We conducted field surveys to provide an up-to-date distribution of the Tropical Mockingbird in the southernmost portion of the species' range, in the state of Rio de Janeiro, supplying an estimate of its current regional population size and conservation status. We surveyed 21 restinga remnants in Rio de Janeiro, covering all major restinga areas in the state. For sites where the species' presence was confirmed through transect line surveys, we estimated the local population size. The species was found at only four sites. The mean local population density was 52 individuals per km^{-2}. The estimated current and historical Extent of Occurrence (EOO) were 256 km^2 and 653 km^2, respectively. Combining the population size and EOO results, we estimated that the population of the state of Rio de Janeiro currently ranges from 2,662 to 13,312 individuals, corresponding to an estimated reduction of 61% to 92% in population size in the last 20 years. The species, therefore, can be considered "Endangered" in the state of Rio de Janeiro. We recommend that a taxonomic study of the southernmost populations is carried out in order to clarify whether they represent a different, likely threatened species. We also recommend that the environmental regulations that protect restingas are used towards the protection of these populations.

KEY WORDS. Brazil; conservation status; IUCN; restinga; Rio de Janeiro.

The Tropical Mockingbird, *Mimus gilvus* (Vieillot, 1808), is widespread in the Neotropics, but some of its populations at the southernmost edge of its distribution are disappearing. The Tropical Mockingbird ranges from Mexico to Brazil, reaching its southernmost limit in the Brazilian state of Rio de Janeiro (MEYER DE SCHAUENSEE 1970, RIDGELY & TUDOR 1989, SIBLEY & MONROE 1990, SICK 2001, CODY 2005), where it is considered threatened (ALVES et al. 2000). There are also records of vagrant individuals found further south, in the Brazilian states of Paraná (SCHERER-NETO et al. 2011, BORNSCHEIN et al. 1997) and Santa Catarina (GHIZONI-JR & AZEVEDO, 2010). The species is absent from the checklist of birds of the state of São Paulo (SILVEIRA & UEZU 2011), although BORNSCHEIN et al. (1997) mentioned a supposed record of the species there. In spite of occupying most of the South American continent, in the eastern coast of Brazil the Tropical Mockingbird only occurs in restinga (SICK 2001),

an ecosystem of the Atlantic Forest biome that is located between the forest and the sea, and is mainly composed of sandy coastal plains (ARAUJO 1992).

There are several described subspecies of *Mimus gilvus* (HELLMAYR 1934, PINTO 1944, CODY 2005, RESTALL et al. 2007), including *M. gilvus antelius* Oberholser, 1819, which occupies the coast of Brazil from the state of Pará in the North, to Rio de Janeiro in the South. *Mimus gilvus antelius* is considered to be endemic to the Atlantic Forest (GONZAGA et al. 2000), and in the state of Rio de Janeiro it is only found in restinga (REIS & GONZAGA 2000). There are reasons to believe that *M. gilvus antelius* is a distinct species (CODY 2005).

According to PARKER et al. (1996), the Tropical Mockingbird is not very sensitive to habitat disturbances, and it is a priority neither for research nor for conservation. In southeastern Brazil, however, the species is highly susceptible to extinction. It was once considered the most common bird in restinga (Sɪᴄᴋ 2001), but in the last decades, it has disappeared from a large number of restinga remnants in the state of Rio de Janeiro (Tᴇɪxᴇɪʀᴀ & Nᴀᴄɪɴᴏᴠɪᴄ 1992, Aʀᴀᴜᴊᴏ & Mᴀᴄɪᴇʟ 1998, Aʀɢᴇʟ-ᴅᴇ-Oʟɪᴠᴇɪʀᴀ & Pᴀᴄʜᴇᴄᴏ 1998, Gᴏɴᴢᴀɢᴀ et al. 2000). Mockingbirds are well-known for their ability to sing, and for this reason human pressure acts on them not only through habitat loss and degradation, but also through illegal trade (Aʀɢᴇʟ-ᴅᴇ-Oʟɪᴠᴇɪʀᴀ & Pᴀᴄʜᴇᴄᴏ 1998, Aʟᴠᴇs et al. 2000).

Currently, the Tropical Mockingbird is listed as "Endangered" in the red list of Rio de Janeiro and Espírito Santo, which, together with Pará, are the only Brazilian states that have such lists within the species' range (Aʟᴠᴇs et al. 2000, Sɪᴍᴏɴ et al. 2007, Pᴀʀᴀ́ 2007). Furthermore, the subspecies M. gilvus antelius is listed as "Almost Threatened" in Brazil (Mᴀᴄʜᴀᴅᴏ et al. 2005). However, none of these lists were based on quantitative data, and only Sɪᴍᴏɴ et al. (2007) provided a justification for the listing, based on an estimated Extent of Occurrence < 5,000 km².

We conducted field surveys of the Tropical Mockingbird to provide up-to-date occurrence records in the southernmost portion of its range, in the state of Rio de Janeiro, and to supply an estimate of the current size of its population in the state. We then used this information to reassess the species' conservation status at the state level, based on two IUCN criteria: Extent of Occurrence and Population Size (IUCN 2001).

MATERIAL AND METHODS

This study was conducted in the state of Rio de Janeiro, Brazil, which represents the southernmost portion of the geographic range of the Tropical Mockingbird. We surveyed 20 areas with restinga remnants within the known geographic range of the Tropical Mockingbird, using line transects and/or casual survey, in order to establish the current distribution of the Tropical Mockingbird in the state (Table I). Both methods were used to confirm the occurrence of the species in each of the visited remnants. The areas were selected based on: I) previously known records of the species, II) restinga remnants indicated by Rᴏᴄʜᴀ et al. (2007), and III) availability of the species' habitat, i.e., restinga with open shrubby vegetation reaching the shoreline (Sɪᴄᴋ 2001). To locate these remnants, we used Google Earth satellite imagery. One additional area was surveyed (area 21 in Table I), outside the species' known range, to investigate a dubious record (Aʀɢᴇʟ-ᴅᴇ-Oʟɪᴠᴇɪʀᴀ & Pᴀᴄʜᴇᴄᴏ 1998) to the south of its known distribution. Surveys were conducted from August 2008 to November 2009, always under stable weather. The same researcher (MSZ) performed all surveys to standardize detectability.

In 15 of the 21 restinga remnants, line transects were applied systematically (Table I). In each of these remnants, one or more line transects were performed, totaling 18. The only restingas where we established more than one line transect were Marambaia (two transects) and Massambaba (three), which are among the largest restinga remnants in Rio de Janeiro. Transects were 2 km long, positioned at the shoreline near the border between the restinga and the beach. These boundaries are at natural sandy mounds, known as the beach ridge ("cordão arenoso" in Portuguese), which represent the highest point of restinga. Mimus gilvus individuals are very conspicuous. They frequently vocalize during the day, and always choose exposed portions of the highest perches to do it (V.C. Tomaz, pers. comm.). For these reasons, they are very easy to spot, even from the beach line. The surveys were always performed in the morning, beginning around six hours. They lasted, on average, 1h30min, with a little variation among the areas due to differences in topography and number of individuals recorded. Binoculars (10 x 50, Olympus or Nikon) were used to search for the birds. Observations were made from the line transect to the interior of the restinga, excluding bare sandy beach. Both visual and auditory contacts were considered. The survey effort was complemented with seven casual expeditions (Table I). These expeditions were also based on the line transects and occurred after the systematic survey, between 7:00 a.m. and 3:00 p.m., in "minor" restinga remnants next to our main study areas. The only methodological difference is that these expeditions were not made during periods of greater birds activity, i.e., the beginning of the morning, although M. gilvus is a very active bird all day long (V.C. Tomaz, pers. comm.). Sites were considered to belong to "minor" restinga remnants when they: 1) had damaged and/or very reduced areas of restinga vegetation, 2) did not have open shrubby vegetation or, if open shrubby vegetation was present, it did not reach the shoreline, and 3) there were no previous records of the Tropical Mockingbird (excepting Enseada do Bananal) in these areas. At all these sites, transects of at least 2 km at the shoreline or on main trails in the natural vegetation were used to search for tropical mockingbirds. In areas where the presence of the Tropical Mockingbird was confirmed through transect line survey, a point count (Hᴜᴛᴛᴏ et al. 1986, Mᴀᴛᴛᴏs et al. 2009) was used to estimate local population size. Ten points were positioned linearly, at 200 m intervals, parallel to the transect line, but 100 m further inside the vegetation than the transect line. The surveys, which began at 6:00 a.m., and lasted 10 minutes on each point, recorded all visual or auditory contacts with tropical mockingbirds into a 100 m-radius circle from the point. Contacts that occurred when the researcher was between two consecutive point counts were not recorded. Point counts were performed just once in each site. The point count surveys ran from October 2008 to August 2009, always when the weather was good (i.e., without rain).

Local population size was used to calculate population density (D) for each area, individually, where the point count method was applied. Population density was defined as the number of individuals counted divided by the area of the site surveyed. We employed the formula $D = \Sigma \pi r^2$, where N is the sum of individuals registered at all 10 point counts and $\Sigma \pi r^2$ is the sum of the ten 100 m-radius circles areas ($\Sigma \pi r^2 = 0,314$ km²).

Table I. Areas surveyed for the Tropical Mockingbird in the state of Rio de Janeiro, Brazil. Surveyed Areas: numbers correspond to areas depicted in Figs. 1-3, and letters in parenthesis to the source of previous record for the species in the area (areas with no source were selected from Rocha et al. (2007) and/or satellite imagery). Geographic Coordinates: refer to the midpoint for the area. Survey Methods: (LT) Line Transect, (CS) Casual Survey, (PC) Point Count.

Surveyed Areas	Geographic coordinates	Size (ha)*	Municipality	Species Present	Survey Method
1. Barra de Itabapoana	21°20'33"S/40°57'48"W	569	São Francisco do Itabapoana	No	LT
2. Grussaí [a]	21°45'20"S/41°01'22"W	302	São João da Barra	No	LT
3. Jurubatiba [a]	22°16'43"S/41°38'58"W	25,141	Carapebus	Yes	LT, PC
4. Itapebussus [a]	22°29'39"S/41°53'58"W	750	Rio das Ostras	Yes	LT, PC
5. Peró [a,b]	22°51'28"S/41°59'13"W	427	Cabo Frio	No	LT
6. Foguete/Dunas [a]	22°54'49"S/42°02'09"W	N/A	Cabo Frio	No	LT
7. Massambaba [a,d]	22°56'38"S/42°10'58"W	7,360	Araruama	Yes	LT, PC, CS
8. Jacarepiá [a]	22°56'04"S/42°26'45"W	508	Saquarema	No	LT
9. Jaconé [a]	22°56'15"S/42°38'47"W	40	Maricá	No	CS
10. Ponta Negra [a]	22°57'29"S/42°41'42"W	126	Maricá	No	CS
11. Barra de Maricá [a]	22°57'43"S/42°50'41"W	272	Maricá	No	LT
12. Itaipuaçu [a]	22°58'09"S/43°00'04"W	1,020	Maricá	No	LT
13. Itacoatiara/Banana [a,c]	22°58'28"S/43°02'00"W	N/A	Maricá	No	CS
14. Itaipu [a]	22°58'20"S/43°02'44"W	23	Maricá	No	LT
15. Marapendi	23°00'46"S/43°23'54"W	621	Rio de Janeiro	No	CS
16. Recreio dos Bandeirantes	23°01'47"S/43°28'05"W	N/A	Rio de Janeiro	No	LT
17. Chico Mendes	23°01'27"S/43°28'18"W	57.3	Rio de Janeiro	No	CS
18. Prainha	23°02'27"S/43°30'20"W	1	Rio de Janeiro	No	CS
19. Grumari	23°02'53"S/43°31'28"W	158	Rio de Janeiro	No	LT
20. Marambaia [a]	23°04'37"S/43°53'50"W	4,940	Rio de Janeiro	Yes	LT, PC
21. Praia do Sul [a]**	23°10'41"S/44°16'31"W	397	Angra dos Reis	No	LT

[a]Argel-de-Oliveira & Pacheco (1998), [b] M.B. Vecchi (pers. comm. 2007), [c] J.A.L. Pontes (pers. comm. 2009), [d] F.A. Galvão (pers. comm. 2009), * Complete list of references, is available on request from the senior author. ** Outside the known range of the species.

We calculated both the historical and current extent of occurrence of the Tropical Mockingbird in the state of Rio de Janeiro. The extent of occurrence (EOO) is 'the area contained within the shortest continuous imaginary boundary which can be drawn to encompass all the known, inferred or projected sites of present occurrence of a taxon, excluding cases of vagrancy' (IUCN 2001). We calculated the EOO of the Tropical Mockingbird using a buffer of 1,100 m from the coastline. In the state of Rio de Janeiro, the sandy portion of beaches do not exceed 100 m, and the Tropical Mockingbird has been mainly recorded up to 1,000 m within the restinga vegetation (M.A.S. Alves pers. obs. 2010), hence a EOO of 1,100 m from the coastline. The historical EOO was based on the scientific literature produced between the 19th century, when foreign naturalists registered their encounters with the Tropical Mockingbird during expeditions in Brazil, and the 1990's, when the first reports on the disappearance of the Tropical Mockingbird took place (a complete list of sources is available on request from the senior author). From that literature we inferred that the Tropical Mockingbird was a common and typical restinga species (Novaes 1950, Sick 2001), and that in Rio de Janeiro it

originally occurred from Marambaia to the Barra do Itabapoana (easternmost historical limit, i.e., the State's boundary). Its historical range, therefore, was most likely continuous throughout the shoreline. The historical EOO was built using the buffer in the entire shoreline of the state of Rio de Janeiro, and the current EOO was the historical EOO minus the portions of restinga were the species was considered absent in our survey and the literature.

To estimate the historical and current regional population size in the state of Rio de Janeiro, we multiplied the mean population density derived from point counts by the historical and current EOO, respectively. This is the maximum regional population size, because it considers that the species occurs throughout its EOO. Nevertheless, because we know that a species does not necessarily occupy the entire region within its EOO (IUCN 2001), we estimated a minimum regional population size by multiplying the maximum regional population size by the probability of finding the species in a given area, i.e., the percent of the 20 areas within the species' original range in which we found the species to be present (following Vale et al. 2007). This is a minimum population size since the probability of finding the species is most likely underestimated because of detection problems. Hence, the species' regional population should be somewhere between the estimated maximum and minimum population sizes.

We used the current estimated EOO and population size to determine the species' conservation status in the state of Rio de Janeiro, applying the two-step process (sensu GÄRDENFORS et al. 2001, IUCN 2001, MILLER et al. 2007).

RESULTS

We did not find the Tropical Mockingbird at Praia do Sul, Ilha Grande (area 21, Table I), a relatively well preserved area according to ROCHA et al. (2007). This confirms that Marambaia is the westernmost limit of the species in the state of Rio de Janeiro and the southernmost limit of the species in South America, except for vagrants. Tropical Mockingbird populations were found at only four of the 20 areas that were surveyed by transect lines or casual expeditions within the known range of the species in the state of Rio de Janeiro (Table I, Figs. 1-3), translating into a probability of finding the species of only 20%.

From our literature survey, the distribution of the Tropical Mockingbird in Rio de Janeiro originally ranged from Marambaia to the Barra do Itabapoana in the state's boundary (Fig. 1). Currently the species' range includes only four isolated areas that were originally part of its historical distribution (Figs. 2 and 3). The westernmost limit of the Tropical Mockingbird in the state of Rio de Janeiro is still Marambaia, but its easternmost limit has retreated to the Restinga de Jurubatiba, and gaps have appeared in the middle of its distribution (Fig. 3). The four currently known populations in the state of Rio de Janeiro occupy only about 250 km of shoreline, and are 211 km apart from the closest population, at Parque Estadual Paulo César Vinha, municipality of Guarapari, state of Espírito Santo (NARCISO 2012, L.C. de Araújo pers. comm. 2014).

In the four sites where point-count was performed, there was great variation in the number of individuals recorded, ranging from three to 38 individuals, which translates into a local population density between 10 and 121 individuals km^{-2} (Table II). The species' estimated EOO in the state of Rio de Janeiro

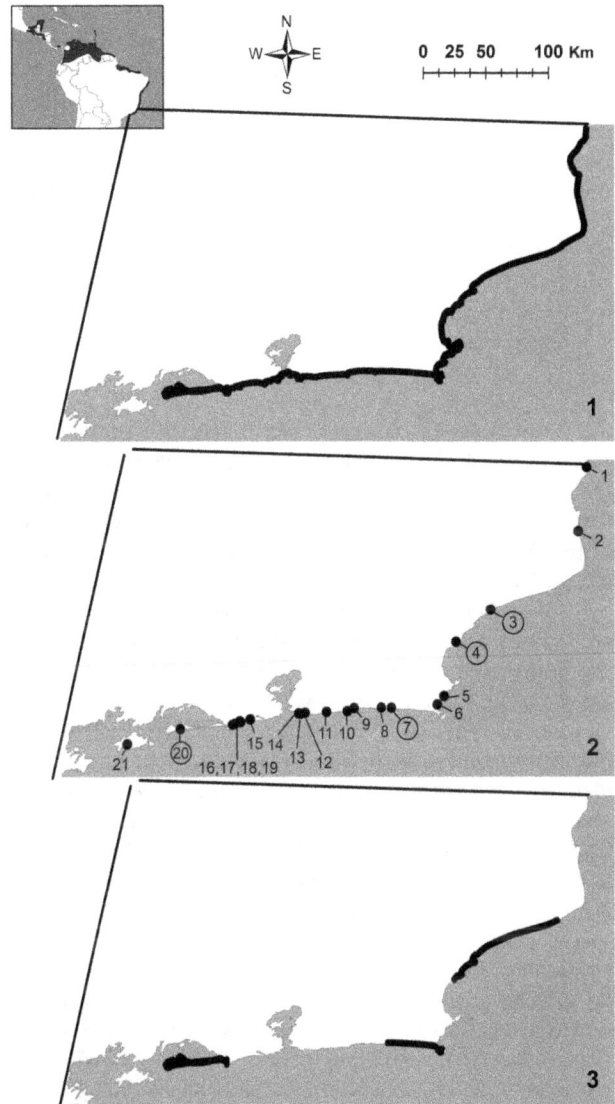

Figures 1-3. Distribution of the Tropical Mockingbird in the state of Rio de Janeiro, Brazil. The inset on the top left shows the global distribution of the species (from NatureServe.org). (1) Original area of occupancy (dark black line). (2) Areas surveyed in the study (numbers correspond to areas in Table I). Numbers in circles show the areas where the species was found. (3) Current area of occupancy (dark black line), depicting the four restinga remnants where the species is still found in the state of Rio de Janeiro.

Table II. Tropical Mockingbird population size and density based on point counts, in the state of Rio de Janeiro, Brazil.

Surveyed Area	Relative size (individuals)	Density (individuals/km²)
Jurubatiba	38	121
Marambaia	8	25
Massambaba *	0	0
Itapebussus	3	10
Average	16.3	52.0

*Although the species was recorded by the transect line method at this site, it was not observed during the survey by point counts. Therefore, its density was not taken into account for the average density computation.

was 256 km², representing only 39% of its original EOO of 653 km². The current EOO of the Tropical Mockingbird in the state of Rio de Janeiro (256 km²), mean population density (52 individuals km²) and the probability of finding the species (20%) translates into a population size ranging from 2,662 to 13,312 individuals in the state. Computing the original EOO (653 km²), we estimate an original population size of 33,904 individuals in the state of Rio de Janeiro, i.e., an estimated reduction of 61% to 92% in population size in the last 20 years.

After the two-step process (sensu GÄRDENFORS et al. 2001, IUCN 2001, MILLER et al. 2007), the Tropical Mockingbird was evaluated as regionally "Endangered", confirming previous non-quantitative assessments for the state of Rio de Janeiro (ALVES et al. 2000). The species has an estimated EOO of 256 km², meeting the 5,000-km² cut-off for the "Endangered" status (IUCN 2001), although being quite close to the 100-km cut-off for the "Critically Endagered" status.

DISCUSSION

Our study of the conservation status of the Tropical Mockingbird shows that the species is Endangered in the state of Rio de Janeiro, confirming previous qualitative assessments (ALVES et al. 2000). Actually, the current Extent of Occurrence of the Tropical Mockingbird approaches the 100 km² threshold for for the Critically Endangered species.

We estimated a population reduction between 30.5% and 46% in 10 years, meeting the threshold of > 30% reduction for the "Vulnerable" status, and an estimated population between 2,662 and 13,312 individuals, which is above the 10,000-individuals cut-off for the "Vulnerable" status. Because IUCN requires that species be listed at the highest threat category identified, we suggest the Tropical Mockingbird be listed as "Endangered" in Rio de Janeiro, 'facing a very high risk of extinction in the wild' (IUCN 2001). The proposed representation of its assignment is EN B1ab (I, III, IV, V), meaning that its EOO is less than 5,000 km²; our estimates indicate that the species exists in no more than five locations; and there is a continuing observed decline in its Extent of Occurrence, habitat size and quality, number of locations and number of mature individuals are in constant decline. We disregard a possible rescue effect from outside populations (GÄRDENFORS et al. 2001, MILLER et al. 2007), because the Tropical Mockingbird is also listed as "Endangered" in the neighboring state of Espírito Santo (SIMON et al. 2007). It was not found in the easternmost restingas of the state of Rio de Janeiro, and it apparently does not occur anymore in Neves, the westernmost restinga in Espírito Santo (M.A.S. Alves pers. obs. 2014).

This study surveyed the restingas of Rio de Janeiro very well. All nine major restinga sections identified by ARAUJO & MACIEL (1998) were surveyed, including 18 of the 21 remnants identified by ROCHA et al. (2007). We confirmed the local extinctions of the Tropical Mockingbird in Barra de Maricá, Niterói, and in the municipality of Rio de Janeiro (excluding the Restinga of Marambaia), previously mentioned in the literature (TEIXEIRA & NACINOVIC 1992, ARGEL-DE-OLIVEIRA & PACHECO 1998, ARAUJO & MACIEL 1998, GONZAGA et al. 2000), and recorded new local extinctions in the eastern portion of the state, above the Jurubatiba National Park (at Grussaí and Barra do Itabapoana remnants, in

the extreme-west of Rio de Janeiro). From the results of the present study, obtained by a systematic survey (and some occasional visits), we also consider the species to be probably extinct in the municipality of Cabo Frio, as suggested by ARGEL-DE-OLIVEIRA & PACHECO (1998). In 2007 two individuals were recorded at Praia Peró, Cabo Frio (M.B. Vecchi, pers.comm.), but were likely vagrants according to the source of the record itself and because our systematic survey at the site did not confirm the species' presence. The WikiAves citizen science website, however, has more recent records of the species at unspecified locations in Cabo Frio (http://www.wikiaves.com). Further scientific investigation should be carried out in order to determine whether these records are from vagrants or if the species is indeed returning to the region.

There are a number of possible explanations for the extinction of most populations of M. gilvus in Rio de Janeiro. One important factor might be the species' association with open shrubby restinga vegetation (SICK 2001), which is the first vegetation to disappear with occupation of the coastline (ROCHA et al. 2007). In our study, we found that all remnants where the species was present had open shrubby restinga vegetation (although the inverse was not true), and that the species was absent from Grumari and Praia do Sul despite the good conservation status of these areas. These two restingas, however, are composed of dense and close vegetation. Another important factor in determining the species' presence might be the amount of available habitat. The Metropolitan Region of Rio de Janeiro and Niterói, where the species is no longer present, is extremely urbanized and degraded (except for Restinga of Marambaia, Rio de Janeiro) (FUNDAÇÃO CIDE 2003). In addition, Maricá and Cabo Frio, which, together with the Metropolitan Region, correspond to the current gap in the middle of the distribution range of M. gilvus, have either been almost completely converted into pastures and agricultural fields, or have lost expressive vegetation cover (FUNDAÇÃO CIDE 2003). Similarly, the boundary with the state of Espírito Santo, which once represented the species' westernmost limit in Rio de Janeiro, is currently predominantly taken by agriculture (FUNDAÇÃO CIDE 2003). At the same time, even relatively large and well preserved areas, such as Restinga of Grussaí (ROCHA et al. 2007), have witnessed the local extinction of M. gilvus. This may result from more subtle pressures, such as the disruption of metapopulation dynamics. The severe fragmentation of the restingas of the state of Rio de Janeiro (ROCHA et al. 2007) might prevent the migration of individuals from well preserved remnants, such as the Jurubatiba National Park. Populations from these areas could act as a source of individuals for sink populations (PULLIAM 1988). Finally, another important factor in the extinction of M. gilvus populations is illegal capture, which is intense because of the remarkable singing abilities of this bird. According to local people, illegal capture does happen in the Jurubatiba National Park, Massambaba and Itapebussus, and there is also evidence that this also occurs in the restingas of

Cabo Frio (J.F. Pacheco, pers. comm. 2008), Maricá, and the municipality of Rio de Janeiro (Argel-de-Oliveira & Pacheco 1998).

Three of the four areas where the Tropical Mockingbird persists in the state of Rio de Janeiro are protected, to some degree, by natural reserves. Jurubatiba, which houses the largest population in our study, is within the Restinga de Jurubatiba National Park. The park is still relatively intact, despite insufficient law enforcement, land parcels that have not yet been expropriated, and open access to cars and people. Marambaia, which has a medium size mockingbird population, is not within a protected area, but is within a military area. The military prevent access to the major part of Marambaia, conserve noteworthy natural areas, and protect fauna against hunting. Therefore, despite not being within a protected area, the conservation status of the Marambaia is good (Rocha et al. 2007). Massambaba and Itapebussus shelter the smaller populations of the Tropical Mokingbird in the state of Rio de Janeiro. Massambaba was been recently incorporated into Costa do Sol State Park, and Itapebussus is within a municipally protected area. Both areas are immersed in a landscape that suffers from unplanned urbanization (including illegal occupation and excessive land parceling) and illegal captures of tropical mockingbirds.

The restinga ecosystem in general is protected by legislation at the national, state and municipal levels. At the national level, restingas are protected by the Forest Code (Brasil 2012) and the Atlantic Forest Law (Brasil 2006). The latter establishes, among other rules, that new facilities should be constructed preferentially in areas that have been already modified or degraded. In practice, however, restingas are under great urbanization pressure such as the construction of summer houses, resorts, ports, and the implementation of roads, pastures, and plantations at the shoreline (Araujo & Lacerda 1987, Diegues 1999, Alves et al. 2000, Rocha et al. 2007, Mattos et al. 2009). In the state of Rio de Janeiro this is worsened by the fact that most restingas are not within protected areas (Rocha et al. 2007).

The Tropical Mockingbird has a very large global distribution, ranging from Mexico to southern Brazil, and therefore is considered of "Least Concern" by IUCN (2011). However, there are enough arguments to take its local disappearance seriously. First, in the southernmost portion of its distribution (excluding records of vagrants), the species is naturally restricted to the coast and is specialized in open restinga habitat. This southern form is described as *M. gilvus antelius* (Hellmayr 1934, Pinto 1944, Cody 2005, Restall et al. 2007), and possibly deserves to be considered as a separate species (Cody 2005). If that is the case, this taxon will likely be threatened, given that it was listed as threatened in Rio de Janeiro and Espírito Santo, and is subjected to the very same human pressures in the remaining of its distribution. There is no red list of threatened species in the remaining ten states within its historical distribution, with the exception of the state of Pará in the Amazon,

where the species is not listed (Pará 2007). Second, the geographical range of a species often begins to shrink at the border of its historical geographic distribution (Chanell & Lomolino 2000, Anjos et al. 2009). Populations at the border of the species' distribution are more vulnerable to extinction because they are often in marginal habitats, and seem to need greater areas to guarantee their continuity (Anjos et al. 2009). Furthermore, these populations usually have fewer available migrants from nearby populations that could rescue the population in case of a local extinction (Anjos et al. 2009). Thus, a small decrease in the range of a species is a good predictor of a much larger population reduction in the future (Chanell & Lomolino 2000). After all, population extinctions precede species extinction (Ceballos & Ehrlich 2002), and *Mimus gilvus antelius* has endured many local extinctions in the last decades. Although conservation tends to focus on extinctions at the species level rather than at the population level, local populations are the ones that play ecological roles in ecosystems, often providing services and goods to humankind (Ehrlich & Daily 1993, Hughes et al. 1997, Ceballos & Ehrlich 2002). For instance, the Tropical Mockingbird disperses restinga nurse plant species, suggesting that it can be a keystone species in the ecological succession of restinga bushes (Gomes et al. 2008, 2010) in areas like Jurubatiba, where it is one of the most abundant resident species (Alves et al. 2004).

"Typical of restingas" and "the restinga's most commonly spotted bird" are expressions that used to be employed in reference to the Tropical Mockingbird (Novaes 1950, Sick 2001). Ironically, the Tropical Mockingbird may currently be one of the most threatened restinga species, after the Restinga Antwren, *Formicivora littoralis* Gonzaga & Pacheco, 1990, (Mattos et al. 2009). We strongly recommend a taxonomic study of the southernmost populations of the Tropical Mockingbird in order to reveal its true taxonomic status. If it represents a different species, as suggested in the literature, then it is likely threatened with extinction. We also recommend enhanced environmental education and environmental regulations that aim to protect restinga areas (Brasil 2006, 2012, Develey & Pongiluppi 2010), in order to avoid further local extinctions.

ACKNOWLEDGEMENTS

We thank Edvandro Ribeiro for his extensive support with field work; Clinton Jenkins for English review; Instituto Biomas and UERJ for transportation; Idea Wild for equipment; INEA, SMAC, SEMAP and Base Aérea de Santa Cruz for research permits; CNPq for a MSc fellowship to MSZ (process 135861/2008 0) and a research grant to MASA (308792/2009-2); FAPERJ for a research grant to MASA (E-26/102837/2012); and PROBIO-II/MCTI/JBRJ/MMA/GEF, PPBio/Rede BioM.A./MCTI/CAPES, and the Brazilian Research Network on Global Climate Change/Rede Clima (Grant No. 01.0405.01) for support to MMV.

REFERENCES

Alves MAS, Pacheco JF, Gonzaga LAP, Cavalcanti RB, Raposo MA, Yamashita C, Maciel NC, Castanheira M (2000) Aves, p. 113-124. In: Bergallo HG, Rocha CFD, Alves MAS, Van Sluys M (Eds) **A fauna ameaçada de extinção do estado do Rio de Janeiro.** Rio de Janeiro, EdUerj, 168p.

Alves MAS, Storni A, Almeida EM, Gomes VSM, Oliveira CHP, Marques RV, Vecchi MB (2004) A comunidade de aves na Restinga de Jurubatiba, p. 199-214. In: Rocha CFD, Esteves FA, Scarano FR (Eds) **Pesquisas de longa duração na Restinga de Jurubatiba: Ecologia, História Natural e Conservação.** São Carlos, RiMA, 376p.

Anjos L, Holt RD, Robinson S (2009) Position in the distributional range and sensitivity to forest fragmentation in birds: a case history from the Atlantic forest, Brazil. **Bird Conservation International** 20(4): 1-8. doi: 10.10.1017/s095927090999025

Araujo DSD (1992) Vegetation types of sandy coastal plains of Tropical Brazil: a first approximation, 337-348p. In: Seeliger U (Ed.) **Coastal plant communities of Latin America.** New York, Academic Press, XX+392p.

Araujo DSD, Lacerda LD (1987) A natureza das restingas. **Ciência Hoje** 6(33): 42-48.

Araujo DSD, Maciel NC (1998) Restingas fluminenses: biodiversidade e preservação. **Boletim FBCN 25:** 25-51.

Argel-de-Oliveira MM, Pacheco JF (1998) Um resumo da situação: *Mimus saturninus* e *M. gilvus* no litoral sudeste brasileiro. **Boletim FBCN 25:** 53-69.

Bornschein MR, Reinert BR, Pichorim R (1997) Notas sobre algumas aves novas ou poucas conhecidas do sul do Brasil. **Ararajuba** 5(1): 53-59.

Brasil (2006) Lei Federal n° 11.428 de 22 de dezembro de 2006. **Diário Oficial da União 246:** 1-4.

Brasil (2012) Lei Federal N° 12.651 de 25 de maio de 2012. **Diário Oficial da União 102:** 1-8.

Ceballos G, Ehrlich PR (2002) Mammal population losses and the extinction crisis. **Science** 296(5569): 904-907. doi: 10.1126/science.1069349

Chanell R, Lomolino MV (2000) Trajectories to extinction: spatial dynamics of the contraction of geographical ranges. **Journal of Biogeography** 27(1): 169-179. doi: 10.1046/j.1365-2699.2000.00382.x

Cody M (2005) Family Mimidae (Mockingbirds and Thrashers), p. 448-495. In: Hoyo J Del, Elliot A, Christie D (Eds) **Handbook of the birds of the world.** Barcelona, Lynx Editions, vol. 10, 895p.

Develey PF, Pongiluppi T (2010) Impactos potenciais na avifauna decorrentes das alterações propostas para o Código Florestal Brasileiro. **Biota Neotropica** 10(4): 43-45. doi: 10.1590/S1676-06032010000400005

Diegues AC (1999) Human populations and coastal wetlands: conservation and management in Brazil. **Ocean & Coastal Management** 42(2-4): 187-210. doi: 10.1016/S0964-5691(98)00053-2

Ehrlich PR, Daily GC (1993) Population extinction and saving biodiversity. **AMBIO 22:** 64-68.

Fundação Cide (2003) **Índice de Qualidade dos Municípios: Verde II.** Rio de Janeiro, Fundação Centro de Informações e Dados do Rio de Janeiro, CD-ROM.

Gärdenfors U, Hilton-Taylor C, Mace GM, Rodríguez JP (2001) The application of IUCN Red List criteria at regional levels. **Conservation Biology** 15(5): 1206-1212. doi: 10.1111/j.1523-1739.2001.00112.x

Ghizoni-Jr IR, Azevedo MAG (2010) Registro de algumas aves raras ou com distribuição pouco conhecida em Santa Catarina, sul do Brasil, e relatos de três novas espécies para o estado. **Atualidades Ornitológicas 154:** 33-46.

Gomes VSM, Correia MCR, Lima HA, Alves MAS (2008) Potential role of frugivorous birds (Passeriformes) on seed dispersal of six plant species of a restinga habitat, southeastern Brazil. **Revista de Biología Tropical** 56(1): 205-216.

Gomes VSM, Buckeridge MS, Silva CO, Scarano FR, Araujo DSD, Alves MAS (2010) Availability peak of caloric fruits coincides with energy-demanding seasons for resident and non-breeding birds in restinga, an ecosystem related to the Atlantic forest, Brazil. **Flora 205** (10): 647-655. doi: 10.1016/j.flora.2010.04.014

Gonzaga LP, Castiglioni GDA, Reis HBR (2000) Avifauna das restingas do Sudeste: estado do conhecimento e potencial para futuros estudos, p. 151-163. In: Esteves FA, Lacerda LD (Eds) **Ecologia de restingas e lagoas costeiras.** Macaé, NUPEM/UFRJ, 446p.

Hellmayr CE (1934) **Catalogue of the birds of the Americas.** Chicago, Field Museum of Natural History, Zoology Series, vol. 13, Part 7, 531p.

Hughes JB, Daily GC, Ehrlich PR (1997) Population diversity: its extent and extinction. **Science 278**(5338): 689-692. doi: 10.1126/science.278.5338.689

Hutto RL, Pletschet SM, Hendricks PP (1986) A fixed-radius point count method for nonbreeding and breeding season use. **The Auk 103:** 593-602.

IUCN (2001) **IUCN Red List Categories and Criteria.** Gland, International Union for Conservation of Nature, v. 3.1, IV+32p.

IUCN (2011) **IUCN Red List of Threatened Species.** International Union for Conservation of Nature, v. 2011.2. Available online at: http://www.iucnredlist.org [Accessed: 2 March 2012]

Machado ABM, Martins CS, Drummond GM (2005) **Lista da fauna brasileira ameaçada de extinção: incluindo as listas das espécies quase ameaças e deficientes em dados.** Belo Horizonte, Fundação Biodiversitas, 160p.

Mattos JCF, Vale MM, Vecchi MB, Alves MAS (2009) Abundance, distribution and conservation of the Restinga Antwren *Formicivora littoralis*. **Bird Conservation International** 19(4): 392-400. doi: 10.1017/S0959270909008697

MEYER DE SCHAUENSEE R (1970) **A guide to the birds of South America.** Philadelphia, Academy of Natural Sciences of Philadelphia, 498p.

MILLER RM, RODRÍGUEZ JP, FOWLER TA, BAMBARADENIYA C, BOLES R, EATON MA, GÄRDENFORS U, KELLER V, MOLUR S, WALKER S, POLLOCK C (2007) National threatened species listing based on the IUCN criteria and regional guidelines: current status and future perspectives. **Conservation Biology** 21(3): 684-696. doi: 10.1111/j.1523-1739.2007.00656.x

NARCISO LC (2012) **Parque Estadual Paulo César Vinha: preservando o nosso quintal.** Cariacica, Espírito Santo, IEMA.

NOVAES FC (1950) Sobre as aves de Sernambetiba, Distrito Federal, Brasil. **Revista Brasileira de Biologia** 10: 199-208.

PARÁ (2007) **Resolução 054 de 2007.** Available online at: http://www.sectam.pa.gov.br/relacao_especies.htm. [Accessed: 25 November 2009]

PARKER T, STOTZ D, FITZPATRICK J (1996) Database A: zoogeography and ecological attributes of bird species breeding in the Neotropics, p. 250. In: Stoltz D, Fitzpatrick J, Parker III T, Moskovits D (Eds) **Neotropical birds: ecology and conservation.** Chicago, University of Chicago Press, XX+502p.

PINTO O (1944) **Catálogo das aves do Brasil: Parte 2.** São Paulo, Secretaria de Agricultura, 700p.

PULLIAM RH (1988) Sources, sinks, and population regulation. **The American Naturalist** 132(5): 652-661. doi: 10.1086/284880

REIS HBR, GONZAGA LP (2000) Análise geográfica das aves das restingas do Estado do Rio de Janeiro, p. 165-178. In: ESTEVES FA, LACERDA LD (Eds) **Ecologia de restingas e lagoas costeiras.** Macaé, NUPEM/UFRJ, 446p.

RESTALL R, RODNER C, LENTINI M (2007) **Birds of northern South America.** New Haven, Yale University Press, vol. 2, XX+656p.

RIDGELY RS, TUDOR G (1989) **The birds of South America.** Austin, University of Texas Press, vol. 1, XVI+516p.

ROCHA CFD, BERGALLO HG, VAN SLUYS M, ALVES MAS, JAMEL CE (2007) The remnants of restinga habitats in the Brazilian Atlantic Forest of Rio de Janeiro State, Brazil: habitat loss and risk of disappearance. **Brazilian Journal of Biology** 67(2): 263-273. doi: 10.1590/S1519-69842007000200011

SCHERER-NETO P, STRAUBE FC, CARRANO E, URBEN-FILHO A (2011) **Lista das aves do Paraná: edição comemorativa do "Centenário da Ornitologia do Paraná".** Curitiba, Hori Consultoria Ambiental, 130p.

SIBLEY CG, MONROE BL (1990) **Distribution and taxonomy of the birds of the world.** New Haven, Yale University Press, XXIV+1111p.

SICK H (2001) **Ornitologia Brasileira.** Rio de Janeiro, Nova Fronteira, 912p.

SILVEIRA LF, UEZU A (2011) Checklist das aves do Estado de São Paulo, Brasil. **Biota Neotropica** 11(Suppl. 1): 83-110. doi: 10.1590/S1676-06032011000500006

SIMON JE, ANTAS PTZ, PACHECO JF, EFÉ MA, RIBON R, RAPOSO MA, LAPS RR, MUSSO C, PASSAMANI JA, PACCAGNELLA SG (2007) As aves ameaçadas de extinção no estado do Espírito Santo, p. 47-63. In: PASSAMANI M, MENDES SL (Eds) **Espécies da fauna ameaçadas de extinção no Estado do Espírito Santo.** Vitória, Instituto de Pesquisas da Mata Atlântica, 140p.

TEIXEIRA D, NACINOVIC JB (1992) Aves da Barra da Tijuca, p. 133-145. In: CARVALHO HOSKEN (Ed.) **Parque da Gleba E, Rio de Janeiro: a entidade.** Rio de Janeiro, Carvalho Hosken S.A., 152p.

VALE MM, BELL JB, ALVES MAS, PIMM SL (2007) Abundance, distribution and conservation of Rio Branco Antbird *Cercomacra carbonaria* and Hoary-throated Spinetail *Synallaxis kollari*. **Bird Conservation International** 17(3): 245-257. doi: 10.1017/S0959270907000743

11

Annual male reproductive activity and stages of the seminiferous epithelium cycle of the large fruit-eating *Artibeus lituratus* (Chiroptera: Phyllostomidae)

Alice A. Notini[1], Talita O. Farias[1], Sônia A. Talamoni[1,*] & Hugo P. Godinho[1]

[1]*Programa de Pós-graduação em Zoologia de Vertebrados, Departamento de Ciências Biológicas, Pontifícia Universidade Católica de Minas Gerais, Avenida Dom José Gaspar, 500, 30535-610, Belo Horizonte, Minas Gerais, Brazil.*
*Corresponding author. E-mail: talamoni@pucminas.br

ABSTRACT. The large fruit-eating phyllostomid bat, *Artibeus lituratus* (Olfers, 1818), forearm 69-75 mm, body mass 66-82 g, has a diversified geographic distribution in the Neotropical region. Therefore it is subjected to different climatic conditions that affect its reproduction, leading to different reproductive strategies such as continuous reproduction, seasonal monoestry or seasonal bimodal polyestry. In this study we used morphometric and histological methods to analyze the annual reproductive activity of *A. lituratus* males in a population living in the Atlantic Forest, Southeastern Brazil. Testis mass, epididymis mass, gonadosomatic index, seminiferous tubule diameter, and Leydig cell nucleus diameter showed no significant differences (p > 0.05) in the two seasons (wet: October to March; dry: April to September). Additionally, the cauda epididymis was packed with sperm throughout the period of study. Our data indicate that in this population spermatogenic activity was continuous throughout the year. Slight variations in accumulated frequency of pre-meiotic, meiotic and post-meiotic stages of the seminiferous epithelium cycle were observed when compared to other bat species, probably due to species-specific characteristics.

KEY WORDS. Morphometry; Neotropical bat; seminiferous epithelium; spermatogenesis; reproduction.

The large fruit-eating phyllostomid bat, *Artibeus lituratus* (Olfers, 1818), is widely distributed in the Neotropical region; it occurs from Central Mexico, the Lesser Antilles, Trinidad and Tobago, south to North-central Argentina. It occupies a variety of habitats from the sea level to at least 2,620 m (Marques-Aguiar 2007). This bat features a large body (mass 66-82 g), with forearm length exceeding 70 mm (Vizotto & Taddei 1973), white facial stripes, and brown body color, although grayish individuals occur in some regions (Reis et al. 2007).

Differences in the reproductive traits of *A. lituratus*, reported by various authors, have been attributed to its wide area of occurrence. A bimodal pattern of reproduction with a minor birth peak during the dry season and a main birth peak in the rainy season have been observed in Panama and Costa Rica (Fleming et al. 1972), and Brazil (Reis 1989). Based on analysis of seasonal testis activity in Southeastern Brazil, Oliveira et al. (2009) reported that the reproductive period of this bat coincides with the rainy season, and is followed by testicular regression before a new cycle of testicular recrudescence. However, in other studies, males have been considered fertile year-round, since they exhibit continuous spermatogenesis (Tamsitt & Valdivieso 1963, Duarte & Talamoni 2010).

The well-defined rainy season of Southeastern Brazil is characterized by abundant rainfall associated with high temperatures and humidity (Sá Júnior et al. 2012). Rains are essential for seed germination and for the fruits that large fruit-eating bats depend on. The dry season is characterized by shortage in rainfall, flowers and food. Despite apparently unfavorable environmental conditions, females of some species reproduce during the dry season and the birth of their offspring coincides with fruiting in the next rainy season (Racey 1982).

In mammals, germ cells are arranged in typical cell associations known as stages (França et al. 2005, Hess & França 2007). These stages follow one another in time, characterizing the seminiferous epithelium cycle. Based on the tubular morphology method, eight stages have been described for the seminiferous epithelium cycle of mammals (Berndtson 1977 França & Russell 1998), including Neotropical bats (Beguelini et al. 2009, Oliveira et al. 2009, Morais et al. 2013).

In this paper, we evaluated data obtained on testis and epididymis mass, gonadosomatic index, diameter of seminiferous tubules, and Leydig cell nucleus diameter in order to determine whether there are variations in the annual reproductive activity of large fruit-eating bat males. The data were grouped into rainy and dry seasons, since these are two major seasonal variables that affect bat reproduction in the Neotropics (Racey & Entwistle 2000). We registered the occurrence of sperm in the cauda epididymis during the period of study. Finally, we characterized the stages of the seminiferous epithelium cycle and their relative frequency of occurrence, which were then analyzed and compared with respect to the rainy-dry seasons.

MATERIAL AND METHODS

The study was conducted in the Special Protection Area of Fechos (SPA Fechos), an area with 1,076 ha, 1,400 m above sea level in the state of Minas Gerais, Southeastern Brazil (20°04'S, 43°57'W). SPA Fechos consists of semideciduous forest fragments, typical of the Atlantic Forest. As in the remaining of Southeastern Brazil, SPA Fechos is subject to a well defined rainy season (from October to March) and a dry season (from April to September) (Sá Júnior et al. 2012). Rainfall in the area during the period of study was 1,457.8 mm. November and January had the highest rainfall values, 321.7 mm and 276.8 mm, respectively. During the same period, the monthly average temperatures ranged from 10.8 to 29.2°C. December to March was the hottest period, with mean temperature 21.3°C, and July to September was the coolest period, with mean temperature 16.9°C.

The bats were captured monthly on two consecutive nights from December 2001 to January 2003, using mist nets left in the forest from 18:00 to 24:00 h, following the procedures described in Duarte & Talamoni (2010). Thirty-three adult males were captured and killed with an overdose of ether followed by cervical dislocation. They were categorized as adults when they displayed ossified epiphyseal plates in the metacarpal and phalangeal bones (Anthony 1988). At the time of capture, the body mass and position of the testis in relation to the scrotum were registered.

Scrotal testes and respective epididymides were removed and weighed. The gonadosomatic index (GSI = mass of testis x 200/body mass) was calculated for each bat. Since the left and right testis and the epididymis did not present significant mass differences (p > 0.05), the left testis and epididymis were cut into fragments and fixed in Bouin solution for 8-12 hours. The fragments were subjected to routine histological techniques for embedding in plastic resin and 5 μm-thick histological sections stained in toluidine blue were obtained. Two-hundred near circular cross-sections of seminiferous tubules from 24 bats were digitally photographed in an optical microscope and had their diameter measured using Image J (Rasband 2012). They were also used to categorize the stages of the seminiferous epithelium cycle, using the method of tubular morphology (Clermont 1972, Berndtson 1977, França & Godinho 2003), based on the following characteristics: shape of spermatid nuclei, occurrence of meiotic divisions, and position of spermatids in the seminiferous epithelium. We also measured the diameter of 10 Leydig cell nuclei close to the seminiferous tubules in stage 1 of the seminiferous epithelium cycle from each of 15 bats.

Bat capture and handling were conducted according to guidelines of the Brazilian National Council for the Control of Animal Experimentation (CONCEA), which are aligned with standard international norms. This study was carried out under license (# 206/2001) granted by the Brazilian Institute of Natural Environment and Renewable Resources (IBAMA). The bats are deposited in the collection of the Pontifical Catholic University of Minas Gerais, Belo Horizonte, MG, Brazil.

Data are presented as mean ± SD (standard deviation). They were initially tested for normality. Variations in testis parameters obtained in the rainy and dry seasons were evaluated by the Student's t test (p < 0.05). Since GSI did not present normal distribution, the non-parametric Mann-Whitney test was used. The relative frequencies of the stages of the seminiferous epithelium cycle were subjected to chi-square test (χ^2) for comparison between the rainy and the dry seasons (Sokal & Rohlf 1995).

RESULTS

Corporal and testis parameters

We registered no statistically significant variation in the parameters analyzed between the rainy and the dry seasons (Table 1). All captured bats had functional spermatogenic testis and cauda epididymis packed with spermatozoa (Fig. 1).

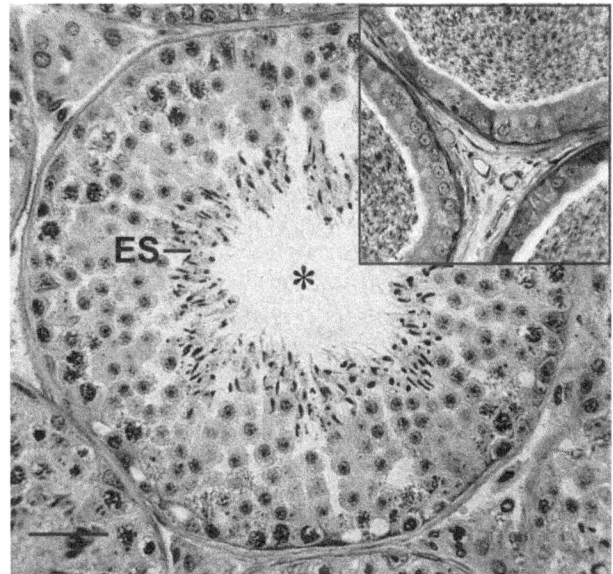

Figure 1. Transverse section of seminiferous tubule of *A. lituratus* in activity (diameter, 194 ± 8 μm). Note elongated spermatids (ES) and abundant spermatozoa in the lumen (*). Insert shows transverse section of cauda epididymis packed with spermatozoa. H-E stain. Scale bars: 69 μm.

Stages of the seminiferous epithelium cycle

We characterized eight stages of the seminiferous epithelium cycle (SEC) of the large fruit-eating bat (Figs. 2-9) according to the tubular morphology method. Most cross-sections

Table 1. Mean values (± 1 Standard Deviation) of corporal mass (CM, g), testis mass (TM, g), epididymis mass (EM, g), gonadosomatic index (GSI), diameter of the seminiferous tubules (DST, μm) and diameter of the Leydig cell nucleus (DLC, μm) of the large fruit-eating bat in rainy and dry seasons (respective sample sizes in parenthesis) of December 2001 to January 2003.

Season	CM (13, 15)	TM (13, 15)	EM (11, 12)	GSI (13, 15)	DST (12, 12)	DLC (8, 8)
Rainy	67.3 ± 4.6	0.13 ± 0.05	0.02 ± 0.01	0.37 ± 0.15	197 ± 10	6.37 ± 0.51
Dry	66.8 ± 6.0	0.12 ± 0.07	0.03 ± 0.02	0.36 ± 0.20	190 ± 5	6.37 ± 0.51
Student's t-test	0.27	0.24	-0.36	83.5*	0.72	0.00
Significance level	ns	ns	ns	ns	ns	ns

ns = p > 0.05; * = Mann-Whitney test.

of seminiferous tubules contained cell associations of only one stage. On rare occasions when two stages were present only the one occupying the largest cross-section area was registered. Composition and spatial arrangement of cells in each stage of the SEC are presented below.

Stage 1 was characterized by the presence of one generation of round spermatids in the upper part of the seminiferous epithelium bordering the lumen (R, Fig. 2). Two generations of primary spermatocytes were registered. The most advanced one was pachytene spermatocytes placed deeper in the seminiferous epithelium and close to round spermatids (P, Fig. 2), whereas preleptotene spermatocytes (Pl, Fig. 2) were located next to the basement membrane. Type-A spermatogonia (A, Fig. 2) rested on the basement membrane. Sertoli cell nuclei (S, Fig. 2) were situated close to basement membrane in this stage as well as in the remaining stages of the SEC. Leydig cells (Ly, Fig. 2) occupied the intertubular space.

Stage 2. Spermatids (E, Fig. 3) in process of nuclear elongation characterized this stage. It also exhibited type-A spermatogonia (not shown in Fig. 3) resting on basement membrane, as well leptotene spermatocytes (L, Fig. 3) and one or two layers of pachytene spermatocytes (P, Fig. 3). Sertoli cell nuclei (S, Fig. 3) and Leydig cells (Ly, Fig. 3) showed characteristics similar to those of the previous stage.

Stage 3. Elongated spermatid nuclei were arranged in bundles (ES, Fig. 4) and occupied the intermediate portion of the seminiferous epithelium. Two generations of primary spermatocytes were still seen in this stage: one generation at zygotene (Z, Fig. 4), and the other at pachytene or diplotene (D, Fig. 4). Type-A spermatogonia and Sertoli cell nuclei (not shown in Fig. 4) resting on basement membrane completed the cell composition of this stage.

Stage 4. The presence of meiotic divisions (M, Fig. 5) characterized this stage: primary spermatocytes divided into secondary spermatocytes (D/II) which then divided to give rise to round spermatids. Zygotene (Z, Fig. 5) and diplotene spermatocytes (not shown in Fig. 5), and bundles of elongated spermatid nuclei (ES, Fig. 5) in the upper region of the seminiferous epithelium were also registered (Fig. 5). Sertoli cell nuclei (S, Fig. 5) and type A-spermatogonia (not shown in Fig. 5) had similar position and morphology to those of previous stages.

Stage 5 was characterized by the presence of newly-formed round spermatids (R, Fig. 6), and by an older generation of elongated spermatids with nuclei arranged in bundles (ES, Fig. 6). Spermatocytes at the transition zygotene/pachytene (Z/P, Fig. 6) were also observed. Sertoli cell nuclei (S, Fig. 6) and type-A spermatogonia (not indicated in this Fig. 6) had similar morphology and occupied same positions as in previous stages.

Stage 6. Most bundles of elongated spermatid nuclei were now next to the lumen of the seminiferous epithelium, whereas the remaining ones were located deeper in the seminiferous epithelium. Type-B spermatogonia (not indicated in this Fig. 7) were first seen in this stage along with type-A spermatogonia (A, Fig. 7). Spermatocytes at pachytene (P, Fig. 7) and layers of round spermatids (R, Fig. 7) were also present.

Stage 7. Loosen bundles of elongated spermatid nuclei (ES, Fig. 8) next to the lumen of the seminiferous tubule characterized this stage. Various layers of round spermatids (R, Fig. 8) were registered interspersed among pachytene spermatocytes (P, Fig. 8). Type-A and type-B spermatogonia (not shown in this Fig. 8), and Sertoli cell nuclei (S, Fig. 8) appeared close to basement membrane.

Stage 8. This stage was characterized by elongated spermatid nuclei (ES, Fig. 9) bordering the seminiferous epithelium. Several residual bodies (RB, Fig. 9) near elongated spermatid nuclei also characterized this stage. Type-A and type-B spermatogonia (not shown in this Fig. 9) rested on basement membrane. One layer of preleptotene spermatocytes (Pl, Fig. 9) which originated from type B spermatogonia, pachytene spermatocytes (P, Fig. 9), and round spermatids (R, Fig. 9) were also present.

Frequency of stages of the seminiferous epithelium cycle

The relative frequencies of stages of the SEC are shown in Fig. 10; no significant differences were found between stage frequencies in the rainy and the dry seasons (χ^2 = 4.307, p > 0.05, n = 8 and 8, respectively). Stages 1 and 7 were the most frequent, and stages 2 and 5 the least frequent. The accumulated relative frequency of pre-meiotic stages (stages 1 to 3) was 40 ± 7.7% and that of post-meiotic stages (stages 5-8) was 49 ± 5.6%. The meiotic stage, corresponding to stage 4, was 12 ± 4.2%.

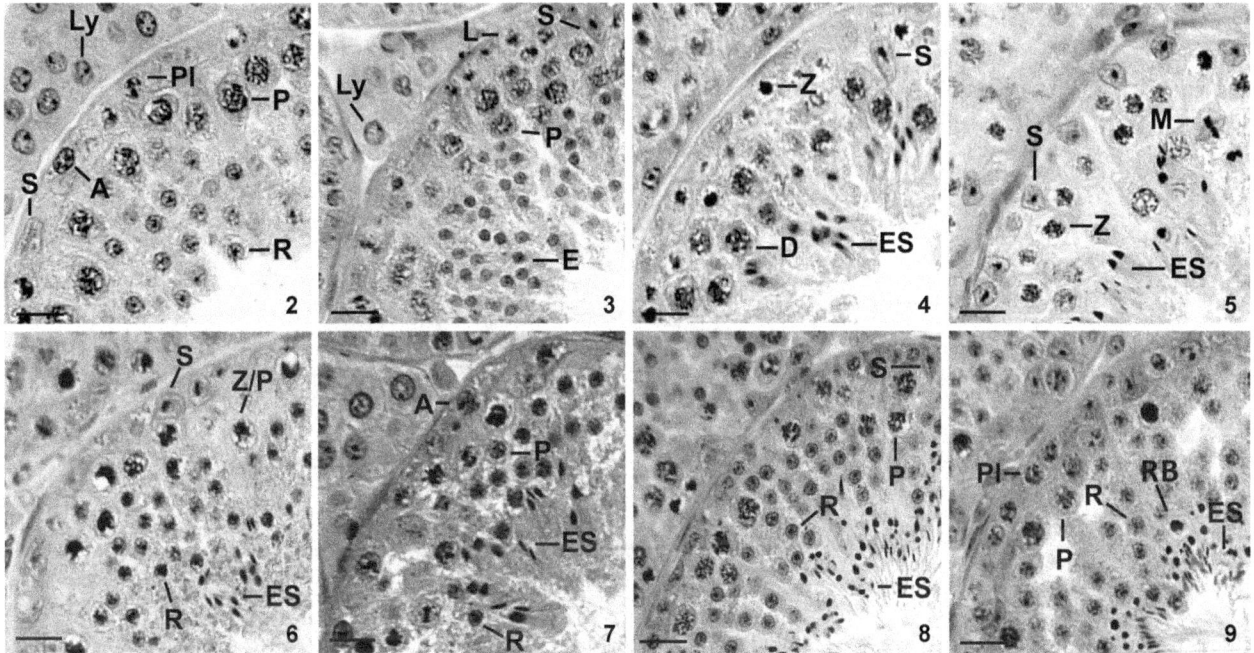

Figures 2-9. Stages 1-4 of the seminiferous epithelium cycle of the large fruit-eating bat based on the tubular morphology method. (2) Stage 1 contains Sertoli cell (S), type-A spermatogonium (A), preleptotene primary spermatocyte (Pl), pachytene primary spermatocyte (P), round spermatid (R), besides Leydig cells (Ly) in the intertubular space. (3) Stage 2 presents Sertoli cell (S), leptotene primary spermatocyte (L), pachytene primary spermatocyte (P) and elongating spermatid (E), besides Leydig cell (Ly). (4) Stage 3 contains Sertoli cell (S), zygotene primary spermatocyte (Z), diplotene primary spermatocyte (D) and elongated spermatid (ES). (5) Stage 4 contains Sertoli cell (S), zygotene primary spermatocyte (Z), diplotene primary spermatocyte (D), meiotic figure (M) and elongated spermatid (ES). (6) Stage 5 contains Sertoli cell (S), zygotene/pachytene primary spermatocyte (Z/P), round spermatid (R) and elongated spermatid (ES). (7) Stage 6 presents type-A spermatogonium (A), pachytene primary spermatocyte (P), round spermatid (R) and elongated spermatid (ES). (8) Stage 7 presents Sertoli cell (S), pachytene spermatocyte (P), round spermatid (R) and elongated spermatid (ES). (9) Stage 8 contains preleptotene primary spermatocyte (Pl), pachytene spermatocyte (P), round spermatid (R), elongated spermatid (ES) and residual bodies (RB). H-E stain. Scale bars: 61 µm.

DISCUSSION

None of the morphometric parameters of the large fruit-eating bat, i.e. testis and epididymis masses, seminiferous tubule diameter, and diameter of Leydig cell nucleus changed significantly between the rainy and the dry seasons. This confirms that continuous spermatogenic activity is taking place, as first described by Duarte & Talamoni (2010) for this species in the Atlantic Forest, Southeastern Brazil. The cauda epididymis packed with spermatozoa throughout the study period was also evidence of male continuous reproductive capacity. Testis regression and recrudescence during the annual reproductive cycle (Oliveira et al. 2009) were not registered in our study. A recent study on the reproductive cycle of the congeneric *Artibeus planirostris* (Spix, 1823) in Brazil also indicates continuous spermatogenic activity based on constant production of spermatozoa and their retention/storage in the cauda epididymis (Beguelini et al. 2013). Similar results have been shown

for other phyllostomid fruit-eating bats, i.e. *Artibeus jamaicensis* Leach, 1821, *Artibeus intermedius* Allen, 1897, and *Dermanura phaeotis* (Miller, 1902), from Mexico, in which males present reproductive activity throughout the year while females tend to be polyestric (Montiel et al. 2011).

The organization of large fruit-eating bat colonies into a harem system (Morrison 1980), also reported for *A. planirostris* (Beguelini et al. 2013), results in polygynous behavior (Wilson 1979). Mammals exhibiting this reproductive behavior have usually low testis mass (Kenagy & Trombulak 1986) which would explain the low GSI value recorded in the present study. Contrarily, other microchiropterans organized in large social groups face increased risk of sperm competition, which results in greater investment in spermatogenesis (Hosken 1997), and consequently larger GSI. Beguelini et al. (2013) showed that GSI values also accompany the two pronounced annual peaks of spermatogenesis in *A. planirostris*. According to latter authors, despite the annual continuous spermatogenic production, GSI

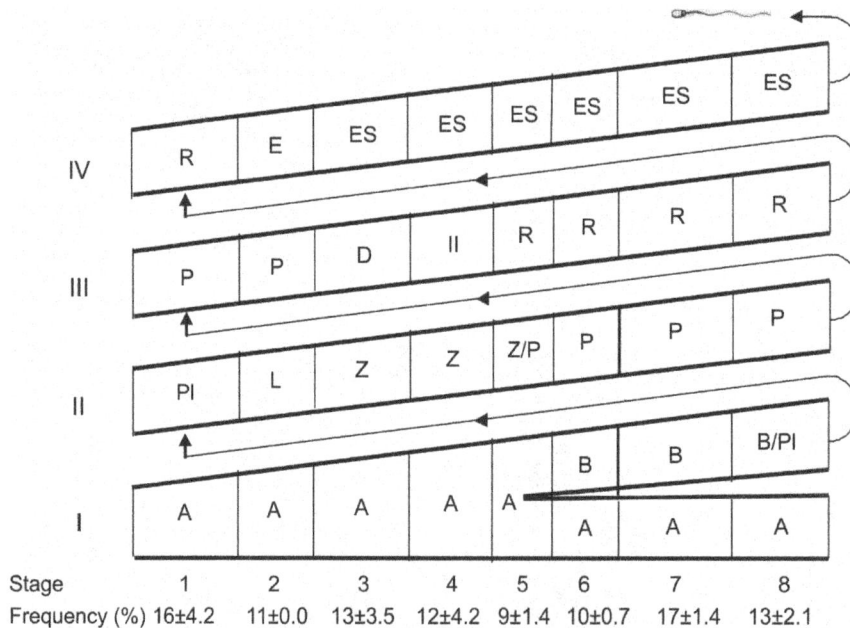

Figure 10. Cell composition and frequency (%) of the stages (1-8) of the cycle (I-IV) of the seminiferous epithelium of the large fruit-eating bat; column width, representing each stage, is proportional to its frequency. (A) Type-A spermatogonium, (B) type-B spermatogonium, (PI) primary spermatocytes in preleptotene, (L) primary spermatocytes in leptotene, (Z) primary spermatocytes in zygotene, (P) primary spermatocytes in pachytene, (D) primary spermatocytes in diplotene, (II) secondary spermatocytes, (R) round spermatids, (E) elongating spermatids, (ES) elongated spermatids.

peaks possibly express a spermatic production needed to synchronize with the bimodal reproductive cycle of females. Thus, monoestry categorization, as suggested for this bat population (DUARTE & TALAMONI 2010), or bimodal polyestry, as observed for other phyllostomids, should be only ascribed after evaluating the female's reproductive cycle.

Three basic types of mammalian spermatogonium recognized under routine microscopy, namely type-A, intermediate, and type-B, are also registered in bats (BEGUELINI et al. 2009, OLIVEIRA et al. 2009). However, most likely due to technical reasons, we did not identify intermediate types of spermatogonium in the large fruit-eating bats of the present study.

As in non-primate mammals (LEAL & FRANÇA 2006), cross-sections of seminiferous tubules of large fruit-eating bats show cell associations belonging to a single stage of the SEC. The eight stages of the SEC previously described for this species (BEGUELINI et al. 2009) have been confirmed in the present study. In general, the cellular composition of these stages is similar to that of other mammals, including other Neotropical phyllostomids (BEGUELINI et al. 2009, OLIVEIRA et al. 2009, MORAIS et al. 2013). One exception is the vespertilionid *Myotis nigricans*, in which different cell associations occur in cross-sections of the seminiferous tubule (BEGUELINI et al. 2009). The species-constant relative frequencies of stages of the SEC (HESS & FRANÇA 2007) were confirmed by our study and, similarly to the re-

sults of a study on *Sturnira lilium* (E. Geoffroy, 1810) (MORAIS et al. 2013), no differences were detected between the rainy and the dry seasons. The smaller accumulated frequency of premeiotic stages with respect to post-meiotic stages, obtained in the present study, was similar to the results of BEGUELINI (2009).

In conclusion, the spermatogenic activity of large fruit-eating bats from the Atlantic Forest in Southeastern Brazil was continuous throughout rainy and dry seasons. Cell associations and the relative frequency of the eight stages of the seminiferous epithelium cycle, categorized by the tubular morphology method, presented no significant differences between the rainy and the dry seasons.

ACKNOWLEDGMENTS

We thank Fundação de Amparo à Pesquisa do Estado de Minas Gerais (FAPEMIG) and Fundo de Incentivo à Pesquisa (FIP PUC Minas) for funding the project, and FIP PUC Minas for scholarships granted to A.A.N. and T.O.F. We also thank Ana Paula G. Duarte for participation in collecting data, to the Brazilian Institute of Natural Environment and Renewable Resources (IBAMA) for providing the license to capture the animals and Companhia de Saneamento do Estado de Minas Gerais (COPASA) for permission to work in the Special Protection Area of Fechos.

REFERENCES

ANTHONY ELP (1988) Age determination in bats, p. 47-57. In: KUNZ TH (Ed). **Ecological and behavioral methods for the study of bats.** Washington, Smithsonian Institution Press.

BEGUELINI MR, MOREIRA PRL, FARIA KC, MARCHESIN SRC, MORIELLE-VERSUTE E (2009) Morphological characterization of the testicular cells and seminiferous epithelium cycle in six species of Neotropical bats. **Journal of Morphology 270**: 943-953. doi: 10.1002/jmor.10731

BEGUELINI MR, PUGA CCL, TABOGA SR, MORIELLE-VERSUTE E (2013) Annual reproductive cycle of males of the flat-faced fruit-eating bat, *Artibeus planirostris* (Chiroptera: Phyllostomidae). **General Comparative Endocrinology 185**: 80-89. doi: 10.1016/j.ygcen.2012.12.009

BERNDTSON WE (1977) Methods for quantifying mammalian spermatogenesis: a review. **Journal of Animal Science 44**: 818-833. doi: 10.2134/jas1977.445818x

CLERMONT Y (1972) Kinetics of spermatogenesis in mammals: seminiferous epithelium cycle and spermatogonial renewal. **Physiological Reviews 52**: 198-236.

DUARTE APG, TALAMONI SA (2010) Reproduction of the large fruit-eating bat *Artibeus lituratus* (Chiroptera: Phyllostomidae) in a Brazilian Atlantic forest area. **Mammalian Biology 75**: 320-325. doi: 10.1016/j.mambio.2009.04.004

FLEMING TH, HOOPER ET, WILSON DE (1972) Three Central American bat communities: structure, reproductive cycles, and movement patterns. **Ecology 53**: 555-569. doi: 10.2307/1934771

FRANÇA LR, GODINHO CL (2003) Testis morphometry, seminiferous epithelium cycle length, and daily sperm production in domestic cats (*Felis catus*). **Biology of Reproduction 68**: 1554-1561. doi: 10.1095/biolreprod.102.010652

FRANÇA LR, RUSSELL LD (1998) The testis of domestic animals, p. 197-219. In: MARTÍNEZ-GARCÍA F, REGADERA J (Eds.). **Male reproduction, a multidisciplinary overview.** Madrid, Churchill Livingstone.

FRANÇA LR, AVELAR GF, ALMEIDA FFL (2005) Spermatogenesis and sperm transit through the epididymis in mammals with emphasis on pigs. **Theriogenology 63**: 300-318. doi: 10.1016/j.theriogenology.2004.09.014

HESS RA, FRANÇA LR (2007) Spermatogenesis and cycle of the seminiferous epithelium, p. 1-15. In: Cheng CY (Ed.). **Molecular mechanisms in spermatogenesis.** Urbana, Landes Bioscience, Springer Science.

HOSKEN DJ (1997) Sperm competition in bats. **Proceedings of the Royal Society of London B: Biological Sciences 264**: 385-392. doi: 10.1098/rspb.1997.0055

KENAGY GJ, TROMBULAK SC (1986) Size and function of mammalian testis in relation to body size. **Journal of Mammalogy 67**: 1-22. doi: 10.2307/1380997

LEAL MC, FRANÇA LR (2006) The seminiferous epithelium cycle length in the black tufted-ear marmoset (*Callithrix penicillata*) is similar to humans. **Biology of Reproduction 74**: 616-624. doi: 10.1095/biolreprod.105.048074

MARQUES-AGUIAR SA (2007) Genus *Artibeus* Leach 1821, p. 301-321. In: GARDNER AL (Ed.). **Mammals of South America. Marsupials, Xenarthrans, Shrews, and Bats.** Chicago, University of Chicago Press.

MONTIEL S, ESTRADA A, LEON P (2011) Reproductive seasonality of fruit-eating bats in northwestern Yucatan, Mexico. **Acta Chiropterologica 13**: 139-145. doi: 10.3161/150811011X578688

MORAIS DB, PAULA TAR, BARROS MS, BALARINI MK, FREITAS MBD, MATTA SLP (2013) Stages and duration of the seminiferous epithelium cycle in the bat *Sturnira lilium*. **Journal of Anatomy 222**: 372-379. doi: 10.1111/joa.12016

MORRISON, D.W. 1980. Foraging and day-roosting dynamics of canopy fruit-bats in Panama. **Journal of Mammalogy 61**: 20-29. doi: 10.2307/1379953

OLIVEIRA RL, OLIVEIRA AG, MAHECHA GAB, NOGUEIRA JC, OLIVEIRA CA (2009) Distribution of estrogen receptors (Era and Erb) and androgen receptor in the testis of big fruit-eating bat *Artibeus lituratus* is cell-and stage-specific and increases during gonadal regression. **General and Comparative Endocrinology 161**: 283-292. doi: 10.1016/j.ygcen.2009.01.019

RACEY PA (1982) Ecology of bat reproduction, p. 57-104. In: KUNZ TH (Ed.) **Ecology of bats.** New York, Plenum Press.

RACEY PA, ENTWISTLE AC (2000) Life-history and reproductive strategies of bats, p. 363-414. In: CRICHTON E, KRUTZSCH P (Eds.). **Reproductive biology of bats.** San Diego, Academic Press.

RASBAND, W.S. 2012 **Image J.** Bethesda, U.S. National Institutes of Health. Available online at: http://imagej.nih.gov/ij [Accessed: 03/07/2012]

REIS SF (1989) Reproductive biology of *Artibeus lituratus* (Olfers, 1818) (Chiroptera: Phyllostomidae). **Revista Brasileira de Biologia 49**: 369-372.

REIS NR, AL PERACCHI, WA PEDRO, LIMA IP (2007) **Morcegos do Brasil.** Londrina, Editora da Universidade de Londrina.

SÁ JÚNIOR A, CARVALHO LG, SILVA FF, ALVES MC (2012) Application of the Köppen classification of climatic zoning in the state of Minas Gerais, Brazil. **Theoretical and Applied Climatology 108**: 1-7. doi: 10.1007/s00704-011-0507-8

SOKAL RR, ROHLF FJ (1995) **Biometry: the principles and practice of statistics in biological research.** New York, WH Freeman and Co., 3rd ed.

TAMSITT JR, VALDIVIESO D (1963) Reproductive cycle of the big fruit-eating bat, *Artibeus lituratus* Olfers. **Nature 198**: 104.

VIZOTTO LD, TADDEI VA (1973) Chave para determinação de quirópteros brasileiros. **Boletim de Ciências da Faculdade de Filosofia, Ciências e Letras, São José do Rio Preto**: 1-72.

WILSON DE (1979) Reproductive patterns, p. 317-378. In: BAKER RJ, JONES JR JK, CARTER DC (Eds.). **Biology of bats of the world. Family Phyllostomidae.** Lubbock, Texas Tech University Press Special Publications of Museum Texas Tech University, Part 3.

12

Environmental parameters affecting the structure of leaf-litter frog (Amphibia: Anura) communities in tropical forests

Carla C. Siqueira[1,2], Davor Vrcibradic[3], Paulo Nogueira-Costa[4], Angele R. Martins[4], Leonardo Dantas[2], Vagner L. R. Gomes[2], Helena G. Bergallo[2] & Carlos Frederico D. Rocha[2]

[1] Programa de Pós-Graduação em Ecologia, Instituto de Biologia, Universidade Federal do Rio de Janeiro. Avenida Carlos Chagas Filho 373, Bloco A, Cidade Universitária, 21941-902 Rio de Janeiro, RJ, Brazil. E-mail: carlacsiqueira@yahoo.com.br
[2] Departamento de Ecologia, Universidade do Estado do Rio de Janeiro. Rua São Francisco Xavier 524, Maracanã, 20550-013 Rio de Janeiro, RJ, Brazil.
[3] Departamento de Zoologia, Universidade Federal do Estado do Rio de Janeiro. Avenida Pasteur 458, Urca, 22290-240 Rio de Janeiro, RJ, Brazil.
[4] Departamento de Vertebrados, Museu Nacional, Universidade Federal do Rio de Janeiro. Quinta da Boa Vista, 20940-040 Rio de Janeiro, RJ, Brazil.

ABSTRACT. Despite a recent increase of information on leaf litter frog communities from Atlantic rainforests, few studies have analyzed the relationship between environmental parameters and community structure of these animals. We analyzed the effects of some environmental factors on a leaf litter frog community at an Atlantic Rainforest area in southeastern Brazil. Data collection lasted ten consecutive days in January 2010, at elevations ranging between 300 and 520 m above sea level. We established 50 quadrats of 5 x 5 m on the forest floor, totaling 1,250 m² of sampled area, and recorded the mean leaf-litter depth and the number of trees within the plot, as well as altitude. We found 307 individuals belonging to ten frog species within the plots. The overall density of leaf-litter frogs estimated from the plots was 24.6 ind/100m², with *Euparkerella brasiliensis* (Parker, 1926), *Ischnocnema guentheri* (Steindachner, 1864), *Ischnocnema parva* (Girard, 1853) and *Haddadus binotatus* (Spix, 1824) presenting the highest estimated densities. Among the environmental variables analyzed, only altitude influenced the parameters of anuran community. Our results indicate that the study area has a very high density of forest floor leaf litter frogs at altitudes of 300-500 m. Future estimates of litter frog density might benefit from taking the local altitudinal variation into consideration. Neglecting such variation might result in underestimated/overestimated values if they are extrapolated to the whole area.

KEY WORDS. Frog richness; density estimates; environmental factors.

Community ecology aims to understand the patterns of distribution, abundance and interactions among organisms (LEIBOLD *et al.* 2004, KELLER *et al.* 2009). However, it is not always clear which factors are responsible for community structure, and whether species assemblies follow any general rules. While ecotones and physical factors are important in structuring amphibian communities, competitive interactions seem to exert comparatively little influence (HOFER *et al.* 1999, 2000, 2004, KELLER *et al.* 2009).

The Brazilian Atlantic rainforest possibly harbors the world's greatest diversity of frog species, most of which are endemic (DUELLMAN 1999, HADDAD & PRADO 2005). Since the mid-1990s, more information on leaf litter frog communities from Atlantic rainforest areas has become available (e.g., GIARETTA *et al.* 1997, 1999, ROCHA *et al.* 2000, 2001, 2007, 2011, 2013, VAN SLUYS *et al.* 2007, ALMEIDA-GOMES *et al.* 2008, SIQUEIRA *et al.* 2009,

SANTOS-PEREIRA *et al.* 2011). Nevertheless, few studies have analyzed the relationship between environmental parameters and community structure of leaf litter frogs (GIARETTA *et al.* 1997, 1999, VAN SLUYS *et al.* 2007, SANTOS-PEREIRA *et al.* 2011). Furthermore, the available information is still insufficient to identify patterns. Consequently, the main factors explaining the local density and richness of frogs in Atlantic Rainforest areas are still poorly understood.

A short-term inventory of the leaf litter frogs of an Atlantic Rain Forest Reserve, Reserva Ecológica de Guapiaçu (REGUA, state of Rio de Janeiro, southeastern Brazil), has provided data on community composition, abundance and density of those frogs (ROCHA *et al.* 2007). In the present study, we analyzed the extent to which certain environmental factors affect the richness and abundance of leaf litter frogs in the same reserve. We also provide new estimates of leaf litter frog density and richness for the area. Our aim was to investigate the factors explaining the local ecological parameters of tropical frog communities.

MATERIAL AND METHODS

The study was carried out at the Reserva Ecológica de Guapiaçu (hereafter REGUA, 22°24'S, 42°44'W), municipality of Cachoeiras de Macacu, Rio de Janeiro State, southeastern Brazil. The area is inserted within one of the largest remnants of Atlantic forest in the state (over 60,000 ha), most of which is encompassed by the Três Picos State Park. Most of the area is covered with Atlantic Rainforest in different levels of conservation, with remnants of undisturbed forests occurring in the higher and less accessible areas of the reserve. The climate is wet and warm, with mean annual rainfall of about 2,600 mm and daily temperatures ranging from 14 to 37°C (BERNARDO et al. 2011).

We sampled during ten consecutive days (from 20 to 29 January 2010), at elevations ranging between 300 and 520 m above sea level. We established 50 plots (quadrats) of 5 x 5 m (25 m²) (see JAEGER & INGER 1994) on the forest floor (five per day), totaling 1250 m² of sampled area. Each quadrat was an independent sampling unit and was at least 100 m apart from the nearest one. While setting the plots, we also maintained a minimum distance of 10 m from streams, in order to minimize the effect of the proximity of water bodies on the frogs (except in one case, where the plot was set ca. 6 m from a stream). During the day, we delimited quadrats by completely enclosing them with a 80 cm high soft mesh fence. The bottom of the fence was buried or attached to the ground with strings and sticks, to prevent frogs from escaping the plot (ROCHA et al. 2001). After sunset, each plot was carefully searched for frogs by a crew of four people wearing head lamps. During searches, each crew member walked the entire plot on hands and knees, side-by-side, using hand rakes. Inside each plot, crew members checked for frogs on and underneath leaves and pieces of dead wood, rock crevices and spaces between tree roots. Additionally, they overturned fallen branches and stones. Searches in each plot lasted about 20-30 minutes. All anurans encountered within a given plot were identified to species and released after the crew had finished searching. Although no individuals were collected in the present study (except for a caecilian), voucher specimens of all frog species recorded herein, collected at the study area during previous fieldwork, are deposited at the Museu Nacional, Rio de Janeiro (see Appendix). For each plot, we recorded the altitude, estimated the mean leaf-litter depth (by measuring the litter depth at each corner of the plot and calculating the mean value), and counted the number of trees with a trunk diameter (DBH) wider than 50 mm.

The evaluation of collection effectiveness was undertaken by a species accumulation curve (collector curve) and by estimators (Bootstrap and Chao 1), using the program EstimateS 8.2.0 (COLWELL 2005). To analyze the relative importance of some

environmental parameters such as leaf litter depth, altitude and number of trees affecting the abundance and richness of leaf litter frogs, we performed multiple regression analyses with frog abundance/richness as the dependent variable and environmental parameters as the independent variables.

The influence of the environmental variables altitude, mean leaf-litter depth and number of trees on anuran species distribution (composition data) was assessed through a Canonical Correspondence Analysis (CCA; TER BRAAK 1986). Data on environmental factors were standardized by centering and normalizing. The statistical significance of the species-environment correlation was evaluated by Monte Carlo test (1000 randomized runs).

RESULTS

A total of 307 individuals belonging to ten frog species were found in the plots (Table I). The number of frogs per plot ranged from zero (5/50 or 10.0% of all plots) to 26 (1/50 or 0.02%) with a median value of 4.5 frogs per plot and a mean value of 6.1 ± 6.3 frogs per plot. The overall density of leaf-litter frogs estimated from the plots was 24.6 ind/100m². *Euparkerella brasiliensis* (Parker, 1926) (7.0 ind/100 m²), *Ischnocnema guentheri* (Steindachner, 1864) (5.7 ind/100 m²), *Ischnocnema parva* (Girard,

Table I. Total number of individuals sampled and estimated density (individuals/100 m², in parenthesis) of each frog species found in the leaf-litter frog community of the Atlantic rainforest of Reserva Ecológica de Guapiaçu, southeastern Brazil, using 5 x 5 plot sampling method. (*) Nine individuals of unidentified species evaded capture.

Frog species	Number of individuals (density)
Brachycephalidae	
Brachycephalus didactylus (Izecksohn, 1971)	15 (1.2)
Ischnocnema guentheri (Steindachner, 1864)	71 (5.7)
Ischnocnema parva (Girard, 1853)	55 (4.4)
Craugastoridae	
Haddadus binotatus (Spix, 1824)	43 (3.4)
Euparkerella brasiliensis (Parker, 1926)	88 (7.0)
Cycloramphidae	
Zachaenus parvulus (Girard, 1853)	11 (0.9)
Hylodidae	
Crossodactylus aeneus Müller, 1924	1 (0.1)
Leptodactylidae	
Adenomera marmorata Steindachner, 1867	7 (0.6)
Physalaemus signifer (Girard, 1853)	6 (0.5)
Microhylidae	
Chiasmocleis carvalhoi Cruz, Caramaschi & Izecksohn, 1997	1 (0.1)
Total (*)	307 (24.6)

1853) (4.4 ind/100 m²) and *Haddadus binotatus* (Spix, 1824) (3.4 ind/100 m²) presented the highest estimated densities (Table I). Those four species accounted for 83.7% of all frogs found in the plots. In addition to frogs, one individual of the caecilian *Siphonops hardyi* Boulenger, 1888 (Gymnophiona, Siphonopidae) was found inside a plot. The species accumulation curve analysis produced a clear asymptotic shape, and the predicted estimates of frog richness (Bootstrap = 10.8 ± 0.3 species; Chao 1 = 11.0 ± 2.3 species) were close to the richness value obtained by us with plot sampling (10 species) (Fig. 1).

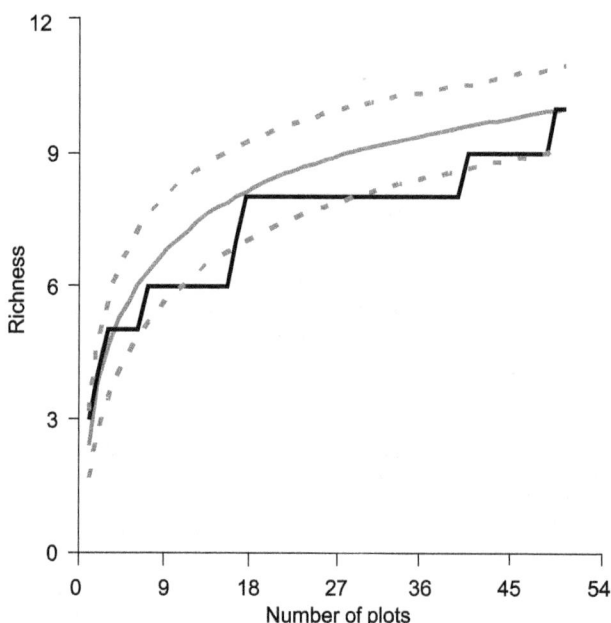

Figure 1. Cumulative (black line) and rarefaction (solid gray line; with the standard deviation represented by dotted gray lines) curves of richness of leaf litter frogs recorded at the Atlantic rainforest of Reserva Ecológica de Guapiaçu, in southeastern Brazil, using 5 x 5 plot sampling method.

The overall abundance of frogs and the three environmental variables (altitude, leaf litter depth and number of trees) were not significantly correlated in the results of the multiple regression analysis (r = 0.39, $F_{3,46}$ = 2.68, p = 0.06). Partial correlations showed a positive and significant relationship between frog abundance and altitude (p < 0.01), but not for the other variables (leaf litter depth, p = 0.60; number of trees, p = 0.63).

Frog richness was not significantly influenced by the combined effect of the three environmental variables tested, according to the multiple regression analysis (r = 0.36, $F_{3,46}$ = 2.22, p = 0.10). Partial correlations showed a positive and significant relationship between frog richness and altitude (p < 0.05), but not for the remaining variables (leaf litter depth, p = 0. 80; number of trees, p = 0.46).

The first three axes of the CCA explained only 9.9% of the variance in species data (6.0, 2.7, and 1.2%, respectively). The biplot scores of the variable "altitude" showed a higher correlation to axis 1 (= -0.345), while "number of trees" had a higher correlation with axis 2 (= 0.216) (Table II). Randomization (Monte Carlo) test did not show statistical significance for species-environment correlations (p = 0.07 for axis 1; P was not reported for axes 2 and 3 because using a simple randomization test for these axes could bias the P values).

Table II. Summary of the CCA performed for three features used to characterize plots and anuran species distribution among them at the Reserva Ecológica de Guapiaçu, southeastern Brazil. The variable scores with the highest influence in the first two axes are shown in boldface.

	Axis 1	Axis 2	Axis 3
Correlations (biplot scores)			
Leaf litter depth	-0.109	0.126	-0.132
Altitude	**-0.345**	-0.055	0.047
Number of trees	-0.024	**0.216**	0.078
Summary statistics for ordination axes			
Eigenvalues	0.137	0.061	0.027
Species-environment			
Correlations	0.665	0.437	0.311

DISCUSSION

With this study, four species, namely *Brachycephalus didactylus* (Izecksohn, 1971), *Crossodactylus aeneus* Müller, 1924, *I. parva*, and *Zachaenus parvulus* (Girard, 1853), are added to the list of leaf litter frogs of the REGUA, previously compiled by ROCHA *et al.* (2007), and which featured 12 species. However, these previous records include the hylid *Scinax* aff. *x-signatus*, a scansorial species that is considered accidental in the leaf litter, and *Euparkerella brasiliensis*, which was erroneously identified as *E. cochranae* Izeksohn, 1988 in that work (L.A. Fusinatto, pers. comm.). Thus, there are currently 15 species known to comprise the leaf litter frog community of the REGUA (excluding *Scinax* aff. *x-signatus* which we do not consider as a leaf litter inhabitant). Most of those species are typical constituents of anuran communities in the leaf litter floor of Atlantic Forest areas in southeastern Brazil (e.g., ROCHA *et al.* 2000, 2007, 2011, ALMEIDA-GOMES *et al.* 2008, SIQUEIRA *et al.* 2009, ARAÚJO *et al.* 2009). However, the hylodid *C. aeneus* is typically found in association with streams (JORDÃO-NOGUEIRA *et al.* 2006), and the only individual recorded during our study was found in a plot set ca. 6 m from a small stream.

Despite the fact that our focus was on leaf litter anurans, it is worth mentioning that we also recorded the caecilian *S. hardyi* during our fieldwork. This finding represents another new record of a forest floor amphibian for the REGUA.

In the present study, the estimated density of forest floor leaf litter frogs at the REGUA (24.6 ind/100m^2) was about three times higher than that of a previous plot study carried out in the same locality (8.4 ind/100 m^2, Rocha *et al.* 2007). It is possible that the different densities found in the two studies are, at least in part, related to the altitude: from 300-520 m a.s.l. in our study *versus* 100-400 m a.s.l. (but predominantly at the lower sites) in the study of Rocha *et al.* (2007). Another plot study carried out between 500 and 800 m a.s.l., in a site only ca. 15 km distant from the REGUA, also yielded relatively high litter frog density estimates (17.1 ind/100 m^2) (Siqueira *et al.* 2009). It is also possible that the months of sampling may have been responsible for part of the differences between our study (January – peak of the rainy season) and that of Rocha *et al.* (2007) (October – beginning of the rainy season), since Atlantic Forest leaf litter frog composition and abundances may vary seasonally (e.g., Santos-Pereira *et al.* 2011).

The litter frog density estimated for the REGUA in our study was not only higher than previously reported for the same region, but also higher than in all other Atlantic Forest areas for which similar data are available (see Rocha *et al.* 2013). Comparing our data with other studies in different tropical forest regions in South America, the estimated density for the REGUA was higher than in some Amazon rainforest areas in Brazil (3.0-6.0 ind/100 m^2; Allmon 1991) and in Peru (4.4-15.5 ind/100 m^2; Toft 1980a). Compared to rainforest sites in Central America, our density estimate was higher (11.5 ind/100 m^2 – Heinen 1992; 15.7 ind/100 m^2 – Lieberman 1986; 7.5-19.4 ind/100 m^2 – Toft 1980b), lower (30.2 ind/100 m^2 – Heatwole & Sexton 1966), or higher/lower (13.4-62.3 ind/100 m^2 – Scott Jr 1976), depending on the locality and/or time of the year when sampling was performed. On the other hand, the estimated leaf-litter frog density obtained in our study was higher than reported for most Old World tropical forests surveyed so far (e.g., 0.5-2.6 ind/100 m^2 in Thailand – Inger & Colwell 1977; 1.2 ind/100 m^2 in Borneo – Lloyd *et al.* 1968; 1.5-2.2 ind/100 m^2 in Uganda – Vonesh 2001; 1.5 ind/100 m^2 in the southern Western Ghats, India – Vasudevan *et al.* 2008; 9.4 ind/100 m^2 in Cameroon – Scott Jr 1982; 3.5-10.2 ind/100 m^2 in Taiwan – Huang & Hou 2004), with one exception (41.8 ind/100 m^2 in Iriomote island, Japan – Watanabe *et al.* 2005). Hence, our data generally support the idea that higher densities of leaf-litter frogs tend to occur in the Neotropical region compared to the Old World tropics.

We detected a significant effect of altitude on the richness and abundance of the leaf litter anuran community at REGUA, despite the relatively small altitudinal range included in our study (ca. 200 m). The influence of altitude on anuran community parameters could be due to variation in environmental factors such as temperature and humidity, for instance (e.g., Giaretta *et al.* 1999). Moreover, it is possible that habitat quality varies with altitude within the area, with more disturbances at lower sites resulting in comparatively less favorable conditions for leaf litter frogs (e.g., Rocha *et al.* 2013). A longer-term study monitoring environmental parameters and quantifying habitat quality (or disturbance), including a wider altitudinal gradient, may clarify the relationship between altitude and leaf litter frog density.

Altitude, in our analysis, was the only variable that significantly affected the parameters of the leaf litter frog community. Giaretta *et al.* (1999) found that altitude, depth of soil cover, leaf litter mass and fallen trunk area all influence the density of leaf litter frogs at an Atlantic Forest area in the state of São Paulo, Brazil. Van Sluys *et al.* (2007), in a study conducted at the island of Ilha Grande (Rio de Janeiro State), found that mean litter depth and the proportion of leaves in the leaf litter were significantly correlated with abundance and richness of litter frogs (Van Sluys *et al.* 2007). In a "restinga" habitat (a relatively xeric ecosystem within the Atlantic Forest biome) in northeastern Brazil, the amount of leaves, percentage of leaf litter, percentage of terrestrial bromeliads exposed to direct sunlight, number of terrestrial bromeliads and soil moisture were the environmental variables that better explained the composition of the local anuran community (Bastazini *et al.* 2007). At a site in the Brazilian Amazon, Allmon (1991) did not obtain any significant results regarding the influence of leaf litter moisture, depth or average dry mass on the abundance of frogs, but plots with frogs had higher average dry litter mass than plots without frogs. Fauth *et al.* (1989) found a negative effect of elevation and a positive effect of leaf litter depth on the density and richness of amphibians in Costa Rica. Data from the aforementioned studies indicate that the effect of different environmental variables on the richness and densities of tropical forest floor herpetofauna varies among sites, but the amount/depth of leaf litter seems to be a consistently important variable overall. In the present study, however, litter depth was not significantly correlated with frog abundance or richness, contrary to what would be expected based on previous studies in other areas.

The results of the present study indicate that the area of the REGUA has a very high density of forest floor leaf litter frogs at altitudes of 300-500 m. This, together with the significant effect of altitude on litter frog abundance and richness in our analyses (and comparisons with an earlier study done mostly at lower altitudes in the same area), shows that estimates of litter frog density within a given area should consider the local altitudinal variation, and that data obtained at a limited altitudinal range may lead to underestimated/overestimated values if extrapolated to the whole area.

ACKNOWLEDGEMENTS

We thank Nicholas J. Locke of the Reserva Ecológica de Guapiaçu (REGUA) for facilitating this study, giving logistical support and for the permit to work in the area; the Instituto Estadual do Ambiente (INEA) for the license (# 005/2008). Carlos F.D. Rocha received resources from the Conselho Nacional de Desenvolvimento Científico e Tecnológico – CNPq (Processes

307653/2003-0 and 476684/2008-0) and from the Fundação de Amparo à Pesquisa do Estado do Rio de Janeiro – FAPERJ (Process E-26/102.404/2009) through the program "Cientistas do Nosso Estado". This project also beneffited from funding of the "Edital Espécies Ameaçadas" of Fundação Biodiversitas/CEPAN and RAN/ICMBio (Project 0158A/012006). During this study Carla C. Siqueira received a PhD grant from the CNPq (Process 141555/2008-4), and currently receives postdoctoral grants also from CNPq (Process 150151/2012-8). Helena G. Bergallo thanks FAPERJ and CNPq for the research grants.

REFERENCES

ALLMON, W.D. 1991. A plot study of forest floor litter frogs, central Amazon, Brazil. **Journal of Tropical Ecology 7** (4): 503-522. doi: 10.1017/S0266467400005885

ALMEIDA-GOMES, M.; D. VRCIBRADIC; C.C. SIQUEIRA; M.C. KIEFER; T. KLAION; P. ALMEIDA-SANTOS; D. NASCIMENTO; C.V. ARIANI; V.N.T. BORGES-JUNIOR; R.F. FREITAS-FILHO; M. VAN SLUYS & C.F.D. ROCHA. 2008. Herpetofauna of an Atlantic Rainforest area (Morro São João) in Rio de Janeiro State, Brazil. **Anais da Academia Brasileira de Ciências 80** (2): 291-300. doi: 10.1590/S0001-37652008000200007

ARAÚJO, O.G.S.; L.F. TOLEDO; P.C.A. GARCIA & C.F.B. HADDAD. 2009. The amphibians of São Paulo State, Brazil. **Biota Neotropica 9** (4): 197-209. doi: 10.1590/S1676-06032009000400020

BASTAZINI, C.V.; J.F.V. MUNDURUCA; P.L.B. ROCHA & M.F. NAPOLI. 2007. Which environmental variables better explain changes in anuran community composition? A case study in the restinga of Mata de São João, Bahia, Brazil. **Herpetologica 63** (4): 459-471. doi: 10.1655/0018-0831(2007)63[459:wevbec]2.0.co;2

BERNARDO, C.S.S.; H. LLOYD; N. BAYLY & M. GALETTI. 2011. Modelling post-release survival of reintroduced Red-billed Curassows *Crax blumenbachii*. **Ibis 153** (3): 562-572. doi: 10.1111/j.1474-919x.2011.01142.x

COLWELL, R.K. 2005. **Estimates: Statistical estimation of species richness and shared species from samples**. Version 7.5. Available online at: ?? [Accessed: ??/??/20??].

DUELLMAN, W.E. 1999. Distribution Patterns of Amphibians in South America, p. 255-327. *In*: W.E. DUELLMAN (Ed.). **Patterns of Distribution of Amphibians**. Baltimore and London, The Johns Hopkins University Press, VII + 648p.

FAUTH, J.E.; B.I. CROTHER & J.B. SLOWINSKI. 1989. Elevational patterns of species richness, evenness and abundance of the Costa Rican leaf-litter herpetofauna. **Biotropica 21** (2): 178-185.

GIARETTA, A.A.; R.J. SAWAYA; G. MACHADO; M.S. ARAÚJO; K.G. FACURE; H.F. MEDEIROS & R. NUNES. 1997. Diversity and abundance of litter frogs at altitudinal sites at Serra do Japi, Southeastern Brazil. **Revista Brasileira de Zoologia 14** (2): 341-346. doi: 10.1590/S0101-81751997000200008

GIARETTA, A.A.; K.G. FACURE; R.J. SAWAYA; J.H.D. MEYER & N. CHEMIN. 1999. Diversity and abundance of litter frogs in a montane

forest of southeastern Brazil: seasonal and altitudinal changes. **Biotropica 31** (4): 669-674. doi: 10.1111/j.1744-7429.1999.tb00416.x

HADDAD, C.F.B. & C.P.A. PRADO. 2005. Reproductive modes in frogs and their unexpected diversity in the Atlantic Forest of Brazil. **BioScience 55** (3): 207-217. doi: 10.1641/0006-3568(2005)055[0207:RMIFAT]2.0.CO;2

HEATWOLE, H. & O.J. SEXTON. 1966. Herpetofaunal comparisons between two climatic zones in Panama. **American Midland Naturalist 75** (1): 45-60. doi: 10.2307/2423482

HEINEN, J.T. 1992. Comparisons in the leaf litter herpetofaunas in abandoned cacao plantations and primary rain forest in Costa Rica: some implications for faunal restoration. **Biotropica 24** (3): 431-439.

HOFER, U.; L.F. BERSIER & D. BORCARD. 1999. Spatial organization of a herpetofauna on an elevational gradient revealed by null model tests. **Ecology 80** (3): 976-988. doi: 10.1890/0012-9658(1999)080[0976:SOOAHO]2.0.CO;2

HOFER, U.; L.F. BERSIER & D. BORCARD. 2000. Ecotones and gradient as determinants of herpetofaunal community structure in the primary forest of Mount Kupe, Cameroon. **Journal of Tropical Ecology 16** (4): 517-533.

HOFER, U.; L.F. BERSIER & D. BORCARD. 2004. Relating niche and spatial overlap at the community level. **Oikos 106** (2): 366-376. doi: 10.1111/j.0030-1299.2004.12786.x

HUANG, C.H. & P.C.L. HOU. 2004. Density and diversity of litter Amphibians in a monsoon forest of southern Taiwan. **Zoological Studies 43** (4): 795-802.

INGER, R.F. & R.K. COLWELL. 1977. Organization of contiguous communities of amphibians and reptiles in Thailand. **Ecological Monographs 47** (3): 229-253. doi: 10.2307/1942516

JAEGER. R. & R.F. INGER. 1994. Standard techniques for inventory and monitoring: quadrat sampling, p. 97-102. *In*: W.R. HEYER, M.A. DONNELLY, R.W. MCDIARMID, L.A.C. HAYEK & M.S. FOSTER (Eds). **Measuring and monitoring biological diversity: standard methods for amphibians**. Washington, D.C., Smithsonian Institution Press, 364p.

JORDÃO-NOGUEIRA, T.; D. VRCIBRADIC; J.A.L. PONTES; M. VAN SLUYS & C.F.D. ROCHA. 2006. Natural history traits of *Crossodactylus aeneus* (Anura, Leptodactylidae, Hylodinae) from an Atlantic forest area in Rio de Janeiro State, southeastern Brazil. **South American Journal of Herpetology 1** (1): 37-41. doi: 10.2994/1808-9798(2006)1[37:NHTOCA]2.0.CO;2

KELLER, A.; M.O. RÖDEL; K.E. LINSENMAIR & T.U. GRAFE. 2009. The importance of environmental heterogeneity for species diversity and assemblage structure in Bornean stream frogs. **Journal of Animal Ecology 78** (2): 305-314. doi: 10.1111/j.1365-2656.2008.01457.x

LEIBOLD, M.A.; M. HOLYOAK; N. MOUQUET; P. AMARASEKARE; J.M. CHASE; M.F. HOOPES; R.D. HOLT; J.B. SHURIN; R. LAW; D. TILMAN; M. LOREAU & A. GONZALEZ. 2004. The metacommunity concept: a framework for multi-scale community ecology. **Ecology Letters 7** (7): 601-613. doi: 10.1111/j.1461-0248.2004.00608.x

LIEBERMAN, S.S. 1986. Ecology of the leaf litter herpetofaunas of a Neotropical rainforest: La Selva, Costa Rica. **Acta Zoologica Mexicana 15** (1): 1-71.

LLOYD, M.; R.F. INGER & W. KING. 1968. On the diversity of reptile and amphibian species in a Bornean rainforest. **American Naturalist 102** (928): 497-515.

ROCHA, C.F.D.; M. VAN SLUYS; M.A.S. ALVES; H.G. BERGALLO & D. VRCIBRADIC. 2000. Activity of leaf-litter frogs: when should frogs be sampled? **Journal of Herpetology 34** (2): 285-287. doi: 10.2307/1565426

ROCHA, C.F.D.; M. VAN SLUYS; M.A.S. ALVES; H.G. BERGALLO & D. VRCIBRADIC. 2001. Estimates of forest floor litter frog communities: a comparison of two methods. **Austral Ecology 26** (1): 14-21. doi: 10.1111/j.1442-9993.2001.01073.pp.x

ROCHA, C.F.D.; D. VRCIBRADIC; M.C. KIEFER; M. ALMEIDA-GOMES; V.N.T. BORGES-JR; P.C.F. CARNEIRO; R.V. MARRA; P. ALMEIDA-SANTOS; C.C. SIQUEIRA; P. GOYANNES-ARAÚJO; C.G.A. FERNANDES; E.C.N. RUBIÃO & M. VAN SLUYS. 2007. A survey of the leaf-litter frog assembly from an Atlantic Forest area (Reserva Ecológica de Guapiaçu) in Rio de Janeiro State, Brasil, with an estimate of frog densities. **Tropical Zoology 20** (1): 99-108.

ROCHA, C.F.D.; D. VRCIBRADIC; M.C. KIEFER; C.C. SIQUEIRA; M. ALMEIDA-GOMES; V.N.T. BORGES-JR; F.H. HATANO; A.F. FONTES; J.A.L. PONTES; T. KLAION; L.O. GIL & M. VAN SLUYS. 2011. Parameters from the community of leaf-litter frogs from Estação Ecológica Estadual Paraíso, Guapimirim, Rio de Janeiro State, southeastern Brazil. **Anais da Academia Brasileira de Ciências 83** (4): 1259-1268.

ROCHA, C.F.D.; D. VRCIBRADIC; M.C. KIEFER; M. ALMEIDA-GOMES; V.N.T. BORGES-JR; V.A. MENEZES; C.V. ARIANI; J.A.L. PONTES; P. GOYANNES-ARAÚJO; R.V. MARRA; D.M. GUEDES; C.C. SIQUEIRA & M. VAN SLUYS. 2013. The leaf-litter frog community from Reserva Rio das Pedras, Mangaratiba, Rio de Janeiro State, Southeastern Brazil: species richness, composition and densities. **North-Western Journal of Zoology 9** (1): 151-156.

SANTOS-PEREIRA, M.; A. CANDATEN; D. MILANI; F.B. OLIVEIRA; J. GARDELIN & C.F.D. ROCHA. 2011. Seasonal variation in the leaf-litter frog community (Amphibia: Anura) from an Atlantic Forest Area in the Salto Morato Natural Reserve, southern Brazil. **Zoologia 28** (6): 755-761. doi: 10.1590/S1984-46702011000600008

SCOTT JR, N.J. 1976. The abundance and diversity of the herpetofauna of tropical forest litter. **Biotropica 8** (1): 41-58.

SCOTT JR, N.J. 1982. The herpetofauna of forest litter plots from Cameroon, Africa, p. 145-150. *In*: N.R.

SCOTT JR (Ed.). **Herpetological communities: a symposium of the Society for the Study of Amphibians and Reptiles and the Herpetologists' League.** Washington, D.C., US Fish and Wildlife Service, 239p.

SIQUEIRA, C.C.; D. VRCIBRADIC; M. ALMEIDA-GOMES; V.N.T. BORGES-JR; P. ALMEIDA-SANTOS; M. ALMEIDA-SANTOS; C.V. ARIANI; D.M. GUEDES; P. GOYANNES-ARAÚJO; T.A. DORIGO; M. VAN SLUYS & C.F.D. ROCHA. 2009. Density and richness of the leaf litter frogs of an Atlantic Rainforest area in Serra dos Órgãos, Rio de Janeiro State, Brazil. **Zoologia 26** (1): 97-102. doi: 10.1590/S1984-46702009000100015

TER BRAAK, C.J.F. 1986. Canonical correspondence analysis: a new eigenvector technique for multivariate direct gradient analysis. **Ecology 67** (5): 1167-1179. doi: 10.2307/1938672

TOFT, C. 1980a. Feeding ecology of thirteen syntopic species of anurans in a seasonal tropical environment. **Oecologia 45** (1): 131-141. doi: 10.1007/bf00346717

TOFT, C. 1980b. Seasonal variation in populations of Panamanian litter frogs and their prey: a comparison of wetter and drier sites. **Oecologia 47** (1): 34-38. doi: 10.1007/bf00541772

VAN SLUYS, M.; D. VRCIBRADIC; M.A.S. ALVES; H.G. BERGALLO & C.F.D. ROCHA. 2007. Ecological parameters of the leaf-litter frog community of an Atlantic Rainforest area at Ilha Grande, Rio de Janeiro state, Brazil. **Austral Ecology 32** (3): 254-260. doi: 10.1111/j.1442-9993.2007.01682.x

VASUDEVAN, K.; A. KUMAR; B.R. NOON & R. CHELLAM. 2008. Density and diversity of forest floor anurans in the rain forests of southern Western Ghats, India. **Herpetologica 64** (2): 207-215. doi: 10.1655/07-066.1

VONESH, J.R. 2001. Patterns of richness and abundance in tropical Africa herpetofauna. **Biotropica 33** (3): 502-510. doi: 10.1111/j.1744-7429.2001.tb00204.x

WATANABE, S.; N. NAKANISHI & M. IZAWA. 2005. Seasonal abundance in the floor-dwelling frog fauna on Iriomote Island of the Ryuku Archipelago, Japan. **Journal of Tropical Ecology 21** (1): 85-91. doi: 10.1017/S0266467404002068

13

Herpetofauna of Paranapiacaba: expanding our knowledge on a historical region in the Atlantic forest of Southeastern Brazil

Vivian Trevine[1], Maurício C. Forlani[1], Célio F. B. Haddad[2] & Hussam Zaher[1,3]

[1] Museu de Zoologia, Universidade de São Paulo. Avenida Nazaré 481, 04263-000 São Paulo, SP, Brazil.
[2] Departamento de Zoologia, Instituto de Biociências, Universidade Estadual Paulista. Avenida 24-A, 1515, Bela Vista, 13506-900 Rio Claro, SP, Brazil.
[3] Correspondig author. Email: hussam.zaher@gmail.com

ABSTRACT. The largest area of preserved Atlantic forest is located in the southern portion of Brazil. The region of Paranapiacaba is depicted in Brazilian zoological studies as one of the first and most intensely sampled areas of the state of São Paulo.We provide a concise list of reptiles and amphibians from the Paranapiacaba Municipal Park. It represents the first comprehensive survey of the group in the area. We recorded 136 species of reptiles and amphibians from field surveys, museum collections and the literature. The anuran diversity of Paranapiacaba is greater than that of Estação Ecológica de Boracéia, which has been considered the most distinctive areas in São Paulo in terms of amphibian diversity. The rich history of herpetological research in the region, including the occurrence of the two most threatened species in Brazil, converts the area to an important conservation landmark for the Brazilian herpetofauna.

KEY WORDS. Anuran conservation; conservation unit; herpetofaunal diversity; historical records; São Paulo state.

The Atlantic Forest Domain (*sensu* AB'SABER 2003) is formed by a mosaic of environments within several biogeographic sub-units (CÂMARA 2003, SILVA & CASTELETI 2005, CARNAVAL *et al.* 2009). It exhibits great latitudinal, altitudinal and longitudinal variation, featuring a unique diversity of flora and fauna, and high levels of endemism (MYERS *et al.* 2000, LAURANCE 2009, RIBEIRO *et al.* 2011). The Atlantic Forest now is 11.7% of its original extension, and more than half of its remnants are restricted to the Serra do Mar formation, in the states of São Paulo and Paraná (GALINDO-LEAL & CÂMARA 2005, RIBEIRO *et al.* 2009).

The herpetofauna of the state of São Paulo is extremely diverse, representing approximately 30% of all Brazilian diversity (ROSSA-FERES *et al.* 2011, ZAHER *et al.* 2011). The amphibian fauna of the Serra do Mar, Serra da Mantiqueira and Serra de Paranapiacaba is clearly the richest of the state (GARCIA *et al.* 2009a). Moreover, most of the Squamata reptiles of São Paulo are endemic to elevated areas within the Serra do Mar range (MARQUES *et al.* 2004, RODRIGUES 2005, ROSSA-FERES *et al.* 2008).

The geomorphological formation knwon as "Serra de Paranapiacaba" encompasses an interiorized portion of the Serra do Mar, and part of the Atlantic plateau along the southern portion of the state of São Paulo (GUIX *et al.* 2000, CRUZ & FEIO 2007). The district known as "Alto da Serra", "Alto da Serra de Paranapiacaba", "Alto da Serra de Cubatão", or simply as "Paranapiacaba," is located at the edge of the Atlantic Plateau between the metropolitan area of São Paulo and its coast,and it is part of the Serra de Paranapiacaba (BOKERMANN 1966). This district was founded in the nineteenth century, along with the construction of the first railroad of the state (Santos-Jundiaí railway) (LOPES & KIRIZAWA 2009). The region was also the setting point of one of the first conservation units of Brazil, the Reserva Biológica do Alto da Serra de Paranapiacaba (REBIO) (LOPES & KIRIZAWA 2009).

Since its foundation, several scientific expeditions have visited Paranapiacaba, which was one of the first locations in São Paulo to be surveyed by zoologists. Several notable scientists explored the area, such as the naturalists Frederico Carlos Hoehne, Hermann Friedrich von Ihering, Jean Massart, Carl Friedrich Von Martius, Auguste de Saint-Hilaire, Arthur Neiva, Affonso de E. Taunay, Hermann Luederwaldt, and the herpetologists Alípio de Miranda-Ribeiro, Adolpho Lutz, Bertha Lutz, Joaquim Venâncio, Oswaldo Peixoto, and Werner Bokermann (MELO *et al.* 2009, VERDADE *et al.* 2009).

Although Paranapiacaba still represents one of the most intensively surveyed localities of the state of São Paulo, being especially relevant for anurans (DIXO & VERDADE 2006), its herpetofauna remains poorly known. Additionally, only a small fraction of the information available in scientific collections, and which deals with specific aspects of the anuran community, has been published (e.g., BOKERMANN 1968, SAZIMA & BOKERMANN 1978, POMBAL & CRUZ 1999, PATTO & PIE 2001, OLIVEIRA *et al.* 2008, PMSA 2008, VERDADE *et al.* 2009).

Here, we provide a concise list of the reptiles and amphibians from the Parque Natural Municipal Nascentes de Paranapiacaba. Our study represents the first comprehensive herpetofaunal survey for the region.

MATERIAL AND METHODS

The present work was conducted at the Parque Natural Municipal Nascentes de Paranapiacaba (PNMNP), municipality of Santo André. The park occupies 426 ha of mountainous Atlantic Forest formation, between 23°47′4.9″S-23°45′27.9″S and 46°18′19.4″W-46°17′7.8″W, at an altitude range of 850 to approximately 1174 m (PMSA 2008). The park surrounds the Paranapiacaba district, adjoining the REBIO, and also the locality known as Campo Grande da Serra, which corresponds to the deactivated Campo Grande train station (BOKERMANN 1966). It is also contiguous with the Parque Estadual da Serra do Mar, and is located at the end of the Mogi river valley (Fig. 1). Historical factors, for instance extraction of natural resources for railroad maintenance and for the construction of the village of Paranapiacaba, led to a predominantly altered vegetational landscape.

The climate of the PNMNP is classified as altitudinal tropical and mesothermic super humid. Mean annual temperatures vary little, ranging from 14° or 15°C to 21° or 22°C. The region is within the greatest cell of precipitation in Brazil (3,300 mm mean average). There is no hydric deficit in the PNMNP, and humidity is high throughout the year. The high humidity, combined with the orographic effect coming from the Serra do Mar, produce a fog that is distinctive (GUTJAHR & TAVARES 2009).

We conducted 15 field surveys, from November 2009 to March 2011. Regular sampling lasted eight to 10 days each month, totaling 117 sampling days. Three complementary methods were used: visual surveys (CRUMP & SCOTT 1994), pitfall traps with drift-fences (GIBBONS & SEMLITSCH 1982, CORN 1994, CECHIN & MARTINS 2000, BLOMBERG & SHINE 2006), and occasional encounters (CAMPBELL & CHRISTMAN 1982, SAWAYA et al. 2008).

Three sets of pitfall traps were used. Each was composed of two stations with ten 100-liter plastic recipients, totaling 60 recipients installed. Every recipient was connected to another by eight m of drift-fences. The fences were 1 m high, and were buried approximately 20 cm into the ground, passing through the center of each recipient. The pitfall sites were located at least 300 m apart from each other, in order to maintain the spatial independence of the sample units. The sites were selected as to encompass different variables, such as arboreal physiognomy and the influence of water bodies. Pitfall traps were inspected daily totaling 111 days of opened traps (6,600 recipients-days).

In addition to pitfall traps, visual and audio surveys were performed at eight different sites within the PNMNP and also in adjacent areas. Sampling was carried out preferentially during the evening and at night, by a team of two to four researchers. Anuran vocalization was also registered with a portable digital recorder (Sony ICD-P630F) to assist species identification.

Data collected by other means than the methods mentioned above were classified as occasional encounters, and pertain exclusively to random species records. Hence, this sample effort was not considered for data analysis. We collected the following information for each specimen registered by us: location, method of collection, date and time of collection, type of environment, activity patterns, and approximate climate conditions. All specimens collected were deposited in the Coleção Célio F.B. Haddad, Universidade Estadual Paulista "Júlio de Mesquita Filho", Rio Claro, São Paulo (amphibians) and Coleção Herpetológica do Museu de Zoologia da Universidade de São Paulo, São Paulo (amphibians and reptiles).

Secondary data was retrieved from the literature and the examination of specimens from Santo André and adjoining municipalities (Santos, Cubatão, Mogi das Cruzes, Rio Grande da Serra and Suzano), deposited in the following herpetological collections: Coleção Célio F. B. Haddad, Universidade Estadual Paulista "Júlio de Mesquita Filho", Rio Claro, São Paulo (CFBH), Coleção Herpetológica "Alphonse Richard Hoge", Instituto Butantan, São Paulo (IB), Coleção Herpetológica do Museu de Zoologia da Universidade de São Paulo, São Paulo (MZUSP), and Museu de História Natural da Universidade Estadual de Campinas, São Paulo (ZUEC). Specimens from the collection of the Instituto Butantan were only considered when their identification was confirmed by one of us, prior to the 2010 fire (KUMAR 2010).

We adopted the taxonomic classification of FROST et al. (2001), KEARNEY (2003), ZAHER et al. (2009), and CARRASCO et al. (2012) for reptiles, with one exception: species of Liophis Wagler, 1830 were allocated to Erythrolamprus Boie, 1826 (CURCIO et al. 2009, FORLANI et al. 2010, GRAZZIOTIN et al., 2012). The classification of amphibians follows FROST (2013) and CARAMASCHI & CRUZ (2013).

We used the sampling data to analyze species composition and richness (number of species). Relative abundance was estimated as a percentage of the number of individuals from each species over the total number of individuals registered. Relative abundance was calculated exclusively based on data collected by pitfall traps. Efficiency of the sampling method was evaluated using species rarefaction curves (COLWELL & CODDINGTON 1994, GOTELLI & COLWELL 2001, THOMPSON et al. 2003), with 95% confidence interval and 1000 aleatorizations, using the software EstimateS 8.20 (COLWELL 2009). Richness was assessed through a non-parametric first order Jackknife index (HELTSCHE & FORRESTER 1983, HELLMANN & FOWLER 1999). Sample effort was considered as the number of open pitfall traps per sampling day. Specimens collected outside the sampling period were not considered in the analysis.

In order to define patterns of species distribution in Paranapiacaba, we allocated each taxon into one of the following five categories: 1) species distribution restricted to the Paranapiacaba region, also encompassing the Estação Ecológica de Boracéia (23°38′S and 45°52′W) (HEYER et al. 1990) (Par); 2) species endemic to the Serra do Mar range within the state of São Paulo (S); 3) species distributed in the entire Serra do Mar

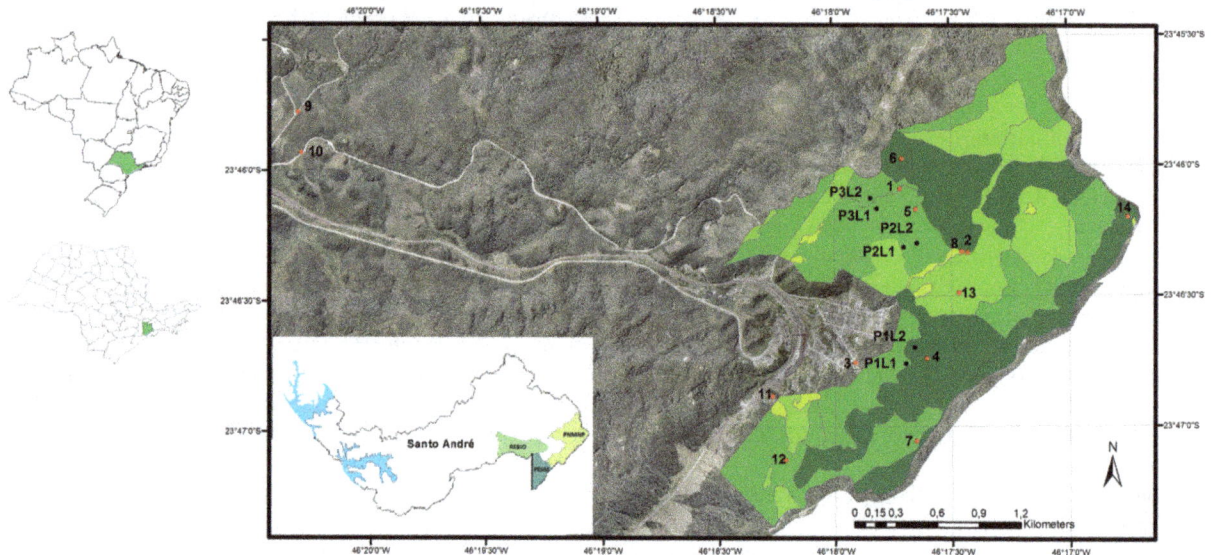

Figure 1. Geographic position of the study area within the state of São Paulo and the municipality of Santo André, including adjacent conservation units. PNMNP: Parque Natural Municipal Nascentes de Paranapiacaba, PESM: Parque Estadual da Serra do Mar, REBIO: Reserva Biológica do Alto da Serra de Paranapiacaba (Adapted from PMSA 2008). Red dots, numbered from 1 to 13, represent localities of visual survey samples, black dots represent localities of pitfall trap samples (P1, P2 and P3).

formation, from south to southeastern Brazil (Se); 4) species broadly distributed within the Atlantic forest (southeastern, northeastern and/or southern Brazil) (Af); 5) species broadly distributed in Brazil, with records outside the Atlantic Forest Domain (Br).

RESULTS

Amphibians

We recorded a total of 80 species of anurans for the PNMNP and Paranapiacaba region, Santo André municipality. Fifty-six species of anurans were catalogued as a result of the field survey, and secondary data (see methods) yielded 73 records. New distributional records for 15 species, including four undescribed species (Table I), resulted for the region. Ten families were registered: Brachycephalidae (11 species), Bufonidae (4), Centrolenidae (1) Craugastoridae (2), Cycloramphidae (5), Hemiphractidae (3), Hylidae (34), Hylodidae (6), Leptodactylidae (13), and Odontophrynidae (1) (Figs 2-61).

Most species are distributed on mountainous areas of the Atlantic forest from south and southeastern Brazil (60.5%). Eight of those are strictly endemic to the Paranapiacaba region (10.5%) and 14 (18.4%) are endemic to the Serra do Mar formation of the state of São Paulo. The remaining species are broadly distributed throughout the Atlantic Forest (23.7%), and 15.8% of the registred anurans occur in more than one biome within Brazil (Table I).

Considering data from pitfall traps alone, we captured 811 specimens from 17 anuran species. The dominant species was *Ischnocnema parva* (314 individuals, 38.7%), followed by another brachycephalid, *I. guentheri* (154 individuals, 18.9%), and by *Physalaemus moreirae* (152 individuals, 18.7%). Eight species exhibited intermediary abundance: *Leptodactylus ajurauana* (33 individuals, 4.1%), *L.* cf. *marmoratus* (31 individuals, 3.8%), *Haddadus binotatus* (26 individuals, 3.2%), *Rhinella icterica* (19 individuals, 2.3%), *R. ornata* (18 individuals, 2.2%), *Brachycephalus* sp. (14 individuals, 1.7%), *Ischnocnema* sp (gr. *lactea*) (13 individuals, 1.6%), and *Paratelmatobius cardosoi* (10 individuals, 1.2%). Six other species were considered rare, comprising less than 1% of the total records: *Bokermannohyla hylax* (seven specimens), *Dendrophryniscus brevipollicatus* (seven), *Cycloramphus eleutherodactylus* (five), *Ichnocnema* cf. *spanios* (four), *I. hoehnei* (two) and *I. juipoca* (two).

Species rarefaction curves from pitfall traps data stabilized for sampled amphibians (Fig. 62). The jackknife index calculated for anurans (Jack1 = 17 ± 0) corresponded fairly to the number of observed species.

Reptiles

We recorded fifty six species of reptiles for the PNMNP and the Paranapiacaba region (Table II). Of these, 39 were snakes classified in four families: Colubridae (six species), Dipsadidae (30), Elapidae (one), Viperidae (one), and Tropidophiidae (one); and 13 were lizards from five families: Anguidae (two species), Gekkonidae (one species), Gymnophtalmidae (five), Leiosauridae

Table I. Species composition of anurans from Parque Municipal Nascentes de Paranapicaba (PNMNP), Santo André, São Paulo. Type of data collected: 1. Primary data from field surveys during the study period; 2. secondary data gathered from herpetological collection catalogs, 3. secondary data from Verdade et al. (2009). Localities of occurrence at site areas: a. Paranapiacaba or Alto da Serra, b. PNMNP (study area), c: Reserva Biológica do Alto da Serra de Paranapicaba, d: Campo Grande da Serra, e: Santo André municipality. Geographical distribution: Par: species exclusive to Paranapiacaba region, S: endemic for the Serra do Mar from São Paulo state, Se: Serra do Mar formation, south and southeastern Brazil, Af: broad distribution within the Atlantic forest, Br: broad distribution in Brazil. * Represents "Paranapiacaba" or "Alto da Serra" as type locality; ** Paranapiacaba as type locality of one of the species synonyms; ***Campo Grande as type locality.

Family	Species	Data	Site	Distribution
Brachycephalidae	Brachycephalus sp.	1,2	b,c	Par
	Brachycephalus hermogenesi (Giaretta & Sawaya, 1998)	3	b,c	Se
	Ischnocnema cf. spanios (Heyer, 1985)	1	b,e	Par
	Ischnocnema gerhti* (Miranda-Ribeiro, 1926)	2,3	a	Par
	Ischnocnema guentheri (Steindachner, 1864)	1,2,3	a,b,c,e	Af
	Ischnocnema hoehnei* (Lutz, 1958)	1,2,3	a,b	Se
	Ischnocnema juipoca (Sazima & Cardoso, 1978)	1,3	b	Br
	Ischnocnema nigriventris*(Lutz, 1958)	1,3	a,b	Par
	Ischnocnema parva (Girard, 1853)	1,2,3	a,b,c	Se
	Ischnocnema sp. (aff. guentheri)	1	b	–
	Ischnocnema sp. (lactea series)	1	b	–
Bufonidae	Dendrophryniscus brevipollicatus** Jiménez de la Espada, 1870	1,2,3	a,b,c	Se
	Dendrophryniscus cf. brevipollicatus	1,2	a,b,c	–
	Rhinella icterica (Spix, 1824)	1,2,3	a,b,c,e	Af
	Rhinella ornata (Spix, 1824)	1,2,3	a,b,c,e	Af
Centrolenidae	Vitreorana uranoscopa (Müller, 1924)	1,2,3	a,b	Af
Craugastoridae	Haddadus binotatus (Spix, 1824)	1,2,3	a,b,c	Af
	Holoaden suarezi Martins & Zaher, 2013	2	a	S
Cycloramphidae	Cycloramphus acangatan Verdade & Rodrigues, 2003	2,3	a,b,c	S
	Cycloramphus dubius* (Miranda-Ribeiro, 1920)	2,3	a,b	S
	Cycloramphus eleutherodactylus* (Miranda-Ribeiro, 1920)	1,2,3	a,b,c	Se
	Cycloramphus semipalmatus* (Miranda-Ribeiro, 1920)	2,3	a,b	S
	Thoropa taophora*(Miranda-Ribeiro, 1923)	1,2,3	a	S
Hemiphractidae	Fritziana fissilis (Miranda-Ribeiro, 1920)	1,2,3	a,b,c	Se
	Fritziana ohausi (Wandolleck, 1907)	1,2,3	a,b	Se
	Gastrotheca fulvorufa*(Andersson, 1911)	1,2	a,b	Se
Hylidae	Aplastodiscus albosignatus* (Lutz and Lutz, 1938)	1,2,3	a,b,c	Af
	Aplastodiscus arildae (Cruz & Peixoto, 1985)	1,2	a,b	Se
	Aplastodicus leucopygius (Cruz & Peixoto, 1985)	1, 2,3	a,b,c,e	Se
	Bokermannohyla astartea* Bokermann, 1967	2,3	a,d	S
	Bokermannohyla circumdata (Cope, 1871)	1,2,3	a,b,c	Af
	Bokermannohyla hylax (Heyer, 1985)	1,2,3	a,b,c,e	Af
	Dendropsophus berthalutzae* (Bokermann, 1962)	1,2,3	a,b,c,d	Af
	Dendropsophus elegans (Wied-Neuwied, 1824)	1	b	Af
	Dendropsophus microps (Peters, 1872)	1,2,3	a,b,d	Af
	Dendropsophus minutus (Peters, 1872)	1,2,3	a,b,c,d	Br
	Dendropsophus nanus (Boulenger, 1889)	2	d	Br
	Dendropsophus sanborni (Schimdt, 1944)	2,3	d	Br
	Hypsiboas albomarginatus (Spix, 1824)	1,2,3	a,b,d	Af
	Hypsiboas albopunctatus (Spix, 1824)	2,3	a,d	Br

Continues

Table I. Continued.

Family	Species	Data	Site	Distribution
	Hypsiboas bandeirantes Caramaschi & Cruz, 2013	1,2,3	a,b,d	Se
	Hypsiboas bischoffi (Boulenger, 1887)	1,2,3	a,b,c,d,c	Af
	*Hypsiboas cymbalum** (Bokermann, 1963)	2,3	d	Par
	Hypsiboas faber (Wied-Neuwied, 1821)	1,2,3	b,c	Af
	Hypsiboas pardalis Spix, 1824	1,2,3	b,d	Se
	Hypsiboas prasinus (Burmeister, 1856)	1, 2,3	b,e	Se
	Phrynomedusa appendiculata (Lutz, 1925)	2	a	Se
	*Phrynomedusa fimbriata** Miranda-Ribeiro, 1923	2,3	a	Par
	Phyllomedusa burmeisteri Boulenger, 1882	2,3	a	Af
	Phyllomedusa rohdei Mertens, 1926	2	d	Se
	Scinax alter (Lutz, 1973)	2,3	a	Af
	Scinax berthae (Barrio, 1962)	2,3	d	Br
	*Scinax brieni** (De Witte, 1930)	1,2,3	a,b	S
	Scinax cf. *perpusillus* Lutz & Lutz, 1939	1,2,3	a,b,c	Se
	Scinax crospedospilus (Lutz, 1925)	1,2,3	a,b,c	Se
	Scinax fuscovarius (Lutz, 1925)	1,2,3	a,b,c,d,e	Br
	Scinax hayii (Barbour, 1909)	1,2,3	a,b,c,d,e	Se
	Scinax hiemalis (Haddad & Pombal, 1987)	1,2	a,b,e	S
	*Scinax rizibilis** (Bokermann, 1964)	1,2,3	b,d	Af
	Scinax squalirostris (Lutz, 1925)	2	d	Br
Hylodidae	*Crossodactylus dispar* Lutz, 1925	2,3	c	S
	*Crossodactylus gaudichaudi*** Duméril & Bibron, 1841	2,3	c,d	Se
	Hylodes asper (Müller, 1924)	1,2,3	a,b	Se
	Hylodes sp. (aff. *phyllodes*)	1	b	–
	Hylodes phyllodes Heyer & Cocroft, 1986	2,3	a	S
	*Megaelosia massarti**(De Witte, 1930)	1,2,3	a,b	Par
Leptodactylidae	*Physalaemus bokermanni**** Cardoso & Haddad, 1985	1,2,3	a,d	S
	Physalaemus cuvieri Fitzinger, 1826	1,2,3	a,b,d	Br
	*Physalaemus maculiventris** (Lutz, 1925)	2,3	a,b	Af
	*Physalaemus moreirae**(Miranda-Ribeiro, 1937)	1,2,3	a,b,c	S
	Physalaemus olfersii (Lichtenstein & Martens, 1856)	2,3	a,c,d	Se
	Leptodactylus ajurauna Berneck, Costa & Garcia, 2008	1,2	a,b	S
	Leptodactylus cf. *marmoratus* (Steindachner, 1867)	1,2,3	a,b,c,e	Af
	Leptodactylus flavopictus Lutz, 1926	2,3	a,b	Af
	*Leptodactylus furnarius** Sazima & Bokermann, 1978	2,3	a,d	Br
	*Leptodactylus jolyi** Sazima & Bokermann, 1978	1,2,3	b,d	Br
	Leptodactylus latrans (Steffen, 1815)	1,2,3	a,b,c,d	Br
	*Paratelmatobius cardosoi** Pombal & Haddad, 1999	1,2,3	a,b	Par
	*Paratelmatobius poecilogaster** Giaretta & Castanho, 1990	1, 2,3	a,b	S
Odontophrynidae	*Proceratophrys melanopogon**(Miranda-Ribeiro, 1926)	1,2,3	a,b,d	Se

(four), and Teiidae (one). There were also two amphisbaenids; and two chelonians (Figs 63-86). Secondary data was responsible for 29 records. Field surveys encountered 18 snakes, 10 lizards and two chelonians, seven of which represent new records for the area according to collection records (see Marques 2009) (Table II).

A significant proportion of reptile species were broadly distributed throughout the Atlantic Forest Domain (41.1%). One snake is endemic to São Paulo (*Atractus serranus*) and 10.7% of the species are restricted to the Serra do Mar formation of southeastern Brazil. Most reptile species recorded in this study,

Figures 2-16. Anuran species sampled for the Parque Natural Municipal Nascentes de Paranapiacaba, Santo André, São Paulo state: (2) *Brachycephalus* sp.; (3) *Ischnocnema* sp. (aff. *guentheri*); (4) *Ischnocnema* sp. (*lactea* series); (5) *Ischnocnema* cf. *spanios*; (6) *Ischnocnema guentheri*; (7) *Ischnocnema hoehnei*; (8) *Ischnocnema juipoca*; (9) *Ischnocnema nigriventris* (10) *Ischnocnema parva*; (11) *Dendrophryniscus brevipollicatus*; (12) *Dendrophryniscus* cf. *brevipollicatus*; (13) *Rhinella icterica*; (14) *Rhinella ornata*; (15) *Vitreorana uranoscopa*; (16) *Haddadus binotatus*; (17) *Cycloramphus dubius* (ZUEC 6859). Photos: Vivian C. Trevine, except: 6, 10 (Juan Camilo Arredondo).

Figures 18-31. Anuran species sampled for the Parque Natural Municipal Nascentes de Paranapiacaba, Santo André, São Paulo state: (18) *Cycloramphus eleutherodactylus*; (19) *Cycloramphus semipalmatus* (ZUEC 2722); (20) *Proceratophrys melanopogon*; (21) *Thoropa taophora*; (22) *Fritziana fissilis*; (23) *Fritziana ohausi*; (24) *Gastrotheca fulvorufa*; (25) *Aplastodiscus albosignatus*; (26) *Aplastodiscus arildae*; (27) *Aplastodiscus leucopygius*; (28) *Bokermannohyla circumdata*; (29) *Bokermannohyla hylax*; (30) *Dendropsophus berthalutzae*; (31) *Dendropsophus elegans*. Photos: Vivian C. Trevine, except: 24 (Juan Camilo Arredondo).

Figures 32-46. Anuran species sampled for the Parque Natural Municipal Nascentes de Paranapiacaba, Santo André, São Paulo state: (32) *Dendropsophus microps*; (33) *Dendropsophus minutus*; (34) *Hypsiboas albomarginatus*; (35) *Hypsiboas bischoffi*; (36) *Hypsiboas cymbalum* (MZUSP 74194); (37) *Hypsiboas faber*; (38) *Hypsiboas pardalis*; (39) *Hypsiboas bandeirantes*; (40) *Hypsiboas prasinus*; (41) *Phyllomedusa rohdei* (MZSUP 81306); (42) *Scinax brieni*; (43) *Scinax* cf. *perpusillus*; (44) *Scinax crospedospilus*; (45) *Scinax fuscovarius*; (46) *Scinax hayii*. Photos: Vivian C. Trevine.

Figures 47-61. Anuran species sampled for the Parque Natural Municipal Nascentes de Paranapiacaba, Santo André, São Paulo state: (47) *Scinax hiemalis*; (48) *Scinax rizibilis*; (49) *Scinax squalirostris* (MZUSP 113525); (50) *Hylodes asper*; (51) *Hylodes* sp. (aff. *phyllodes*); (52) *Megaelosia massarti* (ZUEC 8516); (53) *Physalaemus bokermanni*; (54) *Physalaemus cuvieri*; (55) *Physalaemus moreirae*; (56) *Leptodactylus ajurauna*; (57) *Leptodactylus* cf. *marmoratus*; (58) *Leptodacylus jolyi*; (59) *Leptodactylus latrans*; (60) *Paratelmatobius cardosoi*; (61) *Paratelmatobius poecilogaster*. Photos: Vivian C. Trevine.

however, also occur outside the Atlantic forest (46.4%), in the Cerrado or the Amazon (Table II).

Pitfall trap data yielded 137 specimens from 10 species of Squamata reptiles. The predominant families were Gymnophtalmidae and Dipsadidae. Nonetheless, the dominant species was the leiosaurid *Enyalius* ssp, with 89% of all the records, whereas *E. perditus* (93 individuals, 67.4%) predominated over its congeneric *E. iheringii* (30 individuals, 21.7%). Four species exhibited intermediary abundance, the lizards *Ecpleopus gaudichaudii* (three individuals, 2.1%), *Colobodactylus taunayii* (two individuals, 1.4%), *Placosoma glabellum* (two individuals, 1.4%), and the snakes *Bothrops jararaca* (two individuals, 1.4%) and *Erythrolamprus aesculapii* (two individuals, 1.4%). Other three species (*Placosoma cordilynum*, *Taeniophallus bilineatus*, and *Atractus serranus*) were considered rare, and were represented by only one specimen.

The species accumulation curve for the reptiles sampled by pitfall traps could not be stabilized, even though it showed a slight tendency for stabilization (Fig. 62). The calculated richness index (Jack1 = 14.9 ± 1.9) suggests that the absolute reptile diversity has not yet been sampled by this method.

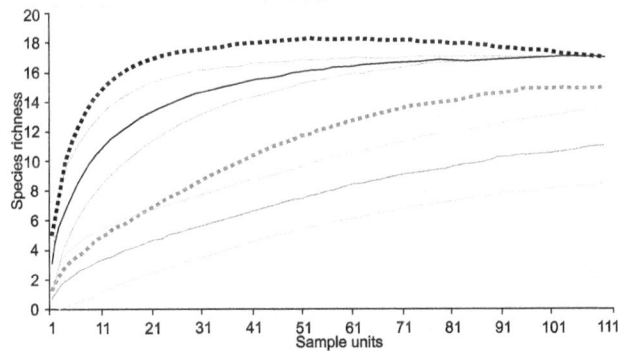

Figure 62. Species accumulation curves (solid lines) and richness index (first order Jackknife) for amphibians (black) and reptiles (grey) sampled by pitfall traps during field surveys in the Parque Natural Municipal Nascentes de Paranapiacaba, Santo André, São Paulo. Dashed lines represent respective standard deviation.

DISCUSSION

Species composition

The amphibian diversity recorded for Paranapiacaba (80 ssp.) in this work is higher than the diversity known for the Estação Ecológica de Boracéia (67 ssp.), which has been considered the most distinctive area in São Paulo in terms of amphibian diversity (HEYER *et al.* 1990, ARAÚJO *et al.* 2009a). The diversity of reptiles (56 ssp) was also very representative for the state (CONDEZ *et al.* 2009, FORLANI *et al.* 2010).

We recorded 15 anuran species for Paranapiacaba that were not present in previous records: *Ischnocnema* cf. *spanios*,

Ischnonema sp. (gr. *lactea*), *Dendrophryniscus* cf. *brevipollicatus*, *Aplastodiscus arildae*, *Dendropsophus elegans*, *D. nanus*, *Phrynomedusa appendiculata*, *Phyllomedusa rohdei*, *Scinax squalirostris*, *S. hiemalis*, *Leptodactylus ajurauna*, and *Holoaden suarezi*; and also the undescribed *Brachycephalus* sp., *Ischnocnema* sp. (aff. *guentheri*), and *Hylodes* sp.(aff. *phyllodes*). Some species cited in VERDADE *et al.* (2009) were not considered here and are mentioned in the taxonomical comments.

Simultaneously, 27 species of reptiles were added to previously published data for the area: 16 snakes, seven lizards, two amphisbeanids, and two chelonians (MARQUES 2009). Contrary to MARQUES (2009), we did not consider the following species: *Leposternon microcephalum*, *Kentropyx paulensis*, *Chironius fuscus*, *C. foveatus*, *Pseudoboa serrana*, *Siphlophis pulcher*, and *Bothrops jararacussu*. Even though their distribution range may include Paranapiacaba, there were no collection records corroborating their presence in the region. However, this may be the result of life habitat, altitudinal range or collection bias, and it is possible that these species will be recorded for the region in the near future. The only exception is the lizard *Kentropyx paulensis*, listed in MARQUES (2009), which is an endemic of Cerrado formations (VALDUJO *et al.* 2009).

Some records were not included in our list, such as *Erythrolamprus typhlus* and *Thamnodynastes strigatus*. Those snakes have been registered for nearby localities. However, as we were not able to match specimens to a precise geographic coordinate, their presence in the PNMNP could not be confirmed. On the other hand, species such as *Amphisbaena alba*, *A. dubia*, *Atractus reticulatus*, *Erythrolamprus jaegeri*, *Oxyrhopus guibei*, and *Sibynomorphus mikanii* were included as probable occurrences. Although they are not believed to occur within the PNMNP boundaries, their distribution is expected for the macro region where Paranapiacaba is inserted (MARQUES *et al.* 2009), with records for Santo André and adjacent municipalities. One single collection record was catalogued for the snake *Echinanthera melanostigma* (IB 1640). However, as the authors could not confirm the identification of the specimen, this species was not included herein.

According to the species accumulation curve, the pitfall trap method was efficient in sampling leaf litter anurans. Reptile diversity was not fully sampled by this method, and we expect that an increased sampling effort would have collected a number of additional species.

Altitudinal factors play an important role in the species composition of the Atlantic Forest Domain (HEYER *et al.* 1990, VASCONCELOS *et al.* 2010), and altitude seems to be a limiting factor for some species at Paranapiacaba, for instance the snakes *Bothrops jararacussu* and *Chironius laevicollis* (MARQUES 2009). Alternatively, some species are restricted to elevated areas, for example *Heterodactylus imbricatus* and *Atractus serranus*. On the other hand, species normally abundant within the Serra do Mar range were scarce in the PNMNP. That appeared to be the case for *Hylodes asper*, otherwise very common in nearby lo-

Table II. Species composition of reptiles from Parque Municipal Nascentes de Paranapicaba, Santo André, São Paulo (PNMNP). Type of data collected: 1. Primary data from field surveys during the study period, 2. secondary data gathered from herpetological collection catalogs, 3. secondary data from Marques (2009). Localities of occurrence at site areas: a. Paranapiacaba or Alto da Serra, b. PNMNP (study area), c: Reserva Biológica do Alto da Serra de Paranapiacaba, d: Campo Grande da Serra, e: Santo André and adjacent municipalities. Geographical distribution: Par: species exclusive to Paranapiacaba region, S: endemic for the Serra do Mar from São Paulo state, Se: Serra do Mar formation, south and southeastern Brazil, Af: broad distribution within the Atlantic forest, Br: broad distribution in Brazil.

Family	Species	Data	Site	Distribution
Squamata				
Amphisbaenidae	*Amphisbaena alba* Linnaeus, 1758	2	e	Br
	Amphisbaena dubia Müller, 1924	2	e	Br
Lacertilia				
Anguidae	*Diploglossus fasciatus* (Gray, 1831)	2,3	a	Br
	Ophiodes cf. *fragilis* Spix, 1824	1,2,3	a,b,c	Br
Gekkonidae	*Hemidactylus mabouia* Moreau de Jonnès, 1818	1, 2	b, e	Br
Gymnophtalmidae	*Colobodactylus taunayi* (Amaral, 1933)	1	b	Af
	Ecpleopus gaudichaudii Duméril & Bibron, 1839	1	b	Af
	Heterodactylus imbricatus Spix, 1825	1	b	Af
	Placosoma cordylinum champsonotus (Tschudi, 1847)	1	b	Af
	Placosoma glabellum Peters, 1870	1	b	Af
Leiosauridae	*Anisolepis grilii* Boulenger, 1891	2,3	e	Af
	Enyalius iheringii Boulenger, 1885	1,2,3	b,c	Af
	Enyalius perditus Jackson, 1978	1,2,3	b,c	Af
	Urostrophus vautieri Duméril & Bibron, 1837	2,3	a	Br
Teidae	*Salvator merianae* (Duméril & Bibron, 1839)	1,2	b, c, e	Br
Serpentes				
Colubridae	*Chironius bicarinatus* Wied, 1820	1,2,3	b,d,e	Af
	Chironius exoletus (Linnaeus, 1758)	2,3	e	Br
	Chironius laevicollis (Wied, 1824)	2,3	e	Br
	Pseustes sulphureus (Wagler, 1824)	2	a	Br
	Spilotes pullatus (Linnaeus, 1758)	2,3	a,b	Br
	Tantilla melanocephala (Linnaeus, 1758)	2	a	Br
Dipsadidae	*Atractus pantostictus* Fernandes & Puorto, 1993	2	a	Br
	Atractus reticulatus (Boulenger, 1885)	2	e	Br
	Atractus serranus Amaral, 1930	1,2,3	a,b	S
	Atractus zebrinus (Jan, 1862)	1	b	Af
	Clelia plumbea (Wied, 1820)	2,3	c	Br
	Dipsas alternans Fischer, 1885	2	a	Af
	Dipsas indica Laurenti, 1768	2	e	Se
	Echinanthera cephalostriata Di Bernardo, 1996	1,2,3	b	Af
	Echinanthera undulata (Wied, 1824)	1,2,3	a,b,d	Af
	Elapomorphus quinquelineatus (Raddi, 1820)	1,2,3	a,b	Af
	Erythrolamprus aesculapii Linnaeus, 1758	1,2,3	b,e	Br
	Erythrolamprus jaegeri (Gunther, 1858)	2	e	Br
	Erythrolamprus miliaris Linnaeus, 1758	1,2,3	b	Br
	Helicops modestus Günther, 1861	2	b,e	Se
	Imantodes cenchoa (Linnaeus, 1758)	2,3	a	Br
	Oxyrhopus clathratus Duméril, Bibron & Duméril, 1854	1,2,3	a,b,d	Af
	Oxyrhopus guibei Hoge & Romano, 1977	2	e	Br
	Philodryas aestiva (Duméril, Bibron & Duméril, 1854)	2,3	b,e	Br

Continues

Table II. Continueed.

Family	Species	Data	Site	Distribution
	Philodryas patagoniensis (Girard, 1858)	1,2	e	Br
	Sibynomorphus mikanii (Schlegel, 1837)	2	e	Br
	Sibynomorphus neuwiedi (Iheringii, 1911)	1,2,3	a,b,d	Af
	Siphlophis longicaudatus (Andersson, 1901)	2,3	a	Af
	Taeniophallus affinis (Günther, 1858)	1,2,3	a,b,e	Af
	Taeniophallus bilineatus (Fischer, 1885)	1,2,3	b	Af
	Thamnodynastes sp. 1	2	a,e	Se
	Tomodon dorsatus Duméril, Bibron & Duméril, 1854	1,2,3	a,b,e	Af
	Tropidodryas serra (Schlegel, 1837)	2	a	Af
	Tropidodryas striaticeps (Cope, 1869)	2,3	a	Af
	Xenodon merremii (Wagler, 1824)	2	a,e	Br
	Xenodon neuwiedii Günther, 1863	1,2,3	a,b	Br
Viperidae	Bothrops jararaca Wied, 1824	1,2,3	a,b,c,e	Af
Elapidae	Micrurus corallinus (Merrem, 1820)	2,3	a	Se
Tropidophiidae	Tropidophis paucisquamis (Müller, 1901)	2	d	Se
Testudines				
Chelidae	Hydromedusa maximiliani Mikah, 1820	1,2	b,e	Se
	Hydromedusa tectifera Cope, 1869	1	b	Br

calities (Patto & Pie 2006). Only five specimens were collected in the area PNMNP area during the entire study period, whereas at least 20 individuals were visualized within few minutes of active daily search in a proximal locality, on the Serra do Mar slope.

Variables such as high humidity and the almost continuous rainy season certainly play an important role in species assemblages in Paranapiacaba, especially for anurans. For instance, the substantial diversity (11 species) of anurans that have direct development (Brachycephalidae) could account for these favorable climatic conditions. The dominance of brachycephalids appears to be a recurrent pattern among Neotropical litter frog faunas (Scott 1976, Duellman 1988, Heinen 1992, Giaretta et al. 1997, 1999, Rocha et al. 2001, 2007, Van Sluys et al. 2007, Almeida-Gomes et al. 2008, Santos-Pereira et al. 2011, Siqueira et al. 2011), and it is possibly associated with higher humidity values in cloud forests of higher altitudes (Giaretta et al. 1999).

The relationship between altitudinal gradients and richness and equitativity of reptile species in the Neotropical region has been debated (Scott 1976, Fauth 1989, Hofer & Bersier 2001, Dixo & Verdade 2006, Vasconcelos et al. 2010) and not fully understood for the Atlantic forest. However, there is an apparent pattern for Atlantic forest localities: frog density seems to be higher in leaf litter assemblages collected in higher altitudes. This pattern has been corroborated for a few localities (Giaretta et al. 1999, Vasconcelos et al. 2010, Siqueira et al. 2011). In the present work, the overall abundance of leaf litter anurans was higher than in lower altitudes (Rocha et al. 2001, 2007, Van Sluys et al. 2007).

Paranapiacaba is the type locality of 23 species of anurans. Historical exploration of the area and the considerable sampling effort made by zoologists during decades could account for such a diverse species catalog. Nevertheless, as one examines anuran diversity in adjacent areas (Estação Ecológica de Boracéia and Parque das Neblinas) (Heyer et al. 1990, Berneck et al. 2008, 2013, Garcia et al. 2009b), a gradient of high diversity and similar species composition can be observed throughout this portion of the Serra do Mar.

Taxonomic comments

The taxonomic status of several species that occur in Paranapiacaba has been intricate (Giaretta & Castanho 1990, Pombal & Cruz 1999). Even though we do not intend to provide a taxonomic review of each group, we do consider necessary to elucidate a few taxonomic aspects of some species.

Verdade et al. (2009) mentioned Brachycephalus ephippium for the REBIO and Paranapiacaba. However, through the analysis of collected specimens and their life habits, we believe that this taxon represents an undescribed species, differentiated from B. ephippium by its darker dorsal coloration and ossification pattern. Brachycephalus sp. is currently being described by Paulo Garcia and collaborators for the Parque das Neblinas, municipality of Mogi das Cruzes. The species seems to be restricted to the most superficial leaf litter layers, and active individuals were always found hidden under leaves, differing from the typical pattern of B. ephippium (Pombal et al. 1994).

Another new brachycephalid sampled, Ischnocnema sp. (aff. guentheri), can be distinguished from the congeneric I. guentheri by a suite of characters that include vocalization and

Figures 63-77. Reptile species sampled for the Parque Natural Municipal Nascentes de Paranapiacaba, Santo André, São Paulo state: (63) *Hydromedusa maximiliani*; (64) *Hydromedusa tectifera*; (65) *Ophiodes* cf. *fragilis*; (66) *Hemidactylus mabouia*; (67) *Colobodactylus taunayi*; (68) *Ecpleopus gaudichaudii*; (69) *Placosoma cordylinum champsonotus*; (70) *Placosoma glabellum*; (71) *Enyalius iheringii*; (72) *Enyalius perditus*; (73) *Chironius bicarinatus*; (74) *Atractus serranus*; (75) *Atractus zebrinus*; (76) *Echinanthera cephalostriata*; (77) *Echinanthera undulata*. Photoss: Vivian C. Trevine, except: 64 (Ingo Grantsau), 66 (Juan Camilo Arredondo).

Figures 78-86. Reptile species sampled for the Parque Natural Municipal Nascentes de Paranapiacaba, Santo André, São Paulo state: (78) *Elapomorphus quinquelineatus*; (79) *Erythrolamprus aesculapii*; (80) *Erythrolamprus miliaris*; (81) *Oxyrhopus clathratus*; (82) *Philodryas patagoniensis*; (83) *Taeniophallus affinis*; (84). *Taeniophallus bilineatus*; (85) *Xenodon neuwiedii*; (86) *Bothrops jararaca*. Photoss: Vivian C. Trevine, except: 78 (Ingo Grantsau), and 84 (Alexandre Missassi).

larger male size (CRC: 30.3 mm) (Clarissa Canedo pers. com.). However, only a more representative sampling of the population and a broader taxonomic revision including the large complex under the name of *I. guentheri*, will provide sufficient grounds to diagnose this new species.

The *Ischnocnema lactea* group represents a complex of 14 species distributed in the Atlantic forest, from Rio de Janeiro to Santa Catarina (HEDGES *et al.* 2008), and Bahia (CANEDO & PIMENTA 2010). During our field surveys, we collected three species from this complex: *Ischnocnema* sp. (gr. *lactea*), *I.* cf. *spanios*, and *I. nigriventris*. *Ischnocnema lactea* is known exclusively from the holotype, and its precarious conservation status is responsible for misleading identifications (CANEDO & PIMENTA 2010). The specimens collected by us match the original description in characteristics such as enhanced apical discs with a bifid pattern,

absence of webbing, and reduced first hand disc (MIRANDA-RIBEIRO 1926). However, we would need the species to be re-described to arrive at a reliable identification. Therefore, we maintain the taxonomic status of the population found in PNMNP as *Ischnocnema* sp., pending further revision of the group.

Ischnocnema cf. *spanios* has been considered restricted to its type locality (Estação Ecológica de Boracéia, São Paulo), but similar specimens were collected from the Atlantic plateau as well, for instance the Parque Estadual de Carlos Botelho and adjacent regions (FORLANI *et al.* 2010). Still, it is possible that such records refer to more than one taxonomically distinct entity under one name, and a reassessment of the group is necessary for further considerations.

Ischnocnema nigriventris is a small species described for the locality known as "Alto da Serra" (Paranapiacaba)

(Bokermann 1966). The species is underrepresented in scientific collections and, because it had an ill-defined type series there used to be a lot of confusion about its identification until it was recently redescribed (Berneck *et al.* 2013). Based on the pattern of dark and light blotches on the venter and inner portions of the thighs, enlarged distal discs on external fingers and toes, dorsal and palpebrous tubercles, and a prominent calcar tubercle (Heyer 1985, Heyer *et al.* 1990, Pombal & Cruz 1999, Berneck *et al.* 2013) we recognize the population collected in the present study as *Ischnocnema nigriventris*.

The bufonid *Dendrophryniscus* cf. *brevipollicatus* resembles its sympatric *D. brevipollicatus*, except for the presence of a white rostral band in *D. leucomystax*. However, morphological and molecular studies on this genus are required for a better taxonomic resolution of the complex.

Scinax perpusillus identified by us might actually correspond to more than one nominal species (Heyer *et al.* 1990, Faivovich *et al.* 2005). Therefore, sampled specimens were listed as *S.* cf. *perpusillus*.

Verdade *et al.* (2009) mentioned the following anurans for the Paranapiacaba region: *Leptodactylus gracilis*, *Rhinella hoogmoedi*, *Scinax fuscomarginatus*, and *Gastrotheca microdiscus*. No additional record was made for *R. hoogmoedi* and *S. fuscomarginatus*, therefore their occurrence there cannot be corroborated. *Leptodactylus gracilis* is known to occur in southeastern and southern Brazil. However, its distributional range does not include the state of São Paulo (Araújo *et al.* 2009b, Rossa-Feres *et al.* 2011). Specimens deposited in the MZUSP correspond to *Leptodactylus jolyi*, and therefore, *L. jolyi* seems to be the taxon that occurs in Paranapiacaba. On the other hand, the only *Gastrotheca* sampled for the PNMNP corresponds to the description available for *G. fulvorufa* (Caramaschi & Rodrigues 2007, Izecksohn & Carvalho-e-Silva 2008). The latter was described for Paranapiacaba, and was recently removed from synonymy with *G. microdiscus* (Caramaschi & Rodrigues 2007).

Among hylodids, *Hylodes* spp. and *Crossodactylus* spp are problematic. The taxonomy of *Crossodactylus* is confusing, and morphological diagnostic characters are misleading (Heyer *et al.* 1990). One specimen resembling a young *Crossodactylus* was collected in a neighboring area, on the Serra do Mar slope. We were not able to provide a precise identification for that specimen. Collection records, however, confirm at least two species occurring in the region, *C. dispar* and *C. gaudichaudii* (see discussion below). Hence, our records for this genus are based solely on secondary data. *Hylodes* sp. (aff. *phyllodes*) is closely related to *H. phyllodes*. The large size of the male (CRC: 33.7-35.9 mm), contrasting with the male of *H. phyllodes* (CRC: 27.6-30.0 mm), as well as the coloration pattern of specimens, might be diagnostic for this new species. However, diagnostic characters still need to be established.

Difficulties identifying cryptic species of the *Leptodactylus marmoratus* complex (*sensu* Heyer 1973) are common (Pombal & Gordo 2004, Berneck *et al.* 2008). The epithet "marmoratus"

does not seem to apply to the populations from Paranapiacaba and Boracéia (Berneck *et al.* 2008), and it is possible that more than one species occur at the PNMNP.

The *Holoaden* specimen recorded for Paranapiacaba (MZUSP 891) was erroneously cited for Campos do Jordão, São Paulo state, and posteriorly defined as the paralectotype of *H. luederwaldti* (Miranda-Ribeiro 1920, Lutz 1958, Caramaschi & Pombal Jr 2006). The mistake was corrected and the specimen was recently described as *H. suarezi* (Martins & Zaher 2013). It is also found in the Estação Ecológica de Boracéia, the Parque Nacional da Serra da Bocaína and the Estação Ecológica do Bananal (Pombal Jr *et al.* 2008, Martins 2010, Martins & Zaher 2013).

Species boundaries are not very clear within the lizard genus *Ophiodes*. The specimens collected from the PNNMP correspond to *Ophiodes fragilis* (Márcio borges-Martins, pers.com.), and we follow that classification until a proper revision is available.

The snake *Dipsas indica* registered by us for Paranapiacaba corresponds to *D. indica bucephala* (Peters 1960). Recent taxonomic changes in the *D. indica* group have resulted in uncertainty about the status of this subspecies (Harvey & Embert 2008). Therefore, we refrain from unsing the subspecies name here.

The *Thamnodynastes* recorded for the Paranapiacaba region corresponds to *Thamnodynastes* "sp. 1" of Franco & Ferreira (2002). According to the authors, this taxon corresponds to the one described by Mikan (1820) as *Coluber nattereri*. Menawhile, as there is no formal redescription available to legitimize the name, the taxon is considered as *Thamnodynastes* "sp. 1".

Conservational status and final comments

The importance of Paranapiacaba for the conservation of Brazilian amphibians is already well established. It is the single known locality for the only anuran considered to be extinct from Brazil (*Phrynomedusa fimbriata*), and for another critically endangered species (*Hypsiboas cymbalum*) (Haddad 2008, Garcia *et al.* 2009a). This alone would be sufficient to establish the importance of the region as a primordial site for conservation. However, declines in anuran populations make the region even more important.

A great deal of attention was given in the 1980's, especially for the REBIO, when research revealed the effects of pollution emerging from the petrochemical pole of Cubatão (Domingos *et al.* 2009, Lopes *et al.* 2009). The original vegetation of the area has been altered as a consequence of pollution (Domingos *et al.* 2000, Moraes *et al.* 2002, 2003), and that has impacted several groups of animals, including amphibians (Lopes *et al.* 2009, Domingos *et al.* 2009, Sugiyama *et al.* 2009, Verdade *et al.* 2011, 2012). According to Verdade *et al.* (2009, 2011), the anurofauna of the REBIO is depauperate, if not in terms of richness, at least in the number of individuals of otherwise common species, for instance *Ischnocnema parva*. Additionally, the authors indicated that the vegetation physiognomy and the water pH of the REBIO are altered, contrasting with the unaltered pH values obtained from the PNMNP. It has been postulated that the mountainous portion of the REBIO is under direct influence of the wind current emerging from the petrochemical pole of Cubatão, whereas the PNMNP is apparently more protected from it (FERREIRA *et al.* 2009, VERDADE *et al.* 2009). These differences are reflected on the anurofauna: species are clearly more abundant in the PNMNP (VERDADE *et al.* 2009). Unfortunately, we were not able to sample the REBIO, and further comparisons are hindered until standardized sampling is performed simultaneously in the two conservation units.

The anuran populations of some species are likely declining in southeastern Brazil (HEYER *et al.* 1988, WEYGOLDT 1989, ETEROVICK *et al.* 2005, VERDADE *et al.* 2011). A few are considered extinct or are under considerable punctual decline. Examples are *Cycloramphus semipalmatus*, *C. boraceiensis*, *Crossodactylus dispar*, *C. gaudichaudii* and *Hylodes phyllodes* for the "Estação Ecológica de Boracéia", state of São Paulo (HEYER *et al.* 1988); and *Cycloramphus fuliginosus* and *Crossodactylus* spp for Santa Teresa, state of Espírito Santo (WEYGOLDT 1989).

Much of the information in the literature is merely speculative. A great diversity of factors might contribute to apparent reductions in population size. Life habitats associated with mountainous streams may be related to environmental sensibility of some of those supposedly declining anurans (HEYER *et al.* 1988, STUART *et al.* 2004, LIPS *et al.* 2005, VERDADE *et al.* 2011). On the other hand, various factors could influence such alterations, such as low population density, stochastic effects caused by natural selection, distributional patterns within the Serra do Mar range, collection biases, or even natural population fluctuations (HEYER *et al.* 1988, HEYER & MAXSON 1983, MAGNUSSON *et al.* 1999).

Nevertheless, even though we strongly recommend caution when evaluating patterns of population decline, we stress that species that now seem to be rarer were frequently collected during campaigns between 1950 and 1980, as documented in collection catalogs from MZUSP and ZUEC (see Appendix 1). Human interference, especially on adjacent areas of the Paranapiacaba village, appears to have played an important role on the current local species composition and loss of biodiversity. As a result from the operation of the railroad, several forested areas have been completely suppressed. Our inability to delimitating the exact path of past collecting expeditions, due to the lack of geographic coordinates for them, makes it difficult to reach more definite conclusions. Regardless, the results presented herein are relevant enough to indicate the need for future studies.

ACKNOWLEDGEMENTS

We are grateful to the Secretaria de Gestão de Recursos Naturais de Paranapiacaba e Parque Andreense da Prefeitura de Santo André and IBAMA/ICMBio for collection permits. Ingo Grantsau for fieldwork assistance, and Carolina Castro-Melo (MZUSP) and Valdir J. Germano (IB) for collection assistance. We are also indebted to Francisco L. Franco (IB), and Felipe Toledo (ZUEC) for permission to access the herpetological collections under their care. To V. Verdade, P. Garcia, I. Martins, B. Berneck, C. Canedo, A. C. Calijorne-Lourenço, D. Baêta, C. Brasileiro, and V. Dill Orrico for information on amphibian taxonomy. VT was supported by a scholarship from Coordenação de Aperfeiçoamento de Pessoal de Nível Superior (CAPES). Funding for this project was provided by Fundação de Amparo à Pesquisa do Estado de São Paulo (BIOTA/FAPESP; grant numbers 2011/50206-9 and 2008/50928-1) to HZ and CFBH, and Conselho Nacional de Desenvolvimento Científico e Tecnológico (CNPq; grant numbers 565046/2010-1; 303545/2010-0) to HZ.

REFERENCES

AB'SABER, A.N. 2003. **Os Domínios de Natureza no Brasil. Potencialidades Paisagísticas.** São Paulo, Ateliê Editorial, 159p.

ALMEIDA-GOMES, M.; D.VRCIBRADIC; C.C.SIQUEIRA; M.C. KIEFER; T. KLAION; P.A. SANTOS; D. NASCIMENTO; C.V. ARIANI; V.N.T. BORGES-JUNIOR; R.F. FREITAS-FILHO; M. VAN SLUYS & C.F.D. ROCHA. 2008. Herpetofauna of an Atlantic rainforest area (Morro São João) in Rio de Janeiro State, Brazil. **Anais da Academia Brasileira de Ciências** 80 (2): 1-10. doi: 10.1590/S0001-37652008000200007

ARAÚJO, C.O.; T.H. CONDEZ & R.J. SAWAYA. 2009a. Anfíbios Anuros do Parque Estadual das Furnas do Bom Jesus, sudeste do Brasil, e suas relações com outras taxocenoses no Brasil. **Biota Neotropica 9** (2): 1-22. doi: 10.1590/S1676-06032009000200007

ARAÚJO, O.G.S.; L.F. TOLEDO; P.C.A. GARCIA & C.F.B. HADDAD. 2009b. The amphibians of São Paulo State, Brazil. **Biota Neotroprica 9** (4): 1-13. h ttp://dx.doi.org/10.1590/S1676-06032009000400020

BERNECK, B.V.M.; C.O.R. COSTA & P.C.A. GARCIA. 2008. A new species of *Leptodactylus* (Anura: Leptodactylidae) from the Atlantic Forest of São Paulo State, Brazil. **Zootaxa 1795**: 46-56.

BERNECK, B.V.M.; M. TARGINO & P.C.A. GARCIA. 2013. Rediscovery and re-description of *Ischnocnema nigriventris* (Lutz, 1925) (Anura: Terrarana: Brachycephalidae). **Zootaxa 3694**: 131-142. doi: 10.11646/zootaxa.3694.2.2

BLOMBERG, S. & R. SHINE. 2006. Reptiles, p. 297-307. In: W.J. SUTHERLAND (Ed.). **Ecological census techniques. A handbook.** Cambridge, Cambridge University Press, 2nd ed., XVI+432p.

BOKERMANN, W.C.A. 1966. **Lista anotada das localidades tipo de anfíbios brasileiros.** São Paulo, RUSP, 183p.

BOKERMANN, W.C.A. 1968. Observações sobre "Hyla pardalis" Spix (Anura, Hylidae). **Revista Brasileira de Biologia 28** (1): 1-5.

CÂMARA, I.G. 2003. Brief history of conservation in the Atlantic forest, p. 31-42. *In*: C. GALINDO-LEAL & I.G. CÂMARA (Eds). **The Atlantic Forest of South America: biodiversity status, threats, and outlook.** Washington, DC, Island Press, 488p.

CAMPBELL, H.W. & S.P. CHRISTMAN. 1982. Field techniques for herpetofaunal community analysis, p. 193-200. *In*: N.J. SCOTT Jr. (Ed.). **Herpetological Communities: a Symposium of the Society for the Study of Amphibians and Reptiles and the Herpetologist's League.** Washington D.C., U.S. Department of the Interior, Fish Wildlife Services, IV+239p.

CANEDO, C. & B.V.S. PIMENTA. 2010. New species of *Ischnocnema* (Anura, Brachycephalidae) from the Atlantic Rainforest of the state of Espírito Santo, Brazil. **South American Journal of Herpetology 5** (3): 199-206. doi: 10.2994/057.005.0305

CARAMASCHI, U. & J.P. POMBAL JR. 2006. Notas sobre a série-tipo de *Holoaden bradei* B. Lutz e *Holoaden luederwaldti* Miranda-Ribeiro (Anura, Brachycephalidae). **Revista Brasileira de Zoologia 23** (4): 1261-1263. doi: 10.1590/S0101-81752006000400039

CARAMASCHI, U. & M.T. RODRIGUES. 2007. Taxonomic status of the species of *Gastrotheca* Fitzinger, 1843 (Amphibia, Anura, Amphignathodontidae) of the Atlantic rain Forest os eastern Brazil, with description of a new species. **Boletim do Museu Nacional, N.S. Zoologia 525**: 1-19.

CARAMASCHI, U. & C.A.G. CRUZ 2013. A new species of the Hypsiboas polytaenius clade from southeastern Brazil (Anura: Hylidae). **South American Journal of Herpetology 8** (2): 121-126.

CARNAVAL, A.C.; M.J. HICKERSON; C.F.B. HADDAD; M.T. RODRIGUES & C.M. MORITZ. 2009. Stability predicts genetic diversity in the Brazilian Atlantic Forest hotspot. **Science 323**:785-789. doi: 10.1126/science.1166955

CARRASCO, P.A.; C.I. MTTONI; G.C. LEYNAUD & G.J. SCROCCHI. 2012. Morphology, phylogeny and taxonomy of South American bothropoid pitvipers (Serpentes, Viperidae). **Zoologica Scripta 41**: 109-124 doi: 10.1111/j.1463-6409.2011.00511.x

CECHIN, S.Z. & M. MARTINS. 2000. Eficiência de armadilhas de queda (pitfall traps) em amostragens de anfíbios e répteis no Brasil. **Revista Brasileira de Zoologia 17**: 729-749. doi: 10.1590/S0101-81752000000300017

COLWELL, R.K. 2009. **EstimateS: Statistical estimation of species richness and shared species from samples.** Version 8.2Available online at: http://purl.oclc.org/estimates [Accessed: January, 2011].

COLWELL, R.K. & J.A. CODDINGTON. 1994. Estimating terrestrial biodiversity through extrapolation. **Philosophical Transactions of the Royal Society of London B 345**: 101-118. doi: 10.1098/rstb.1994.0091

CONDEZ, T.H.; R.J. SAWAYA & M. DIXO. 2009. Herpetofauna of the Atlantic Forest remnants of Tapiraí and Piedade region, São Paulo state, southeastern Brazil. **Biota Neotropica 9** (1): 157-185.

CORN, P.S. 1994. Standard techniques for inventory and monitoring: straight-line drift fences and pitfall traps, p. 109-117. *In*: W.R. HEYER; M.A. DONNELLY; R.W. MCDIARMID, L.C. HAYEK & M.S. FOSTER (Eds). **Measuring and monitoring biological diversity: standard methods for amphibians.** Washington D.C., Smithsoniam Institution Press, 384p.

CRUMP, M.L. & J.R. SCOTT JR. 1994. Standard techniques for inventory and monitoring: visual encounter surveys, p. 84-92. *In*: W.R. HEYER; M.A. DONNELLY; R.W. MCDIARMID; L.C. HAYEK & M.S. FOSTER (Eds). **Measuring and monitoring biological diversity: standard methods for amphibians.** Washington D.C., Smithsoniam Institution Press, 384p.

CRUZ, C.A.G. & R.N. FEIO. 2007. Endemismos em anfíbios em áreas de altitude na Mata Atlântica no sudeste do Brasil, p. 117-126. *In*: L.B. NASCIMENTO & M.E. OLIVEIRA (Eds). **Herpetologia no Brasil II.** Curitiba, Sociedade Brasileira de Herpetologia, 354p.

CURCIO, F.F.; V.Q. PIACENTINI & D.S. FERNANDES. 2009. On the status of the snake genera *Erythrolamprus* Boie, *Liophis* Wagler and *Lygophis* Fitzinger (Serpentes, Xenodontinae). **Zootaxa 2173**: 66-68.

DIXO, M. & V.K. VERDADE. 2006. Herpetofauna de serapilheira da Reserva Florestal do Morro Grande, Cotia (SP). **Biota Neotropica 6** (2): 1-20. doi: 10.1590/S1676-06032006000200009

DOMINGOS, M.; M.I.M.S. LOPES & Y. STRUFFALDI-DE VUONO. 2000. Nutrient cycling disturbance in Atlantic Forest sites affected by air pollution coming from the industrial complex of Cubatão, Southeast Brazil. **Revista Brasileira de Botânica 23** (1): 77-85.

DOMINGOS, M.; A. KLUMPP & G. KLUMPP. 2009. Poluição atmosférica, uma ameaça à Floresta Atlântica da Reserva Biológica de Paranapiacaba, p. 165-184. *In*: M.I.M.S. LOPES; M. KIRIZAWA & M.M.R.F. MELO (Eds). **Patrimônio da Reserva Biológica do Alto da Serra de Paranapiacaba. A antiga Estação Biológica do Alto da Serra.** São Paulo, Instituto de Botânica, 720p.

DUELLMAN, W.E. 1988. Patterns of species diversity in anuran amphibians in American Neotropics. **Annals of the Missouri Botanical Garden 75**: 79-104. doi: 10.2307/2399467

ETEROVICK, P.C.; A.C.O.Q. CARNAVAL; D.M. BORGES-NOJOSA; D.L. SILVANO; M.V. SEGALLA & I. SAZIMA. 2005. Amphibian Declines in Brazil: An Overview. **Biotropica 37** (2): 166-179. doi: 10.1111/j.1744-7429.2005.00024.x

FAIVOVICH, J.; C.F.B. HADDAD; P.C.A. GARCIA; D.R. FROST; J.A. CAMPBELL & W.C. WHEELER. 2005. Systematic review of the frog family Hylidae, with special reference to Hylinae: Phylogenetic analysis and taxonomic revision. **Bulletin of the American Museum of Natural History 294**: 1-240. doi: 10.1206/0003-0090(2005)294[0001:SROTFF]2.0.CO;2

FAUTH, J.E.; B.I. CROTHER & J.B. SLOWINSKI. 1989. Elevational patterns of species richness, evenness and abundance of the Costa Rican leaf-litter herpetofauna. **Biotropica 21** (2): 178-185. doi: 10.2307/2388708

FERREIRA, C.J.; L.K. TOMINAGA; J.M.A. SOBRINHO & M.F. NETO. 2009. Geologia e Geomorfologia, p: 53-71. *In*: M.I.M.S. LOPES; M. KIRIZAWA & M.M.R.F. MELO (Eds). **Patrimônio da Reserva Biológica do Alto da Serra de Paranapiacaba. A antiga Estação Biológica do Alto da Serra.** São Paulo, Instituto de Botânica, 720p.

FORLANI, M.C.; P.H. BERNARDO; C.B.F. HADDAD & H. ZAHER. 2010. Herpetofauna do Parque Estadual Carlos Botelho, São Paulo, Brasil. **Biota Neotropica 10** (3): 265-309. doi.org/10.1590/S1676-06032010000300028

FRANCO, F.L. & T.G. FERREIRA. 2002. Descrição de uma nova espécie de *Thamnodynastes* Wagler, 1830 (Serpentes, Colubridae) do nordeste brasileiro, com comentários sobre o gênero. **Phyllomedusa 1** (2): 57-74.

FROST, D.R. 2013. **Amphibian Species of the World: an Online Reference**. Version 5.6 (9 January 2013). Electronic Database accessible at http://research.amnh.org/herpetology/amphibia/index.html. American Museum of Natural History, New York, USA.

FROST, D.R.; R. ETHERIDGE; D. JANIES & T.A. TITUS. 2001. Total Evidence, Sequence Alignment, Evolution of Polychrotid Lizards, and a Reclassification of the Iguania (Squamata: Iguania). **American Museum Novitates 3343**: 1-39. doi: 10.1206/0003-0082(2001)343<0001:TESAEO>2.0.CO;2

GALINDO-LEAL, C. & I.G. CÂMARA. 2005. Status do hotspot Mata Atlântica: uma síntese, p. 3-11. *In*: C. GALINDO-LEAL & I.G. CÂMARA (Eds). **Mata Atlântica: biodiversidade, ameaças e perspectivas**. São Paulo, Fundação SOS Mata Atlântica, 471p.

GARCIA, P.C.A.; R.J. SAWAYA; I.A. MARTINS; C.A. BRASILEIRO; V.K. VERDADE; J. JIM; M.V. SEGALLA; M. MARTINS; D.C. ROSSA-FERES; C.F.B. HADDAD; L.F. TOLEDO; C.P.A. PRADO; B.M. BERNECK & O.G.S. ARAÚJO. 2009a. Anfíbios, p. 330-347. *In*: P.M. BRESSAN; M.C.M KIERULFF & A.M. SUGIEDA (Eds). **Livro vermelho da fauna ameaçada de extinção do Estado de São Paulo**. São Paulo, SEMA, 645p.

GARCIA, P.C.A.; B.V.M. BERNECK & C.O.R. COSTA 2009b. A new species of *Paratelmatobius* (Amphibia, Anura, Leptodactylidae) from Atlantic Rain Forest of Southeastern Brazil. **South American Journal of Herpetology 4** (3): 217-224. doi: 10.2994/057.004.0303

GIARETTA, A.A. & LM. CASTANHO. 1990. Nova espécie de *Paratelmatobius* (Amphibia, Anura, Leptodactylidae) da Serra do Mar, Brasil. **Papéis Avulsos de Zoologia 37** (8): 133-139.

GIARETTA, A.A.; R.J. SAWAYA; G. MACHADO; M.S. ARAÚJO; K.G. FACURE; H.F. MEDEIROS & R. NUNES. 1997. Diversity and abundance of litter frogs at altitudinal sites at Serra do Japi, southeastern Brazil. **Revista Brasileira de Zoologia 14** (2): 341-346. doi: 10.1590/S0101-81751997000200008

GIARETTA, A.A.; K.J. FACURE; R.J. SAWAYA; J.H.M. MEYER & N. CHEMIN. 1999. Diversity and Abundance of Litter Frogs in a Montane Forest of Southeastern Brazil: Seasonal and Altitudinal Changes. **Biotropica 31** (4): 669-674. doi: 10.1111/j.1744-7429.1999.tb00416.x

GIBBONS, J.W. & R.D. SEMLITSCH. 1982. Terrestrial drift fences with pitfall traps: An effective technique for quantitative sampling of animal populations. **Brimleyana 7**: 1-16.

GOTELLI, N.J. & R.K. COLWELL. 2001. Quantifying biodiversity: procedures and pitfalls in the measurement and comparison of species richness. **Ecological Letters 4**: 379-391. doi: 10.1046/j.1461-0248.2001.00230.x

GRAZZIOTIN, F.G.; H. ZAHER; R.W. MURPHY; G. SCROCCHI; M.A. BENAVIDES; Y. ZHANG & S.L. BONATTO. 2012. Molecular phylogeny of the New World Dipsadidae (Serpentes: Colubroidea): a reappraisal. **Cladistics 1**: 1-23.

GUIX, J.C.; G. LORENTTE; A. MONTORI; M.A. CARRETERO & X. SANTOS. 2000. Una nueva área de elevada riqueza de anuros en el bosque lluvioso atlântico de Brasil. **Boletin de la Asociación Herpetolica Española 11** (2): 100-105.

GUTJAHR, M.R. & R. TAVARES. 2009. Clima, p. 39-52. *In*: M.I.M.S. LOPES; M. KIRIZAWA & M.M.R.F. MELO (Eds). **Patrimônio da Reserva Biológica do Alto da Serra de Paranapiacaba. A antiga Estação Biológica do Alto da Serra**. São Paulo, Instituto de Botânica, 720p.

HADDAD, C.F.B. 2008. Uma Análise da Lista Brasileira de Anfíbios Ameaçados de Extinção, p. 287-320. *In*: A.B.M. MACHADO; G.M. DRUMMOND & A.P. PAGLIA (Eds). **Livro Vermelho da Fauna Brasileira Ameaçada de Extinção**. São Paulo, Ministério do Meio Ambiente, vol. 2, 907p.

HARVEY, M.B. & D. EMBERT. 2008. Review of Bolivian *Dipsas* (Serpentes: Colubridae), with comments on other South American species. **Herpetological Monographs 22**: 54-105. doi: 10.1655/07-023.1

HEDGES, B.S.; W.E. DUELLMAN & M.P. HEINICKE. 2008. New World direct-developing frogs (Anura: Terrarana): Molecular phylogeny, classification, biogeography, and conservation. **Zootaxa 1737**: 1-182.

HEINEN, J.T. 1992. Comparisons of the Leaf Litter Herpetofauna in Abandoned Cacao Plantations and Primary Rain Forest in Costa Rica: Some Implications for Faunal Restoration. **Biotropica 24** (3): 431-439. doi: 10.2307/2388614

HELLMANN, J.J. & G.W. FOWLER. 1999. Bias, precision, and accuracy of four measures of species richness. **Ecological Applications 9** (3): 824-834. doi: 10.1890/1051-0761(1999)009[0824:BPAAOF]2.0.CO;2

HELTSCHE, J. F. & N. E. FORRESTER. 1983. Estimating species richness using the Jackknife procedure. **Biometrics, 39**: 1-11.

HEYER, W.R. 1973. Systematics of the marmoratus group of the frog genus *Leptodactylus* (Amphibia, Leptodactylidae). **Natural History of the Museum of Los Angeles County 251**: 1-50.

HEYER, W.R. 1985. New species of frogs from Boracéia, São Paulo, Brazil. **Proceedings of the Biological Society of Washington 98** (3): 657-671.

HEYER, W.R. & L.R. MAXSON. 1983. Relationships, zoogeography, and speciation mechanisms of frogs of the genus *Cycloramphus* (Amphibia, Leptodactylidae). **Arquivos de Zoologia 30**: 341-373. doi: 10.11606/issn.2176-7793.v30i4p235-339

HEYER, W.R.; A.S. RAND; C.A.G. CRUZ & O.L. PEIXOTO. 1988. Decimations, extinctions, and colonizations of frog populations in southeast Brazil and their evolutionary implications. **Biotropica 20**: 230-235. doi: 10.2307/2388238

HEYER, W.R.; A.S. RAND; C.A.G. CRUZ; O.L. PEIXOTO & C.E. NELSON. 1990. Frogs of Boracéia. **Arquivos de Zoologia 31** (4): 231-410.

HOFER U. & L.F. BERSIER. 2001. Herpetofaunal diversity and abundance in tropical upland forests of Cameroon and Panama. **Biotropica 33**: 142-152.

IUCN. 2010. **IUCN Red List of Threatened Species**. Version 2010.4. Available at: http://www.iucnredlist.org [Accessed: February 16, 2012].

IZECKSOHN, E. & S.P. CARVALHO-E-SILVA. 2008. As espécies de *Gastrotheca* Fitzinger na Serra dos Orgãos, Estado do Rio de Janeiro, Brasil (Amphibia: Anura: Amphignathodontidae). **Revista Brasileira de Zoologia 25** (1): 100-110. doi: 10.1590/S0101-81752008000100014

KEARNEY, M. 2003. Systematics of the Amphisbaenia (Lepidosauria: Squamata) based on morphological evidence from recent and fossil forms. **Herpetological Monographs 17**: 1-74. doi: 10.1655/0733-1347(2003)017[0001:SOTALB]2.0.CO;2

KUMAR, A. 2010. A tragic loss: Fire at Instituto Butantan, Brazil. **Toxicon 56**: 1528-1529. doi: 10.1016/j.toxicon.2010.07.002

LAURANCE, W.F. 2009. Conserving the hottest of the hotspots. **Biological Conservation 142**: 1137. doi: 10.1016/j.biocon.2008.10.011

LOPES, M.I.M.S. & M. KIRIZAWA. 2009. Reserva Biológica de Paranapiacaba, a antiga Estação Biológica do Alto da Serra: história e visitantes ilustres, p. 15-38. *In*: M.I.M.S. LOPES; M. KIRIZAWA & M.M.R.F. MELO (Eds). **Patrimônio da Reserva Biológica do Alto da Serra de Paranapiacaba. A antiga Estação Biológica do Alto da Serra**. São Paulo, Instituto de Botânica, 720p.

LOPES, M.I.M.S.; A.R. SANTOS; R.M. MORAES & M. KIRIZAWA. 2009. Ciclagem de nutrientes e alterações no solo induzidas pela poluição atmosférica, p. 137-164. *In*: M.I.M.S. LOPES; M. KIRIZAWA & M.M.R.F. MELO (Eds). **Patrimônio da Reserva Biológica do Alto da Serra de Paranapiacaba. A antiga Estação Biológica do Alto da Serra**. São Paulo, Instituto de Botânica, 720p.

LIPS, K.R.; P.A. BURROWES; J.R. MENDELSON & G. PARRA-OLEA. 2005. Amphibian declines in Latin America: Widespread population declines, extinctions and impacts. **Biotropica 37** (2): 163-165. doi: 10.1111/j.1744-7429.2005.00023.x

LUTZ, B. 1958. Anfíbios novos e raros das Serras Costeiras do Brasil. **Memórias do Instituto Oswaldo Cruz 56** (2): 373-399. doi: 10.1590/S0074-02761958000200002

MAGNUSSON, W.E.; A.P. LIMA; J.M. HERO & M.C. ARAÚJO. 1999. The rise and fall of a population of *Hyla boans*: Reproduction in a Neotropical gladiator frog. **Journal of Herpetology 33** (4): 647-656. doi: 10.2307/1565582

MARQUES, O.A.V. 2009. A fauna de répteis na região de Paranapiacaba, p. 605-620. *In*: M.I.M.S. LOPES; M. KIRIZAWA & M.M.R.F. MELO (Eds). **Patrimônio da Reserva Biológica do Alto da Serra de Paranapiacaba. A antiga Estação Biológica do Alto da Serra**. São Paulo, Instituto de Botânica, 720p.

MARQUES, O.A.V.; A. ETEROVIC & I. SAZIMA. 2004. **Snakes of the Brazilian Atlantic Forest – An illustrated field guide for the Serra do Mar range**. Ribeirão Preto, Holos Editora, 204p.

MARQUES, O.A.V.; D.N. PEREIRA; F.E. BARBO; V.J. GERMANO & R.J. SAWAYA. 2009. Os Répteis do Município de São Paulo: diversidade e ecologia da fauna pretérita e atual. **Biota Neotropica 9** (2): 139-150. doi.org/10.1590/S1676-06032009000200014

MARTINS, I.A. 2010. Natural history of *Holoaden luederwaldti* (Amphibia: Strabomantidae: Holoadeninae) in southeastern of Brazil. **Zoologia 27** (1): 40-46. doi: 10.1590/S1984-46702010000100007

MARTINS, I.A. & H. ZAHER. 2013. A new species of the highland frog genus *Holoaden* (Amphibia, Strabomantidae) from cloud forests of southeastern Brazil. **Zootaxa 3599** (2): 178-188. doi: 10.11646/zootaxa.3599.2.4

MELO, M.M.R.F.; R.M. MORAES & A.R. SANTOS. 2009. Publicações sobre a reserva Biológica do Alto da Serra de Paranapiacaba, p. 707-720. In: M.I.M.S. LOPES; M. KIRIZAWA & M.M.R.F. MELO (Eds). **Patrimônio da Reserva Biológica do Alto da Serra de Paranapiacaba. A antiga Estação Biológica do Alto da Serra**. São Paulo, Instituto de Botânica, 720p.

MIKAN, J.C. 1820. **Delectus florae et faunae Brasiliensis: jussu et auspiciis Francisci I, Austriae imperatoris investigatae.** Vindobonae [Vienna]:Sumptubus auctoris, 1820 [25]: 43-44.

MIRANDA-RIBEIRO, A. 1920. Algumas considerações sobre *Holoaden luederwaldti* e gêneros correlatos. **Revista do Museu Paulista 12**: 317-320.

MIRANDA-RIBEIRO, A. 1926. Notas para servirem ao estudo dos gymnobatrachios (Anura) Brasileiros. **Arquivos do Museu Nacional 27**: 1-227.

MORAES, R.M.; A. KLUMPP; C.M. FURLAN; G. KLUMPP; M. DOMINGOS; M.C.S. RINALDI & I.F. MODESTO. 2002. Tropical fruit trees as bioindicators of industrial air pollution in southeast Brazil. **Environment International 28**: 367-374. doi: 10.1016/S0160-4120(02)00060-0

MORAES, R.M.; W.B.C. DELITTI & J.A.P.V. MORAES. 2003. Gas exchange, growth, and chemical parameters in a native Atlantic forest tree species in polluted areas of Cubatão, Brazil. **Ecotoxicology and Environmental Safety 54**: 339-345. doi: 10.1016/S0147-6513(02)00067-2

MYERS, N.; R.A. MITTERMEIER; C.G. MITTERMEIER; G.A.B. DA FONSECA & J. KENT. 2000. Biodiversity hotspots for conservation priorities. **Nature 403**: 853-845. doi: 10.1038/35002501

OLIVEIRA, L.E.; R.M.C. OLIVEIRA & A.A. GIARETTA. 2008. *Ischnocnema hoehnei*. Advertisement call. **Herpetological Review 39** (2): 207-208.

PATTO, C.E.G. & M.R. PIE. 2001. Notes on the population dynamics of *Hylodes asper* in Southeastern Brazil (Anura: Leptodactylidae). **Journal of Herpetology 35** (4): 684-686. doi: 10.2307/1565913

PATTO, C.E.G. & M.R. PIE. 2006. Escape behaviour in the Neotropical frog *Hylodes asper* (Anura: Leptodactylidae). **Acta Herpetologica 1** (2): 141-146.

PETERS, J.A. 1960. The Snakes of the Subfamily Dipsadinae. **Miscellaneous Publications Museum of Zoology University of Michigan 114**: 1-228.

POMBAL JR, J.P. & C.A.G CRUZ. 1999. Redescrição de *Eleutherodactylus bolbodactylus* (Lutz, 1925) e a posição taxonômica de *E. gehrti* (Miranda-Ribeiro, 1926) (Anura, Leptodactylidae). **Boletim do Museu Nacional, Nova Série, Zoologia 404**: 1-10.

POMBAL JR, J.P. & M. GORDO. 2004. Anfíbios anuros da Juréia, p. 243-256. *In*: O.A.V. MARQUES & W. DULEBA (Eds). **Estação Ecológica Juréia-Itatins. Ambiente físico, flora e fauna**. Ribeirão Preto, Holos Editora, 384p.

POMBAL JR, J.P.; I. SAZIMA & C.F.B. HADDAD. 1994. Breeding behavior of the pumpkin toadlet, *Brachycephalus ephippium* (Brachycephalidae). **Journal of Herpetology 28** (4): 516-519. doi: 10.2307/1564972

POMBAL JR, J.P.; C.C. SIQUEIRA; T.A. DORIGO; D. VRCIBRADIC & C.F.D. ROCHA. 2008. A third species of the rare frog genus *Holoaden* (Terrarana, Strabomantidae) from a montane rainforest area of southeastern Brazil. **Zootaxa 1938**: 61-68.

PMSA (PREFEITURA DO MUNICÍPIO DE SANTO ANDRÉ). 2008. **Atlas do Parque Natural Municipal Nascentes de Paranapiacaba: revelando o nosso Parque**. São Paulo, Anablume Editora, 2nd ed., 78p.

RIBEIRO, M.C.; J.P. METZGER; A.C. MARTENSEN; F.J. PONZONI & M.M. HIROTA. 2009. The Brazilian Atlantic Forest: how much is left, and how is the remaining forest distributed? Implications for conservation. **Biological Conservation 142**: 1144-1156. doi: 10.1016/j.biocon.2009.02.021

RIBEIRO, M.C.; A.C. MARTENSEN; J.P. METZGER; M. TABARELLI; F. SCARANO & M. FORTIN. 2011. The Brazilian Atlantic Forest: A Shrinking Biodiversity Hotspot, p. 405-434. *In*: F.E. ZACKOS & J.C. HABEL (Eds). **Biodiversity Hotspots. Distribution and Protection of Conservation Priority Areas**. Berlin, Springer Verlag, XVII+546p.

ROCHA, C.F.D.; M. VAN SLUYS; M.A.S. ALVES; M.G. BERGALLO & D. VRCIBRADIC. 2001. Estimates of forest floor litter frog communities: A comparison of two methods. **Austral Ecology 26** (1): 14-21.

ROCHA, C.F.D.; D. VRCIBRADIC; M. KIEFER; C.M. ALMEIDA-GOMES; N.T. BORGES-JUNIOR; P.C.F. CARNEIRO; R.V. MARRA; P. ALMEIDA-SANTOS; C.C. SIQUEIRA; P. GOYANNES-ARAÚJO; C.G.A. FERNANDES; E.C.N. RUBIÃO & M. VAN SLUYS. 2007. A survey of the leaf-litter frog assembly from an Atlantic forest area (Reserva Ecológica de Guapiaçu) in Rio de Janeiro State, Brazil, with an estimate of frog densities. **Tropical Zoology 20** (1): 99-108.

RODRIGUES, M.T. 2005. The Conservation of Brazilian Reptiles: Challenges for a Megadiverse Country. **Conservation Biology 19** (3): 659-664. doi: 10.1111/j.1523-1739.2005.00690.x

ROSSA-FERES, D.C.; M. MARTINS; O.A.V. MARQUES; I.A. MARTINS; R.J. SAWAYA & C.F.B. HADDAD. 2008. Herpetofauna, p. 83-94. *In*: R.R. RODRIGUES; C.A. JOLY; M.C.W. DE BRITO, A. PAESE; J.P. METZGER; L. CASATTI; M.A. NALON; M. MENEZES; N.M. IVANAUSKAS; V. BOLZANI & V.L.R. BONONI (Eds). **Diretrizes para conservação e restauração da biodiversidade no estado de São Paulo**. São Paulo, Imprensa Oficial do Estado de São Paulo, 248p.

ROSSA-FERES, D.C.; R.J. SAWAYA; J. FAIVOVICH; J.G.R. GIOVANELLI; C.A. BRASILEIRO; L. SCHIESARI; J. ALEXANDRINO & C.F.B. HADDAD. 2011.

Anfíbios do Estado de São Paulo, Brasil: conhecimento atual e perspectivas. **Biota Neotropica 11** (1a): 1-19. Available online at: http://www.biotaneotropica.org.br/v11n1a/pt/abstract?inventory+bn0041101a2011 [Accessed: April, 2014].

SANTOS-PEREIRA, M.; A. CANDATEN; D. MILANI; F.B. OLIVEIRA; J. GARDELIN & C.F.D. ROCHA. 2011. Seasonal variation in the leaf-litter frog community (Amphibia: Anura) from an Atlantic Forest Area in the Salto Morato Natural Reserve, southern Brazil. **Zoologia 28** (6): 775-761. doi: 10.1590/S1984-46702011000600008

SAZIMA, I. & W.C.A. BOKERMANN. 1978. Cinco novas espécies de *Leptodactylus* do centro e sudeste brasileiros (Amphibia, Anura, Leptodactylidae). **Revista Brasileira de Biologia 38** (4): 899-912.

SAWAYA, R.J.; O.A.V. MARQUES & M. MARTINS. 2008. Composition and natural history of a Cerrado snake assemblage at Itirapina, São Paulo State, southeastern Brazil. **Biota Neotropica 8** (2): 127-148. doi: 10.1590/S1676-06032008000200015

SCOTT JR., N.J. 1976. The abundance and diversity of the Herpetofaunas of tropical forest litter. **Biotropica 8** (1): 41-58. doi: 10.2307/2387818

SILVA, J.M.C. & C.H.M. CASTELETI. 2005. Estado da biodiversidade da Mata Atlântica brasileira, p. 43-59. *In*: C. GALINDO-LEAL & I.G. CÂMARA (Eds). **Mata Atlântica: biodiversidade, ameaças e perspectivas**. São Paulo, Fundação SOS Mata Atlântica, 471p.

SIQUEIRA, C.C.; D. VRCIBRADIC & C.F.D. ROCHA. 2011. Anurans from two high-elevation areas of Atlantic Forest in the state of Rio de Janeiro, Brazil. **Zoologia 28** (4): 457-464. doi: 10.1590/S1984-46702011000400007

STUART, S.N.; J.S. CHANSON; N.A. COX; B.E. YOUNG; A.S.L. RODRIGUES; D.L. FISCHMAN & R.W. WALLER. 2004. Status and trends of amphibian declines and extinctions worldwide. **Science 306**: 1783-1786. doi: 10.1126/science.1103538

SUGIYAMA, M.; R.P. SANTOS; L.S.J. AGUIAR; M. KIRIZAWA & E.L.M. CATHARINO. 2009. Cararcterização e mapeamento da vegetação, p. 107-117. *In*: M.I.M.S. LOPES; M. KIRIZAWA & M.M.R.F. MELO (Eds). **Patrimônio da Reserva Biológica do Alto da Serra de Paranapiacaba. A antiga Estação Biológica do Alto da Serra**. São Paulo, Instituto de Botânica, 720p.

THOMPSON, G.G.; P.C. WITHERS; E.R. PIANKA & S.A. THOMPSON. 2003. Assessing biodiversity with species accumulation curves, inventories of small reptiles by pittrapping in Western Australia. **Austral Ecology 28**: 361-383. doi: 10.1046/j.1442-9993.2003.01295.x

VALDUJO, P.H.; C.C. NOGUEIRA; L. BAUMGARTEN; F.H.G. RODRIGUES; A. ETEROVIC; M.B. RAMO-NETO & O.A.V. MARQUES. 2009. Squamate Reptiles from Parque Nacional das Emas and surroundings, Cerrado of Central Brazil. **Check List 5** (3): 405-417.

VASCONCELOS, T.S.; T.G. SANTOS; C.F.B. HADDAD & D.C. ROSSA-FERES. 2010. Climatic variables and altitude as predictors os anuran species richness and number of reproductive modes in Brazil. **Journal of Tropical Ecology 26**: 423-432. doi: 10.1017/S0266467410000167

Van Sluys, M.; D. Vrcibradic; M.A.S. Alves; H.G. Bergallo & C.F.D. Rocha. 2007. Ecological parameters of the leaf-litter frog community of an Atlantic Rainforest area at Ilha Grande, Rio de Janeiro state, Brazil. **Austral Ecology 32** (3): 254-260. doi: 10.1111/j.1442-9993.2007.01682.x

Verdade, V.K.; M.T. Rodrigues & D. Pavan. 2009. Anfíbios Anuros da Reserva Biológica de Paranapiacaba e entorno, p. 579-604. *In*: M.I.M.S. Lopes; M. Kirizawa & M.M.R.F. Melo (Eds. **Patrimônio da Reserva Biológica do Alto da Serra de Paranapiacaba. A antiga Estação Biológica do Alto da Serra.** São Paulo, Instituto de Botânica, 720p.

Verdade, V.K.; A.C. Carnaval; M.T. Rodrigues; L. Schiesari; D. Pavan & J. Bertolucci. 2011. Decline of Amphibians in Brazil, p. 85-127. *In*: H. Heatwole; L. Barrio-Amorós & J. Wilkinson (Eds). **Amphibian Biology.** Sydney, Surrey Beatty and Sons, vol. 9, 296p.

Verdade, V.K.; P.H. Valdujo; A.C. Carnaval; L. Schiesari; L.F. Toledo; T. Mott; G.V. Andrade; P.C. Eterovick; M. Menin; B.V.S.

Pimenta; C. Nogueira; C.S. Lisboa; C.D. Paula & D.L. Silvano. 2012. A leap further: the Brazilian Amphibian Conservation Action Plan. **Alytes 29** (1-4): 28-43.

Weygoldt, P. 1989. Changes in the composition of mountain stream frog communities in the Atlantic Mountains of Brazil: frogs as indicators of environmental deteriorations? **Studies on Neotropical Fauna and Environment 243** (4): 249-255.

Zaher, H.; F.G. Grazziotin; J.E. Cadle; R.W. Murphy; J.C. Moura-Leite & S.L. Bonatto. 2009. Molecular phylogeny of advanced snakes (Serpentes, Caenophidia) with an emphasis on South American Xenodontines: A revised classification and descriptions of new taxa. **Papéis Avulsos de Zoologia 49** (11): 115-153. doi: 10.1590/S0031-10492009001100001

Zaher, H.; F.E. Barbo; P.S. Martínez; C. Nogueira; M.T. Rodrigues & R.J. Sawaya. 2011. Reptiles from São Paulo State: current knowledge and perspectives. **Biota Neotropica 11** (1a): 1-14.

14

Size-selective predation of the catfish *Pimelodus pintado* (Siluriformes: Pimelodidae) on the golden mussel *Limnoperna fortunei* (Bivalvia: Mytilidae)

João P. Vieira & Michelle N. Lopes

Laboratório de Ictiologia, Universidade Federal do Rio Grande. Avenida Itália km 8, 96201-900 Rio Grande, RS, Brazil. E-mail: vieira@mikrus.com.br

ABSTRACT. This paper describes the size-selective predation on *Limnoperna fortunei* (Dunker, 1857) by *Pimelodus pintado* (Azpelicueta, Lundberg & Loureiro, 2008) from the time it arrived at the Mirim Lagoon basin (2005). Sampling was carried out using bottom trawl in depths of 3-6 m, from January to November 2005, and from October to November 2008. *Pimelodus pintado* began to prey upon *L. fortunei* soon after its arrival (austral spring of 2005). On the spring of 2008, *L. fortunei* was found to be the most important food item of *P. pintado*. The variation in length of the mussels (0.7-3.2 cm, with a mode of 1.3 cm) indicates that the species is now fully established in the system. Our data indicates that large individuals of *P. pintado* incorporate more mussels in their diets than small individuals. However, regardless of their size, *P. pintado* individuals predate only on small (<1.4 cm) representatives of *L. fortunei*. This prey size corresponds to a phase when the mussel is more mobile and readily available for fish. Larger, more aggregated prey groups that are attached to hard substrates are avoided by fish predators.

KEY WORDS. Diet; freshwater invasion; opportunistic predator; Mirim Lagoon.

The Asian freshwater golden mussel, *Limnoperna fortunei* (Dunker, 1857), was first recorded in the Americas in 1991, from the coast of the Rio de la Plata, by Pastorino *et al.* (1993). *Limnoperna fortunei* has the capability of colonizing a wide range of habitats and has few natural predators, widely colonizing freshwater and estuarine environments in the Neotropical region. Since its introduction, the species has expanded its range to Argentina, Brazil, Paraguay and Uruguay (Darrigran *et al.* 1998, Darrigran & Ezcurra de Drago 2000, Darrigran 2002, Oliveira *et al.* 2006).

In 1998, the Asian freshwater golden mussel was recorded at the northern reaches of the Patos Lagoon drainage basin (Mansur *et al.* 1999, 2003) and, in the next years, it was found in the southern portion of the basin in densities of up to 140,000 ind/m². After that, population densities stabilized at averages 60,000 ind/m² (Mansur *et al.* 2003). The golden mussel invaded the Mirim Lagoon in 2005 through the São Gonçalo Channel, which connects the Mirim and Patos Lagoons (Langone 2005, Burns *et al.* 2006b, Colling *et al.* 2012, Lopes & Vieira 2012).

There is little question now that catastrophic biological events like these can profoundly affect entire ecosystems to the point that the invader species monopolizes a large proportion of the resources available (Sylvester *et al.* 2005, Darrigran & Damborenea 2011). The cascading effects of such trophic web disruptions can be extremely important (Power 1992, Ruetz *et al.* 2002, Thorp & Casper 2003). The high densities of golden mussel and the fact that individuals become fixed to the sub-strate by their byssal threads create a new microenvironment. This microenvironment, in turn provides a new habitat for some epifaunal species and, at the same time, can lead to the displacement of other benthic organisms (Santos *et al.* 2012). Since its invasion of South America, *L. fortunei* has threatened the survival and modified the natural occurrence and abundance of several native macroinvertebrate species (Darrigran *et al.* 1998, Darrigran 2002, Santos *et al.* 2012), including the Anomura crab *Aegla platensis* Schmitt, 1942 in the São Gonçalo Channel (Lopes *et al.* 2009).

The predation strength of fish upon the golden mussel ranges from negligible to efficient, when the total control of the mussel's population growth is achieved (Stewart *et al.* 1998, Bartsch *et al.* 2005). In some cases, mussel predators showed increased productivity and growth as a result of the new food supply (Poddubnyi 1966, Boltovskoy *et al.* 2006, Karatayev *et al.* 2007). Ancillary data on Argentine freshwater fish yields support to the above conclusions, suggesting that *L. fortunei* may have had a positive effect on fish biomass in Neotropical systems (Boltovskoy *et al.* 2006).

Catfish of the genus *Pimelodus* have omnivorous feeding habits (Garcia *et al.* 2006,2007 which are characterized by a generalist feeding behavior and the opportunistical exploitation of eventual peaks in prey abundance (Bonetto *et al.* 1963, Montalto *et al.* 1999, Braga 2000, García & Protogino 2005). The use of *L. fortunei* as a food source by *Pimelodus* spp. (Paraná River and Patos Lagoon basin) had been previously reported

(Montalto *et al.* 1999, Cataldo *et al.* 2002, Darrigran & Damborenea 2011). For this reason, we hypothesized that the catfish *Pimelodus pintado* (Azpelicueta, Lundberg & Loureiro, 2008) at the Mirim Lagoon would incorporate the Asian freshwater golden mussel in its diet.

The objective of this study was to describe the chronological incorporation of *L. fortunei* into the diet of *P. pintado*, from the time the prey arrived in the São Gonçalo Channel, at the Mirim Lagoon basin. Additionally, we show that *P. pintado* selects mussel prey within a certain size range, from the wide range of sizes available in the environment.

MATERIAL AND METHODS

The Mirim Lagoon basin is located between 31 and 34°S and 52 and 54°W in the eastern part of the South American central plains (Fig. 1). The basin area covers 62,250 km², with 29,250 km² in southern Brazil and the remaining 33,000 km² in eastern Uruguay. The main geographical feature of this basin is the Mirim lagoon itself, with an average area of 3.749 km² (Bracco *et al.* 2005). During periods of intense rainfall, water from the Mirim Lagoon and its adjacent wetland system drain through the natural São Gonçalo Channel (75 km long, 200 to 500 m wide, 6 m maximum depth) into the Patos Lagoon, which ultimately connects with the Atlantic Ocean through the Rio Grande channel. The São Gonçalo Channel Dam divides the São Gonçalo Channel into a freshwater environment to the south and an estuarine environment to the North (Burns *et al.* 2006a, Moura *et al.* 2012).

Bottom fauna samples were obtained from the limnetic region of the São Gonçalo Channel between 31°48′S, 52°23′W and 32°7′S, 52°35′W (Fig. 1). Bottom trawl was carried 3 to 6 m deep, from January to November 2005, and from October to November 2008. Every season, four samplings at three different sampling areas (Dam, Tigre and Piratini), totalizing twelve samplings per season, were carried out using a fishermen's wooden boat (10.9 m long with a 60-Hp engine). Five-minute tows (approximately 400 m) were performed using a 10.5 m (head rope) shrimp trawl (1.3 cm bar mesh wings and body with a 0.5 cm bar mesh cod end liner) and a pair of weighted outer doors (Lopes *et al.* 2009, Lopes & Vieira 2012).

Representatives of *P. pintado* collected in each tow were stored in separate plastic bags and fixed in 10% formaldehyde. Voucher specimens were deposited at "Coleção Ictiológica da FURG" number 6,056 (25) 6061 (6). At the laboratory, the total length (TL) of each individual was measured to the nearest millimeter. Stomachs were extracted by cutting out the esophagus and pylorus and fixed in 70% alcohol. Prey items were identified to the lowest taxonomic level, counted and grouped into categories. We determined Frequency of Occurrence (FO%), Percent Area (PA%) and Index of Relative Importance (IRI) of the *P. pintado* diet for samples collected during the austral summer (January and February), autumn (April), winter (July and

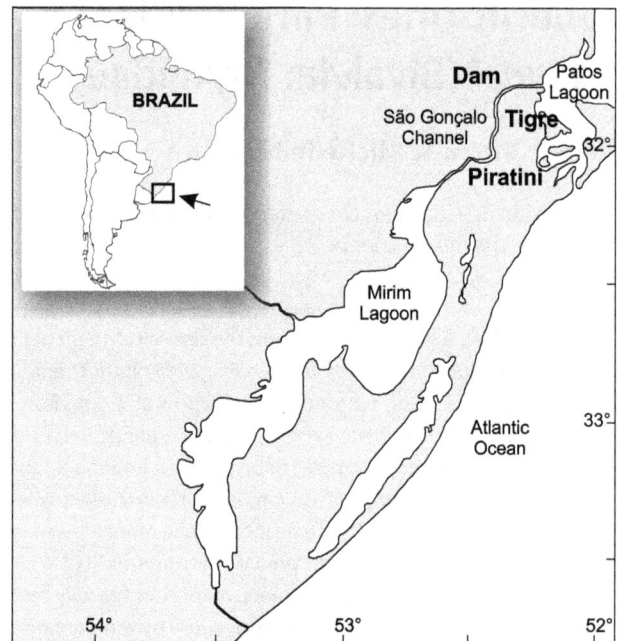

Figure.1 Mirim Lagoon (drainage basin, 62.250 km²) and the São Gonçalo channel that connects it with the Patos Lagoon. The limnetic region is localized southern the dam. The sampling areas are show (Dam, Tigre and Piratini).

August) and the austral spring (September, October and November) of 2005 and the austral spring (October and November) of 2008. IRI was determined as FO% x (PN%+PA%), where PN% are number percent (Hyslop 1980) of each food item. We compared de relative importance of the golden museel in the stomachs of catfish sampled in the spring of 2005 with those sampled in the spring of 2008 (three years after *L. fortunei* occurrence was recorded in the system).

Using an electronic caliper with accuracy of 0.01 mm, the maximum shell length of *L. fortunei* prey was measured using the individuals present in the stomachs of *P. pintado*, as well as using the specimens collected in the environment (bottom trawl, during 2008). The maximum length was considered as the distance from the anterior end, situated just above and ahead of the umbones, until the rear end of the shell (Mansur *et al.* 1987).

After logarithmic transformation ($\log_{10}+1$) the differences between the sizes of golden mussels found in the digestive tract of different sizes classes of *P. pintado* were tested using one-way analysis of variance (size class 10-15 cm TL, N = 6; size class 15-20 cm TL, N = 52; size class 20-25 cm TL, N = 51; size class 25-30 cm TL, N = 38) (Zar 1999). The differences between the sizes of golden mussels found in the digestive tract of all *P. pintado* and those collected in the environment were tested using the nonparametric Kolmogorov-Smirnov test (Sokal & Rohlf 1995).

RESULTS

Asian golden mussels were not observed in the stomachs of catfish collected during the austral summer, autumn or winter of 2005. The first record of the mussel was during the austral spring (September, October and November) of 2005, when it was found in 21.8% of the 180 stomachs of *P. pintado* (10-30 cm TL). In this period, *L. fortunei* represented 1.8% of the Relative Importance Index (IRI) in the diet to *P. pintado* (Table I).

During the austral spring of 2005, *L. fortunei* individuals were only observed in the stomachs of fish larger than 14 cm, and large fish incorporated more mussels in their diets (Fig. 2). The length golden mussel shells (0.3-1.4 mm) found in the digestive tracts of *P. pintado* did not differ significantly among the different predator length class analyzed (Fig. 3, ANOVA, $F_{3,143} = 0.9744$, p = 0.407).

In the austral spring of 2008 (Table I), *L. fortunei* was found in 60.8% of the 30 stomachs of *P. pintado* analyzed and was the most important food item in the diet of the cat fish (IRI = 31.4%). Specimens of *L. fortunei* (N = 7789) collected by the shrimp trawl in 2008 were measured and compared with the 51 specimens found in the stomachs of *P. pintado* (20.7-30.5 cm TL) sampled in the same area. The length range of fortune of the mussel in the environment (n = 7789) was 0.7 to 3.2 cm, with a mode of 1.3 cm TL, and the shell length of mussels found in the in the guts (n = 147) of *P. pintado* ranged from 0.3 to 1.4 cm (Fig. 4) and the nonparametric Kolmogorov-Smirnov test revealed a significant difference (Fig. 5, D = 0.82491, p < 0.001) between them.

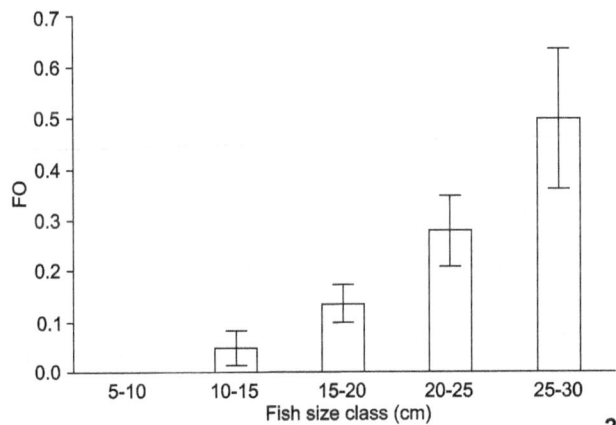

Figures 4-5. (4) Shell size distribution of *Limnoperna fortunei* collected in the environment (ENV), and in the guts (GUT) of *Pimelodus pintado* at the spring of 2008. (5) Mean size and 95% confidence interval of *Limnoperna fortunei*.

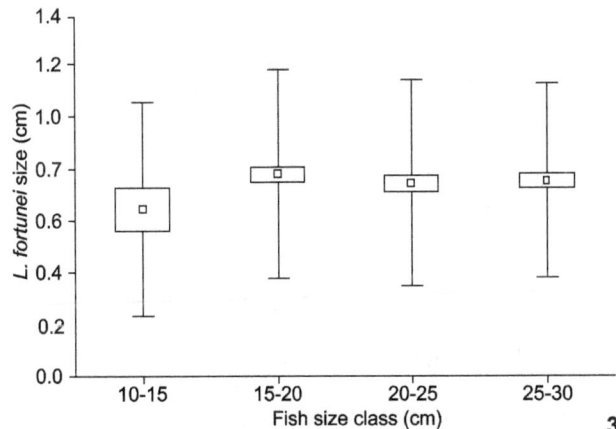

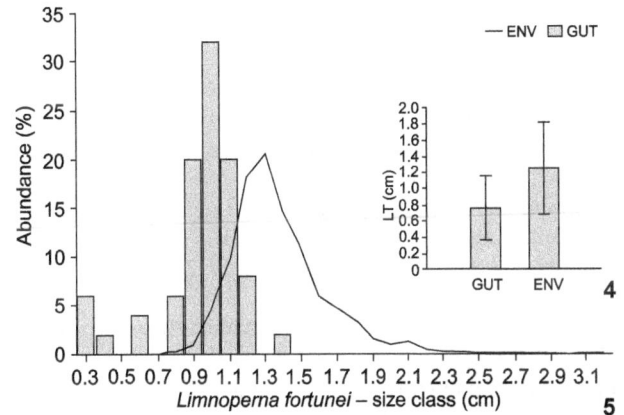

Figures 2-3. (2) Frequency of occurrence of *Limnoperna fortunei* in the digestive tracts of *P. pintado* at different fish size class analyzed in the spring of 2005. The line represent the standard deviation and the bars represent mean. (3) Box plot of *Limnoperna fortunei* shell size distribution by *Pimelodus pintado* size class in the spring of 2005.

DISCUSSION

BOLTOVSKOY & CATALDO (1999) showed that during the first year of the introduction of *L. fortunei*, mussels reached up to 2.0 cm in length, and by the end of their second year, some reached 3.0 cm. The maximum length of mussels in this study was 3.2 cm. Additionally, BURNS *et al.* (2006b) recorded the species in the São Gonçalo Channel for the first time in 2005. Both facts suggest that the species was already fully established in São Gonçalo Channel in 2008 (LOPES & VIEIRA 2012).

The mean shell length of *L. fortunei* found in the digestive tracts of *P. pintado* were similar for the different predator length classes analyzed and reached a maximum size of 1.4 cm (Fig. 3). MONTALTO *et al.* (1999) and LOPES & VIEIRA (2012) showed that Asian golden mussels are predated in different development stages by several fish species in the Neotropical region, but small mussels (< 1.5 cm) were more frequent and abundant in fish guts. CANTANHÊDE *et al.* (2008) analyzed the diet of a

Table I. Seasonal variation of the Frequency of Occurrence (FO%), Percentage by Number (PN%), Percentage by Area (PA%), and Relative Importance Index (IRI%) of items found in the diet of *Pimelodus pintado* during 2005 and 2008.

Items	2006							2008			
	Summer	Autumn	Winter	Spring				Spring			
	FO%	FO%	FO%	FO%	PN%	PA%	IR%	FO%	PN%	PA%	IR%
Small crustaceans	13.51	47.89	65.48	74.15	63.77	40.07	67.76	17.39	7.04	3.64	1.97
Crabs	20.27	2.82	1.19	3.40	0.21	2.34	0.08	21.74	2.18	19.75	5.06
Insects	35.14	60.56	64.29	57.14	15.33	19.27	17.40	65.22	16.26	8.11	16.85
L. fortunei	–	–	–	**21.77**	5.16	4.49	1.85	**60.87**	18.83	34.77	31.37
Corbicula spp.	–	4.23	11.90	2.72	0.25	0.12	0.01	13.04	1.94	1.02	0.41
Heliobia spp.	27.03	12.68	7.14	16.33	3.16	0.87	0.58	47.83	51.70	9.20	30.88
Fish	70.27	49.30	16.67	21.29	4.28	7.42	2.17	4.35	0.24	2.68	0.13
Plant	2.70	35.21	15.48	39.46	3.23	22.98	9.10	52.17	3.88	17.52	11.84
Other	9.48	4.23	2.38	19.73	1.93	2.07	0.69	26.09	1.94	2.98	1.36
Mineral	–	9.86	4.76	13.61	2.70	0.37	0.37	8.70	0.97	0.32	0.12

predator fish larger than *P. pintado* (*Pterodoras granulosus*; 17-55 cm TL) and found similar results, whereas the size range of *L. fortunei* individuals ranged from 0.8 to 1.7 cm in mean, although the mussel is well known to reach more than 3 cm (BOLTOVSKOY & CATALDO 1999, MAROÑAS *et al.* 2003, MANSUR *et al.* 2008).

Young and adult *L. fortunei* individuals have a considerable ability to move to new locations with some taxis, suggesting that this represents an ability to avoid predation by hiding (LOPES & VIEIRA 2012). The distance that mussels are able to move decreases with increasing shell length, and they tend to aggregate after reaching 1.5 cm TL (URYU *et al.* 1996). The observations of the present study suggest that more mobile individuals of *L. fortunei* (smaller than <1.4 cm) are able to crawl over the bottom and are probably more readily available in the São Gonçalo Channel to fish such as *P. pintado* than larger and more aggregated mussels that get attached to the hard substrate. Fish with a generalist feeding behavior are good samplers of prey diversity and can easily detect new sources of food, and are flexible enough to exploit eventual peaks in prey abundance (GLOVA & SAGAR 1989, MENDOZA-CARRANZA & VIEIRA 2007). Then, independently from the size of individuals, the catfish under study consumes only golden mussels that are smaller than 1.4 cm TL, which implies that the predator prefers smaller prey because they are more readily available.

It is important to note that *L. fortunei* was first reported in the limnetic zone of the São Gonçalo Channel in January 2005 (BURNS *et al.* 2006b). Since the diet of *P. pintado* is influenced by the availability of food in the environment, the data presented here suggest that the trend observed in this study, i.e., the absence of *L. fortunei* in the stomachs of *P. pintado* from the austral summer to the winter and the presence of it in the spring 2005, is a good indicator that *L. fortunei* started to be abundant in the limnetic zone of the São Gonçalo Channel

during the austral spring of 2005. At the present time, *L. fortunei* is fully established in the São Gonçalo Channel (COLLING *et al.* 2012, LOPES & VIEIRA 2012).

Little empirical information is available to explain the success of invaders (MANSUR *et al.* 2012). The arrival and spread of *L. fortunei* at the Mirim Lagoon will probably bring rapid changes in the benthic community as well as the displacement of other mollusk species, as described by DARRIGAN (2002) and DARRIGRAN & DAMBORENEA (2005) for other South American fresh water habitats. Currently, unpublished data suggest that *L. fortunei* is much more abundant in the dense vegetated limnetic part of the São Gonçalo Channel (61 km long and 17 m wide) than in the wide open Mirim Lagoon itself (3,749 km² area) which has few hard subtracts for mussel fixation (LOPES *et al.* 2009).

With the invasion of *L. fortunei* in the Neotropical region, dietary changes have been noted in omnivorous fishes, which have switched from a low quality, predominantly plant-based diet, to an energetically rich diet dominated by invasive mollusks (MONTALTO *et al.* 1999, FERRIZ *et al.* 2000, BOLTOVSKOY *et al.* 2006, LOPES & VIEIRA 2012). With the invasion of *L. fortunei*, part of the organic matter in the turbid São Gonçalo Channel will be filtered and modified into a form available to organisms that cannot feed on small particles, like fishes.

LOPES & VIEIRA (2012) shows that 12 of 19 predators in the Mirin/São Gonçalo Channel feed on *L. fortunei*. Regardless of the size and foraging behavior of the predator, individuals smaller than 14 mm on average are preyed upon. The incidence of individuals of *L. fortunei* smaller than 14 mm in the diet of detritivorous fishes like *Rineloricaria microlepdogaster* and *R. strigilata*, which do are not adapted to predate on mollusks, confirms the hypothesis that individuals of golden mussel up to 14 mm TL are more vagile than larger individuals (URYU *et al.* 1996), and frequently move on the bottom of the São

Gonçalo Channel, being more available to fish predation than individuals larger than 14 mm TL, which tend to be clustered or hindered in crevices of the substrate (Lopes & Vieira 2012).

ACKNOWLEDGEMENTS

This work was supported by CNPq (Conselho Nacional de Desenvolvimento Científico e Tecnológico) and PELD Program (Pesquisas Ecológicas de Longa Duração). M.N.L was a former postgraduate student of Biologia de Ambientes Aquaticos Continentais, FURG.

REFERENCES

BARTSCH, M.R.; L.A. BARTSCH & S. GUTREUTER. 2005. Strong effects of predation by fishes on an invasive macroinvertebrate in a large floodplain river. **Journal North American Benthological Society 24**: 168-77.

BOLTOVSKOY, D. & D.H. CATALDO. 1999. Population dynamics of *Limnoperna fortunei*, an invasive fouling mollusc, in the lower Parana River (Argentina). **Biofouling 14**: 255-263.

BOLTOVSKOY, D.; N. CORREA; D. CATALDO & F. SYLVESTER. 2006. Dispersion and ecological impact of the invasive freshwater bivalve *Limnoperna fortunei* in the Río de la Plata watershed and beyond. **Biological Invasions 8**: 947-963.

BONETTO, A.A.; C. PIGNALBERI & E. CORDIVIOLA. 1963. Ecologia alimentaria del amarillo y moncholo, *Pimelodus clarias* (Bloch) y *Pimelodus albicans* (Valenciennes) (Pisces, Pimelodidae). **Physis 24**: 87-94.

BRACCO, R.; L. PUERTO; H. INDA & C. CASTIÑEIRA. 2005. Mid-late Holocene cultural and environmental dynamics in Eastern Uruguay. **Quaternary International 132**: 37-45.

BRAGA, F.M. DE S. 2000. Biologia e Pesca de *Pimelodus maculatus* (Siluriformes, Pimelodidae) no reservatório de Volta Grande, Rio Grande (MG-SP). **Acta Limnologica Brasiliensia 12**: 1-14.

BURNS, M.D.; A.M. GARCIA; J.P. VIEIRA; M.A. BEMVENUTI; D.M.L. MOTTA MARQUES & V. CONDINI. 2006a. Evidence of fragmentation affecting fish movement between Patos and Mirim coastal lagoons in southern Brazil. **Neotropical Icthiology 4**: 69-72.

BURNS, M.D.; R.M. GERALDI; A.M. GARCIA; C.E. BEMVENUTI; R.R. CAPITOLI & J.P. VIEIRA. 2006b. Primeiro registro de ocorrência do mexilhão dourado *Limnoperna fortunei* na Bacia de drenagem da Lagoa Mirim, RS, Brasil. **Biociências 14**: 83-83.

CANTANHÊDE, G.; N.S. HAHN; E.A. GUBIANI & R. FUGI. 2008. Invasive molluscs in the diet of *Pterodoras granulosus* (Valenciennes, 1821) (Pisces, Doradidae) in the Upper Paraná River floodplain, Brazil. **Ecology of Freshwater Fish 17**: 47-53.

CATALDO, D.; D.D. BOLTOVSKOY; V. MARINI & N. CORREA. 2002. Limitantes de *Limnoperna fortunei* en la cuenca del Plata: la predación por peces. *In*: **Tercera jornada sobre conservación de la fauna íctica en el río Uruguay**. Paysandu, Uruguay, La Comisión Administradora de Río Uruguay.

COLLING, L.A.; R.M. PINOTTI & C.E. BEMVENUTI. 2012. *Limnoperna fortunei* na Bacia da Lagoa dos Patos e Lagoa Mirim, p. 187-191. *In*: C.P. SANTOS; D. PEREIRA; I.C.P. PAZ; L.M. ZURITA; M.C.D. MANSUR; M.T. RAYA RODRIGUEZ; M.V. NERHKE & P.A. BERGONCI (Eds). **Moluscos límnicos invasores no Brasil: biologia, prevenção e controle**. Porto Alegre, Redes Editora, 412p.

DARRIGRAN, G. 2002. Potencial impact of filter-feeding invaders on temperate inland freshwater environments. **Biological Invasions 4**: 145-156.

DARRIGRAN, G. & E. DE DRAGO. 2000. Invasion of *Limnoperna fortunei* (Dunker, 1857) (Bivalvia: Mytilidae) in South America. **Revista Nautilus 114**: 69-73.

DARRIGRAN, G. & C. DAMBORONEA. 2005. A bioinvasion history in South America. *Limnoperna fortunei* (Dunker, 1857), the golden mussel. **American Malacological Bulletin 20**: 105-112.

DARRIGRAN, G. & C. DAMBORONEA. 2011. Ecosystem engineering impacts of *Limnoperna fortunei* in South America. **Zoological Science 28**: 1-7.

DARRIGRAN, G.; S.M. MARTIN; B. GULLO & L. ARMENDARIZ. 1998. Macroinvertebrate associated to *Limnoperna fortunei* (Dunker, 1857) (Bivalvia, Mytilidae). Río de La Plata, Argentina. **Hydrobiologia 367**: 223-230.

FERRIZ, R.A.; C.A. VILLAR; D. COLAUTTI & C. BONETTO. 2000. Alimentación de *Pterodoras granulosus* (Valenciennes) (Pisces, Doradidae) en la baja cuenca del Plata. **Revista del Museo Argentino de Ciencias Naturales 2**: 151-156.

GARCIA, A.M.; D.J. HOEINGHAUS; J.P. VIEIRA; K.O. WINEMILLER; D.M.L.M. MARQUES & M.A. BEMVENUTI. 2006. Preliminary examination of food web structure of Nicola Lake (Taim Hydrological System, south Brazil) using dual C and N stable isotope analyses. **Neotropical Ichthyology 4**: 279-284.

GARCIA, A.M.; D.J. HOEINGHAUS; J.P. VIEIRA & K.O. WINEMILLER. 2007. Isotopic variation of fishes in freshwater and estuarine zones of a large subtropical coastal lagoon. **Estuarine Coastal and Shelf Science 73**: 399-408.

GARCÌA, M.L.A. & L.C. PROTOGINO. 2005. Invasive freshwater molluscs are consumed by native fishes in South America. **Journal of Applied Ichthyology 21**: 34-38.

GLOVA, G.J. & P.M. SAGAR. 1989. Prey selection by *Galaxias vulgaris* in the Hawkins River, New Zealand. **New Zealand Journal of Marine and Freshwater Research 23**: 153-161.

HYSLOP, E.J. 1980. Stomach contens analysis- a review of methods and their applications. **Journal Fish Biology 17**: 411-429.

KARATAYEV, A.Y.; D.K. PADILLA; D. MINCHIM; D. BOLTOVSKOY & L.E. BURLAKOVA. 2007. Changes in global economies and trade: the potential spread of exotic freshwater bivalves. **Biological Invasions 9**: 161-180.

LANGONE, J.A. 2005. Notas sobre el mejillón dorado *Limnoperna fortunei* (Dunker, 1857) (bivalvia, mytilidae) en Uruguay. **Publicación extra del Museo Nacional de Historia Natural y Antropologia 1**: 17.

LOPES, M. & J.P. VIEIRA. 2012. Predadores potenciais para o controle do mexilhão-dourado, p. 357-363. *In*: C.P. SANTOS; D.

Pereira; I.C.P. Paz; L.M. Zurita; M.C.D. Mansur; M.T. Raya Rodriguez; M.V Nerhke & P.A. Bergonci (Eds). **Moluscos límnicos invasores no Brasil: biologia, prevenção e controle.** Porto Alegre, Redes Editora, 412p.

Lopes, M.N.; J.P. Vieira & M.D.M. Burns. 2009. Biofouling of the golden mussel *Limnoperna fortunei* (Dunker, 1857) over the Anomura crab *Aegla platensis* Schmitt, 1942. **Pan-American Journal of Aquatic Sciences 4**: 222-225.

Mansur, M.C.D.; C. Schulz & L.M.M.P. Garces. 1987. Moluscos bivalves de água doce: Identificação dos gêneros do sul e leste do Brasil. **Acta Biológica Leopoldencia 9**: 181-202.

Mansur, M.C.D.; L.M.Z. Richinitti & C.P. Santos. 1999. *Limnoperna fortunei* (Dunker, 1857) molusco bivalve invasor na bacia do Guaíba, Rio Grande do Sul, Brasil. **Biociências 7**: 147-149.

Mansur, M.C.D.; C.P. Santos; G. Darrigran; I. Heydricht; C.T. Callil & F.R. Cardoso. 2003. Primeiros dados quail-quantitativos do mexilhão dourado, *Limnoperna fortunei* (Dunker), no Delta do Jacuí, no lago Guaíba e na Laguna dos Patos, Rio Grande do Sul, Brasil e alguns aspectos de sua invasão no novo ambiente. **Revista Brasileira de Zoologia 20**: 75-84.

Mansur, M.C.D.; H. Figueiró; C.P. Santos; L. Glock; P.E.A. Bergonci & D. Pereira. 2008. Variação espacial do comprimento e do peso úmido total de *Limnoperna fortunei* (Dunker, 1857) no delta do rio Jacuí e lago Guaíba (RS, Brasil). **Biotemas 21**: 49-54.

Mansur, M.C.D.; C.P. Santos; D. Pereira; I.C.P. Paz; M.L.L. Zurita; M.T.R. Rodriguez; M.V. Nehrke & P.E.A. Bergonci. 2012. **Moluscos Límnicos Invasores no Brasil: biologia, prevenção, controle.** Porto Alegre, Redes Editora, 412p.

Maroñas, M.E.; G.A. Darrigran; E.D. Sendra & G. Breckon. 2003. Shell growth of the golden mussel, *Limnoperna fortunei* (Dunker, 1857) (Mytilidae), in the Río de la Plata, Argentina. **Hydrobiologia 495**: 41-45.

Mendoza-Carranza, M. & J.P. Vieira. 2007. Whitemouth croaker *Micropogonias furnieri* (Desmarest, 1823) feeding strategies across four southern Brazilian estuaries. **Aquatic Ecology 42** (1): 83-93.

Montalto, L.; O.B. Oliveros; I. E. de Drago & L.D. Demonte. 1999. Peces del rio Parana Medio predadores de una especie invasora: *Limnoperna fortunei* (Bivalvia, Mytilidae). **Revista de la Facultad de Bioquimica y Ciencias Biológicas de la Universidad Nacional del Litoral 3**: 85-101.

Moura, P.M.; J.P. Vieira; A.M. Garcia. 2012. Fish abundance and species richness across an estuarine freshwater ecosystem in the Neotropics. **Hydrobiologia 696**: 107-122.

Oliveira, M.D.; A.M. Takeda; L.F. Barros; S.D. Barbosa; E.K. Rezende. 2006. Invasion by *Limnoperna fortunei* (Dunker, 1857) (Bivalvia, Mytilidae) of the Pantanal wetland, Brazil. **Biological Invasions 8** (1): 97-104.

Pastorino, G.; G. Darrigran; S. Martin & L. Lunaschi. 1993. *Limnoperna fortunei* (Dunker, 1857) (Mytilidae), nuevo bivalvo invasor em águas Del Rio de la Plata. **Neotropica 39**: 101-102.

Poddubny, A.G. 1966. Adaptive response of *Rutilus rutilus* to variable environmental conditions. **Trudy Instiyuta Biologii Vnutrennykh Vod Akademii Nauk 10**: 131-138.

Power, M.E. 1992. Habitat heterogeneity and the functional significance of fish in River Food Webs. **Ecology 73**: 1675-1688.

Ruetz, C.R.; R.M. Newman & B. Vondracek. 2002. Top-down control in a detritus-based food web: fish, shredders and leaf breakdown. **Oecologia 132**: 307-315.

Santos, S.B.; S.C. Thiengo; M.A. Fernandez; I.C. Miyahira; I.C.B. Gonçalves; R.F. Ximenes; M.C.D. Mansur & D. Pereira. 2012. Espécies de moluscos límnicos invasores no Brasil, p. 25-49. *In*: C.P. Santos; D. Pereira; I. C. P. Paz; L.M. Zurita; M.C.D. Mansur; M.T. Raya Rodriguez; M.V Nerhke & P.A. Bergonci (Ed.). **Moluscos límnicos invasores no Brasil: biologia, prevenção e controle.** Porto Alegre, Redes Editora, 412p.

Sokal, R.R. & F.J. Rohlf. 1995. Biometry: **The Principles and Practice of Statistics in Biological Research.** New York, W.H. Freeman, 937p.

Stewart, T.W.; J.G. Miner & R.L. Lowe. 1998. A experimental analysis of crayfish (*Orconectes rusticus*) effects on a Dreissena dominated benthic macroinvertebrate community in western Lake Erie. **Canadian Journal of Fisheries and Aquatic Sciences 55**: 1043-1050.

Sylvester, F.; J. Dorado; D. Boltovskoy; A. Juàrez & D. Cataldo. 2005. Filtration rates of the invasive pest bivalve *Limnoperna fortunei* as a function of size and temperature. **Hydrobiologia 534**: 71-80.

Thorp, J.H. & A.F. Casper. 2003. Importance of biotic interactions in large rivers: an experiment with planktivorous fish, dreissenid mussels and zooplankton in the St. Lawrence River. **River Research and Applications 19**: 265-279.

Uryu, Y.; K. Iwasaky & M. Hinque. 1996. Laboratory experiments on behaviour and movement of a freshwater mussel, *Limnoperna fortunei* (Dunker). **Journal of a Molluscan Studies 62**: 327-341.

Zar, J.H. 1999. **Biostatistical Analysis.** New Jersey, Prentice Hall, 662p.

15

Longer is not always better: The influence of beach seine net haul distance on fish catchability

Pryscilla Moura Lombardi[1,2], Fábio Lameiro Rodrigues[1] & João Paes Vieira[1]

[1] *Laboratório de Ictiologia, Instituto de Oceanografia, Universidade Federal do Rio Grande. Avenida Itália km 8, 96203-900 Rio Grande, RS, Brazil.*
[2] *Corresponding author. E-mail: pryscilla_lombardi@yahoo.com.br*

ABSTRACT. The aim of this study was to compare the influence of different haul distances of a codend beach seine on the catchability of fish in a surf zone. Two different surf zone sites (A and B) at the Cassino Beach (Rio Grande do Sul, Brazil) were sampled at three different distances, parallel to the beach (30, 60, and 90 m). Starting 40 m from the beach line, diagonal distances of approximately 50, 70, and 100 m were swept. The total CPUE and CPUA haul distances are compatible with a declining trend in catch rates with increased haul distance at both sites. However, statistically significant differences were observed only for the short distance CPUA (50 m) in relation to the other haul distances at one of the sites sampled. Two fish size groups were observed (TL $\leqslant$ 40 and > 40 mm), but only small individuals ($\leqslant$ 40 mm) captured in the shorter haul distance at site B showed significant differences in CPUA. This result indicates that size structure for hauls at different distances was equal and that smaller individuals determined the pattern of fish abundance. The net performance indicates that a short haul ($\leqslant$ 50 m) is the best strategy to reduce net avoidance and fish escape when using this type of sampling gear.

KEY WORDS. Fish abundance; fish size; sampling gears; surf zone.

A broad variety of sampling strategies and fishing gears have been developed to collect and record the presence and abundance of different fish species occurring in estuarine and coastal marine habitats (VAN MARLEN 2003, VIEIRA *et al.* 2006, ROTHERHAM *et al.* 2007, QUEIROLO *et al.* 2009). According to KING (1995) and VIEIRA *et al.* (2006), the beach seine net is the most effective fishing gear for sampling in shallow, non-vegetated surf zone areas. VIEIRA *et al.* (2006) recommended the use of a particular beach seine net (a beach seine with a codend) for sampling estuarine environments of the Brazilian coast (Fig. 1). A number of studies have used and approved on this beach seine in estuarine and freshwater habitats in southern Brazil (e.g., BURNS *et al.* 2006, GARCIA *et al.* 2006, ARTIOLI *et al.* 2009). However, only one record of the use of this type of gear exists for the marine surf zone in Brazil, for the southeast coast (MAZZEI *et al.* 2009).

Even when using the same sampling gear, the selected haul distance represent an important factor for comparing catch results. Several authors have used beach seines at surf zones, but different distances are generally applied. For example, LAYMAN (2000) and MAZZEI *et al.* (2009) selected a haul length of 15 m, SILVA *et al.* (2004) selected a length of 30 m, and MONTEIRO-NETO & PRESTRELO (2013) selected 100 m hauls, hindering comparisons among different data sets. Therefore, the present study was conducted to test the performance of a beach seine net with a bag (codend) in a marine surf zone area in southern Brazil (Fig. 3), and to identify the most effective protocol for this net in a wave-dominated environment.

MATERIAL AND METHODS

The codend beach seine used in the current study is made from a multifilament net with the following dimensions: 9 m in length × 2.4 m high; each wing measured 3.25 m in length (Fig. 1), and the codend was 3 m in length; the mesh in the lateral wings was 13 mm, and the mesh in the codend was 5 mm (Fig. 1). The net was pulled by two people, with a third person holding a rope tied to the codend to prevent the bag from rising in the waves, thus hindering the process of dragging.

Starting from a fixed distance perpendicular to the beach (Per_{dist} = 40 m), each haul was performed on a transversal line to the beach, dragging the net in the direction of the current from a depth of 1.2 m up to the shoreline. Three different distances parallel to the beach (Par_{dist}) were previously established (30, 60, and 90 m; Fig. 2), and the haul distance (H) was calculated using the Pythagorean Theorem ($H^2 = Par_{dist}^2 + Per_{dist}^2$) (Fig. 2). Since Per_{dist} was fixed at 40 m and Par_{dist} comprised 30, 60, and 90 m, the H estimates were 50.0, 72.1, and 98.5 m. For simplicity, the H values have been referred to as 50, 70, and 100 m in the text. However, the original H values were retained for the calculation of the area swept. The standard seine width of the net was estimated to be 6 m, and the swept area of the

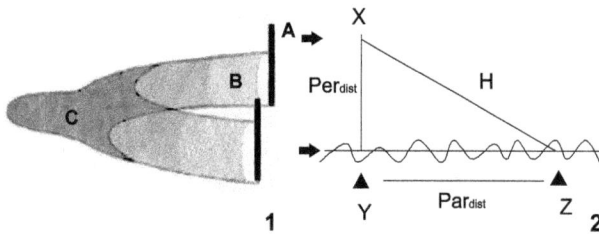

Figures 1-2. Illustrative picture of a bag seine net (1), in which "A" are the poles, "B" are the wings and "C" the center sac; and (2) the sampling design, in which "H" represents the net trajectory.

net was calculated by multiplying H by 6, which resulted in values of 300.0, 432.7, and 590.9 m², respectively.

Sampling was conducted at two different sites (A and B) in the Cassino Beach surf zone (Fig. 3). Cassino Beach is a dissipative beach, exhibiting medium wave energy, a smooth slope with few and inexpressive beach cusps, and fine sandy sediments (CALLIARI 1998, CALLIARI et al. 2005). Sites A (32°12'33.3"S, 052°10'45.3"W) and B (32°09'41.9"S, 052°06'21.8"W) are located 9 km and 500 m south of the west jetty of the Patos Lagoon, respectively. Despite the proximity of these two sites, LIMA & VIEIRA (2009) recorded more wave lines at site B than at site A. This variability was considered advantageous for testing the utility of this fishing gear under different conditions.

Each site was visited three times between March and April of 2009. During each visit, three random hauls were carried out for every parallel distance (Par$_{dist}$ = 30, 60, and 90 m) at each site; thus, there were nine samples per site. Each seine haul was performed immediately adjacent to the end of the previous one, but the order of parallel distances was selected at random. All fish were identified to the lowest taxonomic level using the keys by FIGUEIREDO & MENEZES (1980, 2000) and; MENEZES & FIGUEIREDO (1980, 1985); the specimens captured were counted and measured to the nearest millimeter (total length – TL). For each site, the number of fish caught per haul were independently expressed as catch per unit effort (individuals per sample – CPUE), which represents the number of fish caught in a single seine haul, and as catch per unit area (individuals per square meter – CPUA), which represents the number of fish caught per unit area.

Fishes were classified by size classes of 10 mm intervals and separated into two size groups (≤ 40 mm TL and > 40 mm TL). Based on VIEIRA (2006), CPUE and CPUA by size class (CPUA-SC) were calculated for individuals smaller or equal to 40 mm and larger than 40 mm TL.

Even after log transformation, the CPUE, CPUA, and CPUA-SC values did not meet the assumptions of Analyses of Variance (ANOVA) (normality and variance homogeneity); thus, a non-parametric analysis (Kruskal-Wallis test) was selected to compare the mean CPUE, CPUA, and CPUA-SC per Par$_{dist}$ for each site (at 0.05 signifficant level). The free statistics program PAST ver. 1.81 (HAMMER et al. 2001) was used for these analyses.

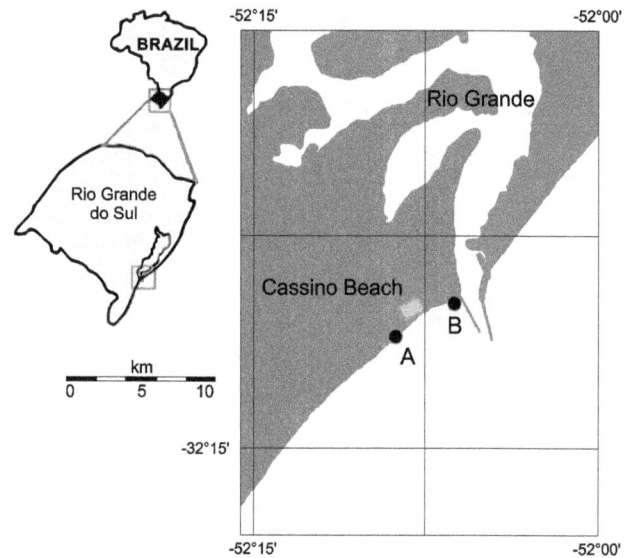

Figure 3. Geographic location of the study area, Cassino Beach, in detail, with the two sampling sites (A and B). Modified by the authors from http://www.aquarius.geomar.de (Online Map Creation).

To compare the faunistic similarity between the different haul distances, we used the minimum percentage of similarity (P$_{min}$) based on CPUE% (ARTIOLI et al. 2009), which was described as P$_{min}$ = Σ$_i$ minimum (p1$_i$ and p2$_i$), where p1$_i$ = the percentage of species i in sample 1 and p2$_i$ = the percentage of species i in sample 2 (KREBS 1989). Based on the CPUE values, we calculated the numerical percentage (CPUE%) and frequency of occurrence (FO%) of each species, for each sampling day and each Par$_{dist}$. Species that presented FO% ≥ mean FO% in each haul distance were considered frequent, while species with FO% < mean FO% were considered rare. A similar method was used for CPUE%, in which species with CPUE% ≥ mean CPUE% in each seine distance were considered abundant, while species with CPUE% < mean CPUE% were considered not abundant. Finally, combinations of FO% and CPUE% allowed us to classify the species into 4 groups: abundant and frequent, frequent but not abundant, abundant but rare, and not abundant and rare (BURNS et al. 2006, GARCIA et al. 2006, ARTIOLI et al. 2009, CENI & VIEIRA 2013).

Vouchers for the species collected in this study are available in the "Coleção Ictiológica da FURG".

RESULTS

The total number of fish caught during the longer haul distance (100 m) was less than during the shorter distances (50 and 70 m), at both sites (Table I). The CPUE and CPUA at both sites showed a tendency to decrease with increased haul distance (Fig. 4). The Kruskal-Wallis post-hoc test did not reveal significant differences (p > 0.05) for the mean CPUE of differ-

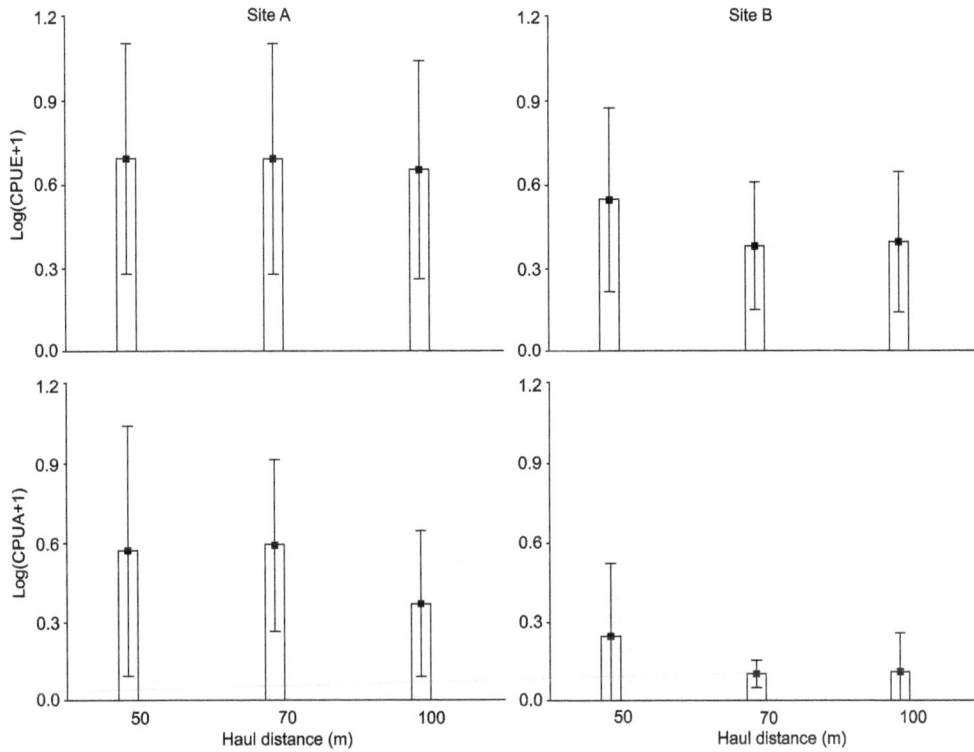

Figure 4. Variations of mean CPUE and CPUA (log$_{10}$ transformed) values by haul distance (50, 70 and 100 m) in each site (A and B). The vertical bars indicate the mean values and lines indicate the standard deviation.

Table I. Total numbers of dominant taxa in each haul distance (50, 70 and 100 m). Taxa were classified as Abundant and Frequent (bold), Frequent and Not-Abundant (underline), Abundant and rare (italic), and the other ones are present but not frequent or abundant.

Taxa	Site A			Site B		
	50 m	70 m	100 m	50 m	70 m	100 m
Mugil curema Valenciènnes, 1836	**12463**	**10678**	**6366**	**1551**	**155**	103
Mugil liza Valenciènnes, 1836	**1539**	**2070**	**1117**	**564**	**256**	1327
Brevoortia pectinata (Jenyns, 1842)	**1228**	679	789	<u>246</u>	<u>55</u>	<u>28</u>
Mugil sp.	<u>922</u>	1387	*867*	<u>141</u>	<u>55</u>	<u>53</u>
Clupeidae	<u>578</u>	<u>517</u>	<u>555</u>	*853*	4	1
Trachinotus marginatus Cuvier, 1832	<u>335</u>	<u>394</u>	<u>412</u>	<u>172</u>	<u>67</u>	**220**
Genidens barbus (Lacepède, 1803)	<u>111</u>	<u>27</u>	<u>41</u>	39	301	62
Atherinella brasiliensis (Quoy & Gaimard, 1825)	<u>14</u>	<u>48</u>	<u>33</u>	83	46	<u>51</u>
Pomatomus saltatrix (Linnaeus, 1766)	<u>11</u>	<u>5</u>	8	1	3	
Engraulidae	10	9	1			
Menticirrhus americanus (Linnaeus, 1758)	7	8	9		1	
Elops saurus Linnaeus, 1766	4					
Odontesthes argentinensis (Valenciennes, 1835)	3	6	3	1	5	5
Menticirrhus littoralis (Holbrook, 1847)	2	1		11	8	<u>18</u>
Epinephelus marginatus (Lowe, 1834)	1		1			
Micropogonias furnieri (Desmarest, 1823)	1	8	10			
Others		3	2	5	2	7
Total number	17229	15840	10214	3667	958	1875
Species richness	16	17	16	13	13	13

ent haul distances for either sampled site (Table II). However, the mean CPUA at site B showed significant differences between 50 and 70 m (p = 0.05), and 50 and 100 m (p = 0.01) distances (Table II).

Table II. The resulting p values from Kruskal-Wallis post-hoc test for CPUE and CPUA between the different haul distances (50, 70, and 100 m), for each sampling site (A and B).

Haul distances comparison	p values			
	CPUE		CPUA	
	Site A	Site B	Site A	Site B
50 x 70 m	0.95	0.78	0.66	0.05
50 x 100 m	0.83	0.73	0.66	0.01
70 x 100 m	0.87	0.92	0.13	0.08

Size distribution was similar between different haul distances at both sites, ranging from 10 mm to 100 mm, with peak abundance in length classes being lower than or equal to 40 mm TL (Fig. 5). At site A, the CPUA-SC of individuals from both size groups ($\leq$ 40 and > 40 mm TL) showed a tendency to decrease with increasing haul distance (Fig. 6); however, the Kruskal-Wallis post-hoc test did not reveal any significant differences (p > 0.05) (Table III). At site B, the CPUA-SC of small individuals ($\leq$ 40 mm TL) was significantly higher at the 50 m haul distance when compared to 70 or 100 m haul distances, but there was no statistical difference between the 70 and 100 m haul distances. There was no significant difference in CPUA-SC for higher size class (> 40 mm TL) at site B (Table III).

Table III. The resulting p values from Kruskal-Wallis post-hoc test for CPUA between the different haul distances (50, 70 and 100 m), for individuals $\leq$ 40 mm and > 40 mm in total length, for each sampling site.

Haul distances comparison	p values site A		p values site B	
	$\leq$ 40 mm	> 40 mm	$\leq$ 40 mm	> 40 mm
50 x 70 m	0.86	0.86	0.01	0.93
50 x 100 m	0.13	0.33	0.01	0.72
70 x 100 m	0.13	0.21	0.60	0.54

The total number of species that were collected per treatment did not differ among haul distances for both sites (Table I). At site A, the similarity among haul distances was always higher than 89%, suggesting that the same proportion of the same species group was captured at all three haul distances (Table I). At site B, the similarity among distances did not exceed 51%. At this site, comparison of the 50 m and 100 m haul distances showed the lowest similarity value (< 34%); this indicates that while the same species were sampled, different proportions of them were captured (Table I).

Six species were identified as abundant and frequent, but only *Mugil liza* Valenciennes, 1836 was abundant and frequent for all haul distances at both sites (Table I). *Mugil curema* Valenciennes, 1836 was abundant and frequent for all haul distances at site A and for distances 50 and 70 m at site B, *Brevoortia pectinata* (Jenyns, 1842) was abundant and frequent for distances 50 and 100 m only at site A, *Mugil* sp. was abundant and frequent for distance 70 m at site A, *Trachinotus marginatus* Cuvier, 1832 was abundant and frequent for distance 100 m at site B, and *Genidens barbus* (Lacepède, 1803) was abundant and frequent for distance 70 m at site B (Table I). This and other species that were abundant or frequent or rare are listed in Table I.

DISCUSSION

Contrary to expectations, more individuals were captured by the shorter haul distances (50 and 70 m) than the longer haul distances (100 m), even though this difference was statistically significant only at site B. The wave action, according to Hahn et al. (2007) may affect the shape of the seine and can temporarily lift lead lines or submerge float lines. Those difficulties may contribute to the differences found between the distances tested, considering that longer hauls passed through more waves than shorter hauls.

Our study showed that, while the three different haul distances (50, 70, and 100 m) caught nearly the same species, with similar abundance and frequencies at site A, species abundances and frequencies were different at site B. Evaluating the three haul distances, only CPUA and CPUA-SC for site B showed differences between the shorter haul distance and the two other for both abundance and size class. However the distribution pattern of individuals among species was similar among the efforts tested. Layman (2000), using a codend seine similar to that used in the present work, but with even smaller haul distances (15 m and 120 m² of swept area), also reported low diversity and few abundant but frequent species.

The same general pattern of size distribution was found at both sampling sites, with most individuals being smaller than or equal to 40 mm TL. The expressive dominance of the $\leq$ 40 mm TL size group in all sampled areas indicates that the pattern of abundance in the surf zone is dominated by small individuals. Using a larger codend seine (26 m) with a smaller mesh (4 mm) in Japan, Suda et al. (2002) found primarily small juveniles (mostly smaller than 50 mm TL), with a few species dominating the catches. These findings are consistent with studies that found that the marine surf zone has low diversity of fish and a few highly dominant species, which comprise small transient or resident individuals that use the surf zone as a nursery area (Godefroid et al. 2003, Monteiro-Neto et al. 2003, Felix et al. 2007, Lima & Vieira 2009, Rodrigues & Vieira 2013).

The shorter haul distance we tested (i.e., 50 m) is similar to haul distances applied in different studies at surf zone areas,

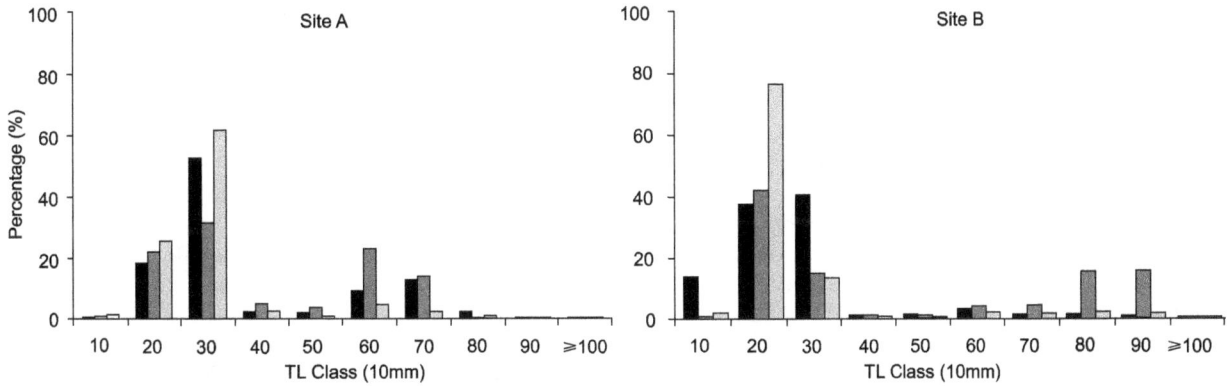

Figure 5. Percentage of individuals captured by length class for each haul distance (50, 70 and 100 m), for each sampling site (A and B). (■) 50 m, (■) 70 m, (■) 100 m.

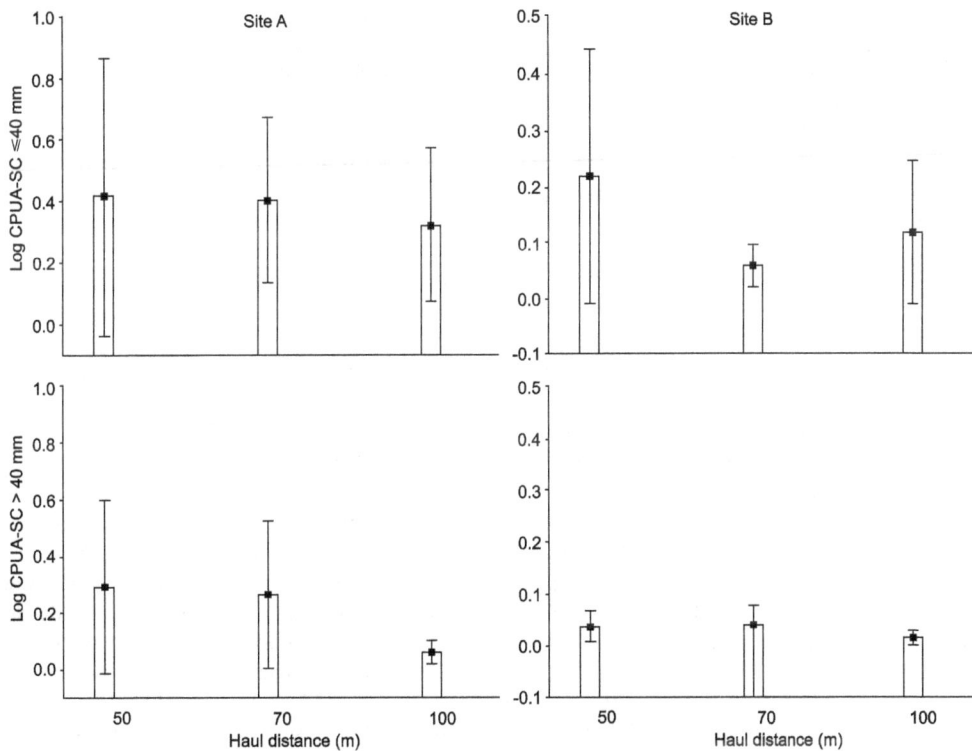

Figure 6. Variations of mean LogCPUA-SC values per haul distance for groups ≤ 40 mm and > 40 mm of total length (TL) in site A and site B. The vertical bars indicate the mean values and the lines indicate the standard deviation.

for instance: Monteiro-Neto & Musick (1994) also used 50 m haul distance perpendicular to the shoreline with a small beach seine; and Kanou et al. (2004) applied a 20 m haul distance parallel to the shoreline, using a small bag seine net. Vieira et al. (2006) suggested that this particular type of 9 m beach seine with codend should be used for short distance hauls, in order to improve the efficiency at the haul.

Based on our results and in the literature we recommend short haul distances (≤ 50 m) as the strategy for the codend net type used in this study, since the size structure of individuals, and species composition and structure of the different haul distances are the same, shorter haul distances also take less time to be performed, becoming an economic and productive approach to surveys of fish assemblages. It is important to register that small beach seines without a codend, although often easier to operate in surf zones with considerable wave action, may let fish easily evade from the seine, and the codend helps to retain more fish.

ACKNOWLEDGMENTS

P.M.L. thanks a master's degree fellowship provided by Coordenação de Aperfeiçoamento de Pessoal de Nível Superior (CAPES). This work is a contribution of the Brazilian Long Term Ecological Research Program (PELD) and SISBIOTA Program from Conselho Nacional de Desenvolvimento Científico e Tecnológico (CNPq – Proc. 403805/2012-0; 563263/2010-5) and Fundação de Amparo à Pesquisa do Estado do Rio Grande do Sul (FAPERGS – Proc. 11/2262-7).

REFERENCES

ARTIOLI, L.G.S.; J.P. VIEIRA; A.M. GARCIA & M.A. BEMVENUTI. 2009. Distribuição, dominância e estrutura de tamanhos da assembléia de peixes da lagoa Mangueira, sul do Brasil. **Iheringia 99** (4): 409-418. doi: 10.1590/S0073-47212009000400011.

BURNS, M.D.M.; A.M. GARCIA; J.P. VIEIRA; A.M. BEMVENUTI; D.M.L.M. MARQUES & V. CONDINI. 2006. Evidence of habitat fragmentation affecting fish movement between the Patos and Mirim coastal lagoons in southern Brazil. **Neotropical Ichthyology 4** (1): 69-72. doi: 10.1590/S1679-62252006000100006.

CALLIARI, L.J. 1998. Características Geomorfológicas, p. 101-104. *In*: U. SEELIGER; C. ODEBRECHT & J.P. CASTELLO (Eds). **Os ecossistemas costeiro e marinho do extremo sul do Brasil.** Rio Grande, Ed. Ecoscientia.

CALLIARI, L.J.; T. HOLAND; M.S. DIAS; S. VINZON; E.B. THORTON & T.P. STANTON. 2005. Experimento Cassino 2005: Uma síntese dos levantamentos efetuados na ante-praia e zona de arrebentação. *In*: **Anais do Congresso da Associação Brasileira de Estudos do Quaternário.** Vitória, ABEQUA, [CD].

CENI, G. & J.P. VIEIRA. 2013. Looking through a dirty glass: how different can the characterization of a fish fauna be when distinct nets are used for sampling. **Zoologia 30** (5): 499-505. doi: 10.1590/S1984-46702013000500005.

FÉLIX, F.C.; H.L. SPACH; P.S. MORO & C.W. HACKRADT. 2007. Ichthyofauna composition across a wave – energy gradient on southern Brazil beaches. **Brazilian Journal of Oceanography 55** (4): 281-292. doi: 10.1590/S1679-87592007000400005.

FIGUEIREDO, J.L. & N.A. MENEZES. 1980. **Manual de peixes marinhos do sudeste do Brasil III. Teleostei (2).** São Paulo, Museu de Zoologia Universidade de São Paulo, 93p.

FIGUEIREDO, J. L. & N.A. MENEZES. 2000. **Manual de peixes marinhos do sudeste do Brasil VI. Teleostei (5).** São Paulo, Museu de Zoologia Universidade de São Paulo, 54p.

GARCIA, A.M.; M.A. BEMVENUTI; J.P. VIEIRA; D.M.L.M. MARQUES; M.D.M. BURNS; A. MORESCO & M.V. CONDINI. 2006. Checklist comparison and dominance patterns of the fish fauna at Taim Wetland, South Brazil. **Neotropical Ichthyology 4** (2): 261-268. doi: 10.1590/S1679-62252006000200012.

GODEFROID, R.S.; H.L. SPACH; R.S. JUNIOR & G.M.L. QUEIROZ. 2003. A fauna de peixes da praia do balneário Atami, Paraná, Brasil. **Atlântica 25** (2): 147-161. doi: 10.5088/atl%C3%A2ntica.v25i2.2302.

HAHN, P.K.J.; R.E. BAILEY & A. RITCHIE. 2007. Beach Seining, p. 267-324. *In*: D.H. JOHNSON; B.M. SHRIER; J.S. O'NEAL; J.A. KNUTZEN; X. AUGEROT; T.A. O'NEIL & T.N. PEARSONS (Eds). **Salmonid Field Protocols Handbook: Techniques for Assessing Status and Trends in Salmon and Trout Populations.** Bethesda, American Fisheries Society in association with state of the salmon.

HAMMER, Ø.; D.A.T. HARPER & P.D. RYAN. 2001. Past: Paleontological Statistics Software Package for Education and Data Analysis. **Palaeontologia Electronica 4** (1): 9p. Available online at: http://palaeo-electronica.org/2001_1/past/past.pdf [Accessed: February 2013].

KANOU, K.; M. SANO & H. KOHNO. 2004. Catch efficiency of small seine for benthic juveniles of the yellowfin goby *Acanthogobius flavimanus* on a tidal mudflat. **Ichthyological Research 51**: 374-376. doi: 10.1007/s10228-004-0231-9.

KING, M. 1995. **Fisheries biology, assessment and management.** Oxford, Fishing News Books, Blackwell Science, 341p.

KREBS, C.J. 1989. **Ecological Methodology.** New York, Harper and Row Publishers, 654p.

LAYMAN, C.A. 2000. Fish Assemblage Structure of the Shallow Ocean Surf-Zone on the Eastern Shore of Virginia Barrier Islands. **Estuarine and Coastal Shelf Science 51** (2): 201-213. doi: 10.1006/ecss.2000.0636.

LIMA, M.S.P. & J.P. VIEIRA. 2009. Variação espaço-temporal da ictiofauna da zona de arrebentação da Praia do Cassino, Rio Grande do Sul, Brasil. **Zoologia 26** (3): 499-510. doi: 10.1590/S1984-46702009000300014.

MAZZEI, E.F.; C.R. PIMENTEL; R.M. MACIEIRA & J.C. JOYEUX. 2009. Resultados preliminares da variação espacial da ictiofauna de praias arenosas sobre influência do estuário dos rios Piraquê-Açê e Piraquê-Mirim, ES. **IX Anais do Congresso de Ecologia do Brasil. São Lourenço: Sociedade de Ecologia do Brasil.** Available online at: http://www.seb-ecologia.org.br/2009/resumos_ixceb/892.pdf [Accessed: 22 January 2011].

MENEZES, N.A. & J.L. FIGUEIREDO. 1980. **Manual de peixes marinhos do sudeste do Brasil IV. Teleostei (3).** São Paulo, Museu de Zoologia Universidade de São Paulo, 98p.

MENEZES, N.A. & J.L. FIGUEIREDO. 1985. **Manual de peixes marinhos do sudeste do Brasil V. Teleostei (4).** São Paulo, Museu de Zoologia Universidade de São Paulo, 107p.

MONTEIRO-NETO, C. & J.A. MUSICK. 1994. Effects of beach seine on the assessment of surf-zone fish communities. **Atlântica 16**: 23-29.

MONTEIRO-NETO, C.; L.P.R. CUNHA & J.A. MUSICK. 2003. Community Structure of Surf-zone Fishes at Cassino Beach, Rio Grande do Sul. **Brazilian Journal of Coastal Research 35**: 492-501.

MONTEIRO-NETO, C. & L. PRESTRELO. 2013. Comparing sampling strategies for surf-zone fish communities. **Marine and Freshwater Research 64**, 102-107. doi: 10.1071/MF12070.

QUEIROLO, D.; H. DELOUCHE & C. HURTADO. 2009. Comparison between dynamic simulation and physical model testing of a new trawl design for Chilean crustacean fisheries. **Fisheries Research 97**: 86-94. doi: 10.1016/j.fishres.2009.01.005.

RODRIGUES, F.L. & J.P. VIEIRA. 2013. Surf zone fish abundance and diversity at two sandy beaches separated by long rocky jetties. **Journal of the Marine Biological of the United Kingdom 93(4)**: 867-875. doi: 10.1017/S0025315412001531.

ROTHERHAM, D.; A.J. UNDERWOOD; M.G. CHAPMAN & C.A. GRAY. 2007. **A strategy for developing scientific sampling tools for fishery-independent surveys of estuarine fish in New South Wales, Australia. ICES Journal of Marine Science 64: 1512**-1516. doi: 10.1093/icesjms/fsm096.

SILVA, M.A.; F.G. ARAÚJO; M.C.C. AZEVEDO & J.N.S. SANTOS. 2004. The nursery function of sandy beaches in a Brazilian tropical bay for 0-group anchovies (Teleostei: Engraulidae): diel, seasonal and spatial patterns. **Journal of Marine Biology Assessment UK 84**: 1229-1232. doi: 10.1017/S0025315404010719h.

SUDA, Y.; T. INOUE & H. UCHIDA. 2002. Fish communities in the surf zone of a protected sandy beach at Doigahama, Yamaguchi Prefecture, Japan. **Estuarine and Coastal Shelf Science 55**: 81-96. doi: 10.1006/ecss.2001.0888.

VAN MARLEN, B. 2003. Improving the selectivity of beam trawls in The Netherlands: The effect of large mesh top panels on the catch rates of sole, plaice, cod and whiting. **Fisheries Research 63**: 155-168. doi: 10.1016/S0165-7836(03)00075-4.

VIEIRA, J.P. 2006. Ecological analogies between estuarine bottom trawl fish assemblages from Patos Lagoon, Rio Grande do Sul, Brazil and York River, Virginia, USA. **Revista Brasileira de Zoologia 23** (1): 234-247. doi: 10.1590/S0101-81752006000100017.

VIEIRA, J.P.; T. GIARRIZZO & H. SPACH. 2006. Necton. *In*: P.C. LANA; A. BIANCHINI; C. RIBEIRO; L.F.H. NIENCHESKI; G. FILMANN & C.S.G. SANTOS (Eds). **Avaliação Ambiental de Estuários Brasileiros: Diretrizes Metodológicas.** Rio de Janeiro, Museu Nacional, vol. 1.

16

Habitat use and seasonal activity of insectivorous bats (Mammalia: Chiroptera) in the grasslands of Southern Brazil

Marília A. S. Barros[1,3], Daniel M. A. Pessoa[1] & Ana Maria Rui[2]

[1] *Departamento de Fisiologia, Centro de Biociências, Universidade Federal do Rio Grande do Norte. Campus Universitário Lagoa Nova, 59078-970 Natal, RN, Brazil.*
[2] *Departamento de Ecologia, Zoologia e Genética, Instituto de Biologia, Universidade Federal de Pelotas. Campus Universitário Capão do Leão, Caixa Postal 354, 96001-970 Pelotas, RS, Brazil.*
[3] *Corresponding author E-mail: barrosmas@gmail.com*

ABSTRACT. In temperate zones, insectivorous bats use some types of habitat more frequently than others, and are more active in the warmest periods of the year. We assessed the spatial and seasonal activity patterns of bats in open areas of the southernmost region of Brazil. We tested the hypothesis that bat activity differs among habitat types, among seasons, and is influenced by weather variables. We monitored four 1,500-m transects monthly, from April 2009 to March 2010. Transects corresponded to the five habitat types that predominate in the region. In each sampling session, we detected and counted bat passes with an ultrasound detector (Pettersson D230) and measured climatic variables at the transects. We recorded 1,183 bat passes, and observed the highest activity at the edge of a eucalyptus stand (0.64 bat passes/min) and along an irrigation channel (0.54 bat passes/min). The second highest activity values (0.31 and 0.20 bat passes/min, respectively) were obtained at the edge of a riparian forest and at the margin of a wetland. The grasslands were used significantly less (0.05 bat passes/min). Bat activity was significantly lower in the winter (0.21 bat passes/min) and showed similar values in the autumn (0.33 bat passes/min), spring (0.26 bat passes/min), and summer (0.29 bat passes/min). Bat activity was correlated with temperature, but it was not correlated with wind speed and relative humidity of the air. Our data suggest that, in the study area, insectivorous bats are active throughout the year, and use mostly forest and watercourses areas. These habitat types should be considered prioritary for the conservation of bats in the southernmost region of Brazil.

KEY WORDS. Acoustic monitoring; activity patterns; Molossidae; South American *Pampas*; Vespertilionidae.

Insectivorous bats represent 70% of all bat species and are widely distributed (SIMMONS 2005). They play an important ecological role in the transfer of nutrients in ecosystems (PIERSON 1998) and in the control of insect populations, including agricultural pests (BOYLES *et al.* 2011). Insectivorous bats occupy high trophic levels, are indicators of habitat quality (JONES *et al.* 2009), and may undergo population decrease in response to environmental disturbances (TUTTLE 1979, GERELL & LUNDBERG 1993, O'DONNELL 2000).

In temperate zones, insectivorous bats use some types of habitat more frequently than others, and tend to respond positively to the presence of trees and water bodies (WALSH & HARRIS 1996). Bats are highly active in forests and forest fragments in rural areas (ERICKSON & WEST 2003, LUMSDEN & BENNETT 2005), mainly in hedgerows and forest edges (RUSS *et al.* 2003, PETTIT & WILKINS 2012), as well as around rivers, lakes, and lagoons (VAUGHAN *et al.* 1997, BROOKS 2009). Furthermore, in temperate regions the activity levels of bats vary seasonally in response to fluctuations in climatic conditions and their own energy

requirements throughout the year (CIECHANOWSKI *et al.* 2010, JOHNSON *et al.* 2011). Some studies highlighted the influence of abiotic factors on bat activity, such as temperature (HAYES 1997, RUSS *et al.* 2003), wind speed (AVERY 1985, JOHNSON *et al.* 2011), and relative humidity of the air (LACKI 1984, ADAM *et al.* 1994).

There is plenty of information available about habitat use and seasonal activity of insectivorous bats, which is mainly based on studies carried out in temperate regions of the northern hemisphere. In Brazil, studies focusing on habitat use by insectivorous bats are few, though there is one study carried out in an urban area in southeastern Brazil (ALMEIDA *et al.* 2007). One of the places where the Brazilian bat fauna is understudied is the state of Rio Grande do Sul, mainly its southern half, where the Pampa biome is located (BERNARD *et al.* 2011). The Pampa biome corresponds to the Brazilian part of the South American Pampas, which extends through Uruguay and Argentina and is characterized by plains covered by grasslands (IBGE 2004). Although the Pampa covers approximately 2% of the Brazilian territory (IBGE 2004), it occupies the third place among the six continental biomes of Brazil in terms of number of endangered species, surpassing even the Amazon, Caatinga, and Pantanal (PAGLIA *et al.* 2008). The main threats to the biodiversity of the Pampa are the expansion of agriculture, silviculture, and exotic grasses, which have been responsible for a considerable loss of natural grasslands in the past three decades (PILLAR *et al.* 2009).

Information about patterns of habitat use is important for bat conservation (FENTON 1997) and, hence, for the conservation of ecological processes associated with bats. In the present study, we describe spatial and seasonal activity patterns of insectivorous bats in the southernmost region of Brazil, in a Pampa area. Our objectives were: a) to compare the main characteristic habitats of this region in terms of bat activity; b) to test for seasonal variations in bat activity; and c) to test whether bat activity is influenced by abiotic factors, such as temperature, wind speed, and relative humidity of the air. We hypothesized that bat activity is higher in habitats with trees and/or water, and also that bats decrease their activity in the colder months of the year.

MATERIAL AND METHODS

We carried out the present study in private properties of the rural area of Santa Vitória do Palmar, in the southernmost region of the state of Rio Grande do Sul, Brazil. The study area is located in the geomorphologic region of Planície Costeira (a coastal lowland), in the Pampa biome, classified as steppe according to the international phytogeographic system of world vegetation (IBGE 2004).

The landscape is wide lowland with sandy grasslands, located between Mirim Lagoon and the Atlantic Ocean. The predominant vegetation in the region is grassland, which is mainly formed by grasses, sedges, and other herbs and subshrubs (RAMBO 2000). In areas under freshwater influence there is paludal and shrubby vegetation, forming small riparian forests along watercourses (RAMBO 2000). In addition, there are lines and stands of introduced eucalyptus isolated in the grassland, planted to serve as windbreaks and shelter for farm livestock.

According to the Köppen system, the climate of the region is classified as Cfa, i.e., temperate without dry season and with hot summer (PEEL *et al.* 2007). According to data obtained between 1990 and 2010 at the meteorological station of Santa Vitória do Palmar (Instituto Nacional de Meteorologia/INMET), the average annual temperature is 17°C. The coldest months are June and July, with an average temperature of 12°C, and the warmest months are January and February, with an average temperature of 22°C. The average monthly rainfall is 109 mm, and the annual total rainfall was 1,153 mm in 2009 and 1,176 in 2010.

The term "activity" has broad meaning; in our study we are referring to the flight/foraging activity of bats, outside day roosts. Monitoring of bat activity was carried out in four 1,500-m long transects, located in the different types of habitats that predominate in the region (Table I, Figs 1-6. 1). We marked 30 fixed points at each transect, 50 m away from each other. We marked the transects TR II and TR IV in structurally homogeneous environments in terms of vegetation, and considered them to contain the same type of habitat. Transects TR I and TR III comprised two types of habitat each. Each point at the transects comprised only one type of habitat, considering the type of vegetation in a radius of 50 m around the point.

Every month from April 2009 to March 2010 we monitored one transect each night and, whenever possible, during four consecutive nights. We used the ultrasound detector Pettersson Elektronik AB D230 (frequency range: 10-120 kHz, bandwidth: 8 kHz ± 4 kHz), in the mode heterodyne, which artificially reduces the ultrasound frequency and makes it audible to humans in real time (PARSONS & SZEWCZAK 2009). The transects were always monitored by the same observer, and the ultrasound detector was used for three minutes in each one of the 30 points. Transect monitoring began 10 minutes after sunset and was carried out at night when it was not raining. At each point, the observer kept the ultrasound detector at approximately 1 m above ground, at 45° of inclination, and turned it 360° covering all directions. For each point, we recorded the number of bat passes, which are defined as a sequence of a bat echolocation calls on the detector from beginning to end (following KUENZI & MORRISON 2003). To increase the chance of recording the highest possible number of species, we constantly altered the frequency of the bat detector between 10 and 120 kHz (following CELUCH & KROPIL 2008).

We considered the activity of the bat assemblage as a whole. Spatial and seasonal activity patterns may vary among species (BROOKS & FORD 2005, CIECHANOWSKI *et al.* 2010), but information on general bat activity is useful for the identification of general tendencies of habitat use and priority areas for bat conservation (ESTRADA *et al.* 2004, WALSH & HARRIS 1996). In the study area, the fauna of insectivorous bats corresponds to molossid and vespertilionid species, and sampling with mist nets confirmed the presence of *Tadarida brasiliensis* (I. Geoffroy, 1824), *Molossus molossus* (Pallas, 1766), *Molossus rufus* É. Geoffroy, 1805, and *Eptesicus brasiliensis* (Desmarest, 1819) in the region (M.A.S. Barros, unpublished data) – although probably more species occur in the area.

The sampling effort was 72 hours on 48 sampling nights. The average time to completely monitor transects was 132,4 ± 9.5 minutes (time with the bat detector switched on plus movement time on foot between points). According to information from temperate zones of North America (KUNZ 1973, HAYES 1997), and also to data from other parts of the region of Planície Costeira in Rio Grande do Sul (Rui & Barros, unpublished data), our survey was carried out in the typical activity peak of insectivorous bats in the first two-three hours after sunset. Because of this, we assume that the timing of activity monitoring did not bias our results, and transects were always surveyed in the same direction (from point 1 to 30).

Figures 1-6. Habitat types monitored for bat activity with acoustic surveys, from April 2009 to March 2010, in the grasslands of southernmost Brazil: (1) Eucalyptus stand (TR I); (2) Grassland (TR I); (3) Channel (TR II); (4) Riparian forest (TR III); (5) Wetland (TR III); (6) Grassland (TR IV).

Table I. Transects and habitat types monitored for bat activity with acoustic surveys, from April 2009 to March 2010, in the grasslands of southernmost Brazil.

Transect	Sites	Coordinates (UTM)	Habitat	Description
TR I	01 to 10	288369/6282442	Eucalyptus stand	Edge of eucalyptus stand, with an area of 600 m² and maximum height of 15 m.
TR I	11 to 30	288914/6281884	Grassland	Grassland with only underbrush vegetation, used as pasture for bovine livestock.
TR II	01 to 30	293353/6284182	Channel	Margin of an artificial managed channel of 15 m in width and 3 m in depth, used for water extraction for rice irrigation.
TR III	01 to 15	293208/6280102	Riparian forest	Edge of dense arboreal-shrubby riparian forest fragment, with maximum height of 5 m, located on the margin of a native wetland.
TR III	16 to 30	293107/6280711	Wetland	Margin of native wetland, composed of a flooded field covered by herbaceous and shrubby vegetation.
TR IV	01 to 30	289340/6287498	Grassland	Grassland with only underbrush vegetation, used as pasture for bovine livestock.

We recorded the following abiotic parameters: temperature, relative humidity of the air, and wind velocity. We collected these data four times while monitoring transects: at point 01 (beginning of the transect), at point 10 (500 m), at point 20 (1,000 m), and at point 30 (end of the transect).

The statistical analysis was carried out in the program PASW Statistics 18 (Statistical Package for the Social Sciences/SPSS Inc.). Since our data were not normally distributed (Kolmogorov-Smirnov test, $\alpha = 0.01$), we used non-parametric tests.

We compared the number of bat passes/3 min among the five habitat types, pooling the 12 sampling months, and compared bat activity between habitats for each season. In both analyses, we used Kruskal-Wallis tests ($\alpha = 0.05$) and Mann-Whitney *post hoc* tests ($\alpha = 0.005$, with Bonferroni correction).

For the assessment of seasonal variations in bat activity, we compared the number of bat passes/3 min among seasons, pooling data of the five types of habitats. We also compared bat activity among seasons for each habitat alone, carrying out five additional analyses, one for each type of habitat. For these analyses, we used the Friedman ANOVA ($\alpha = 0.05$) and the Wilcoxon *post hoc* test ($\alpha = 0.008$ with Bonferroni correction).

To assess the influence of abiotic factors on bat activity, we calculated correlations between the number of bat passes and temperature (°C), relative humidity of the air (%), and wind speed (m/s), with the Spearman coefficient ($\alpha = 0.05$). Since the abiotic factors were measured only four times when we monitored transects, and did not vary much from one point to the next, we extrapolated the values measured in one point to its neighbors. For the analyses, we extrapolated the abiotic data obtained at point 01 to points 02 to 05, the data of point 10 to points 6 to 15, the data of point 20 to points 16 to 25, and the data of point 30 to points 26 to 29.

RESULTS

Use of habitat by bats

We recorded 1,183 bat passes during one year. Bat activity differed significantly among habitat types (H = 311.38, df = 4, p < 0.001). We observed the highest activity in the eucalyptus stand and in the channel, followed by the riparian forest and the wetland (Fig. 7). The grassland was the least used habitat (Fig. 7). All habitats differed from each other (p < 0.001), except for the comparisons "eucalyptus vs. channel" (p = 0.038) and "riparian forest vs. wetland" (p = 0.162) (Fig. 8).

Bat activity differed also among habitat types in the different seasons (Fig. 9). The difference in activity among habitats was significant in the autumn (H = 190.56, df = 4, p < 0.001), the winter (H = 81.65, df = 4, p < 0.001), the spring (H = 71.48, df = 4, p < 0.001), and the summer (H = 90.54, df = 4, p < 0.001). In the autumn, the highest activity levels were recorded in the eucalyptus stand and in the channel; all habitats differed from each other (p ≤ 0.002), except for the com-

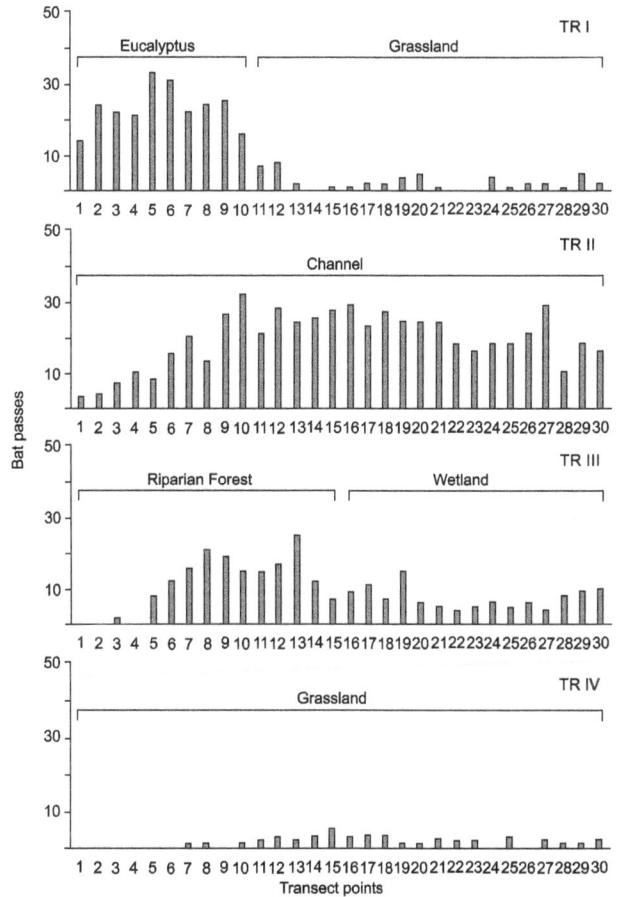

Figure 7. Total number of bat passes recorded with a bat detector in each fixed point on 1,500-m transects monitored for bat activity, from April 2009 to March 2010, in the grasslands of southernmost Brazil.

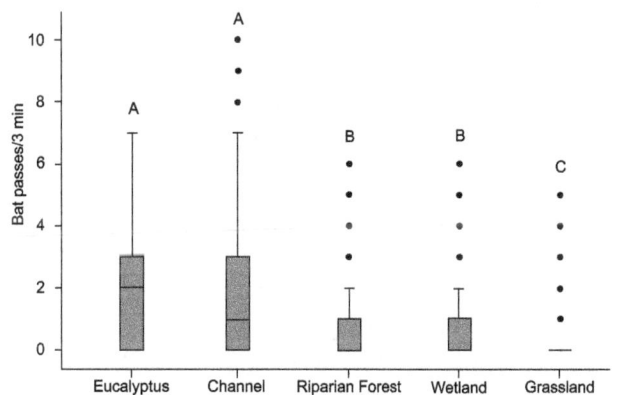

Figure 8. Box plot of the number of bat passes/3 min recorded with a bat detector in each habitat type, from April 2009 to March 2010, in the grasslands of southernmost Brazil. Different letters indicate significant differences at the 0.005 probability level (with Bonferroni correction).

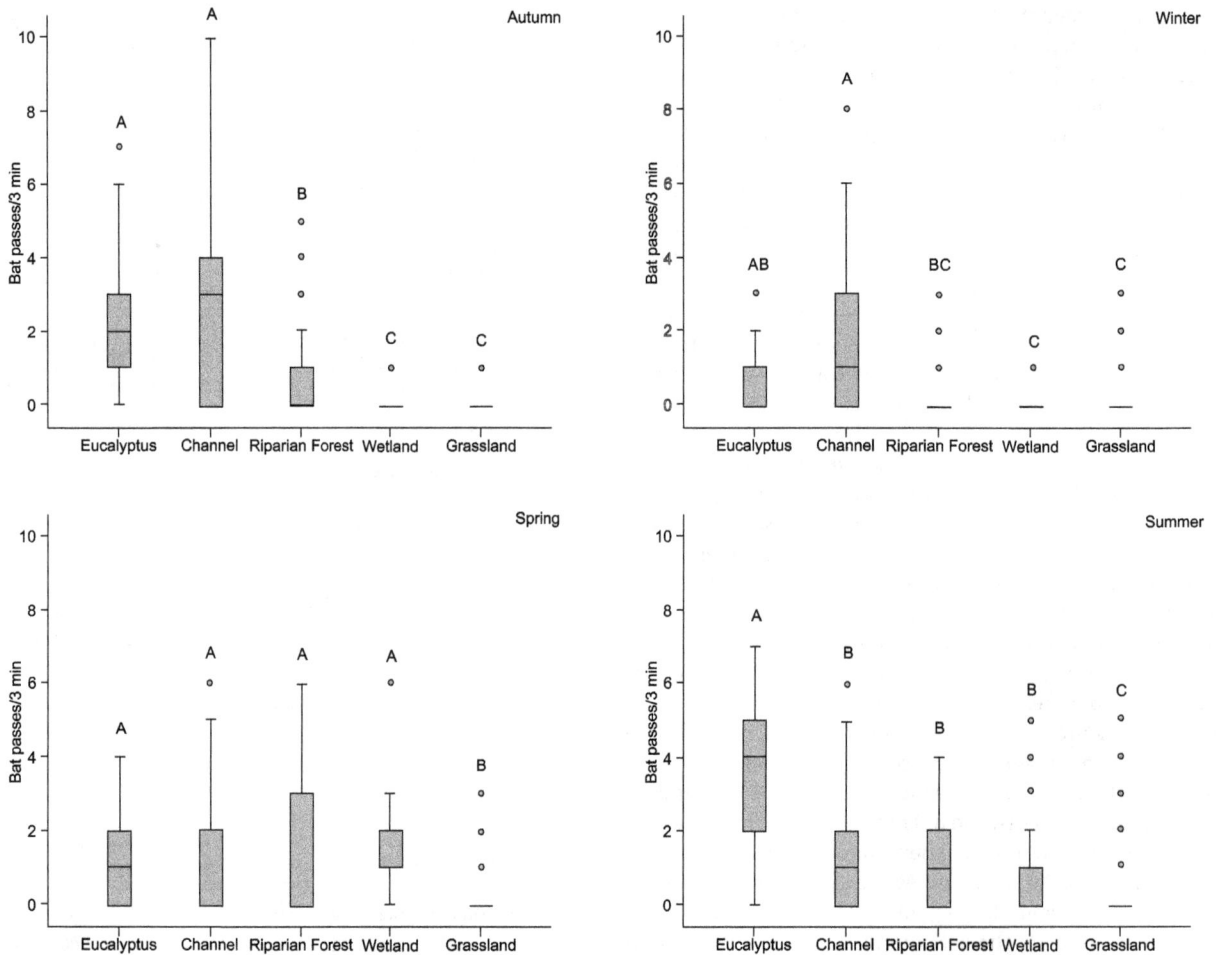

Figure 9. Box plot of the number of bat passes/3 min recorded with a bat detector in each habitat type in different seasons, from April 2009 to March 2010, in the grasslands of southernmost Brazil. Different letters indicate significant differences at the 0.005 probability level (with Bonferroni correction).

parisons "eucalyptus vs. channel" (p = 0.834) and "grasslands vs. wetland" (p = 0.113). In the winter, activity was higher in the channel than in the other habitats (p < 0.001), except for the eucalyptus stand (p = 0.018). In the spring, bat activity was lower in the grassland than in the other habitats (p < 0.001), which did not differ from each other (p ≥ 0.010). In the summer, the eucalyptus stand showed the highest activity, differing from all other types of habitats (p < 0.001).

Seasonal variations in bat activity

Bat activity varied among seasons (F_r = 10.34, df = 3, p = 0.016). There was no significant difference in the number of bat passes between the autumn (356 bat passes in total), summer (316), and spring (282) (p ≥ 0.116). The number of bat passes in the winter (229) was significantly smaller than in the autumn and summer (p ≤ 0.004), and statistically similar to the number of bat passes in the spring (p = 0.106).

We also observed seasonal variations in bat activity for each habitat alone, in the eucalyptus stand (F_r = 36.42, df = 3, p < 0.001), the channel (F_r = 32.46, df = 3, p < 0.001), the riparian forest (F_r = 12.12, df = 3, p = 0.007), the wetland (F_r = 61.94, df = 3, p < 0.001), and the grassland (F_r = 28.14, df = 3, p < 0.001). In the eucalyptus stand, riparian forest, wetland and grassland, the highest bat activity was observed in the spring and summer (Fig. 10). However, in the channel, bat activity was higher in the autumn and winter; bat activity in the autumn was significantly higher than in the other seasons (p ≤ 0.006) (Fig. 10).

Influence of climatic factors in bat activity

In the winter we recorded the lowest average temperature, and also the highest humidity and highest wind speed (Table II). We observed the highest bat activity (≥ 7 bat passes/3 min) in the points where the temperature was higher than 15°C, and we observed null or very low values (≤ 1 bat pass/

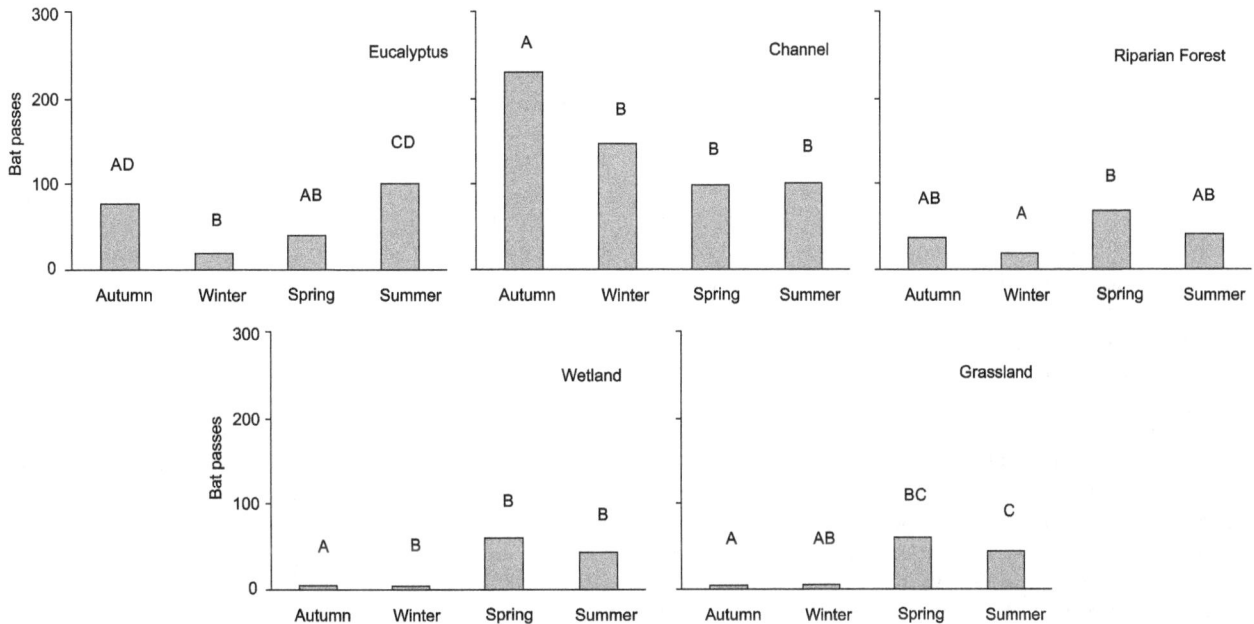

Figure 10. Total number of bat passes recorded with a bat detector in each season for the five habitat types, from April 2009 to March 2010, in the grasslands of southernmost Brazil. Different letters indicate significant differences at the 0.008 probability level (with Bonferroni correction).

Table II. Average values and standard deviation of the climatic factors measured in the transects monitored for bat activity, from April 2009 to March 2010, in the grasslands of southernmost Brazil.

Season	Temperature (°C)	Humidity (%)	Wind speed (m/s)
Autumn	15.0 ± 4.7	80.5 ± 9.5	1.5 ± 1.6
Winter	11.9 ± 2.5	92.7 ± 5.3	2.6 ± 1.5
Spring	14.7 ± 4.5	88.7 ± 8.9	2.1 ± 1.3
Summer	17.8 ± 2.3	85.2 ± 11.8	1.2 ± 0.9

3 min) when temperatures were between 5°C and 10°C. The number of bat passes was correlated with temperature ($r_s = 0.138$, $p < 0.001$). We recorded high levels of activity (≥ 7 bat passes/3 min) in nights with different values of wind speed and relative humidity of the air. Bat passes were not significantly correlated with relative humidity of the air ($r_s = 0.009$, $p = 0.722$) or wind speed ($r_s = -0.030$, $p = 0.258$).

DISCUSSION

Use of habitat

The hypothesis that bats mostly use areas with tall trees and water bodies was corroborated. The association of insectivorous bats with forest edges has been broadly documented for other regions (LUMSDEN & BENNETT 2005, KOFOKY et al. 2007, MORRIS et al. 2010). This is also true for aquatic habitats (LUNDE & HARESTAD 1986, VAUGHAN et al. 1997, BROOKS 2009). The use of the eucalyptus stand and the channel were probably due to the availability of aerial insects, which are frequently abundant in forest edges (LEWIS 1970) and water bodies (BARCLAY 1991).

The canopy of the eucalyptus stand is approximately three times higher than the canopy of the riparian forest. Since linear vegetation edges serve as landmarks and also offer protection against the wind and predators (LIMPENS & KAPTEYN 1991), eucalyptus probably are more efficient for spatial orientation and protection than riparian forest. In addition, the eucalyptus trees have a large diameter, a reasonable space between each other, and trunks with cavities, cracks and loose bark, which favors their use as diurnal roost by bats, and also for flights inside the stand. The vegetation of the riparian forest, on the contrary, harbors trees with smaller diameter and high density, a factor that is negatively correlated with bat activity (ERICKSON & WEST 2003).

In southern Brazil, there are records of association between insectivorous bats and tree species introduced for commercial use (Pinus spp.), which are used by vespertilionid bats to roost during the day (BARROS & RUI 2011). Exotic trees tends to acquire particular significance when the original vegetation has been altered and the availability of tall native trees is reduced. In the study area, the natural vegetation is known as Butiazal, which is composed of jelly palm (Butia capitata) clusters that occur along the coastal lowland of southern Brazil (RAMBO 2000). These formations are currently rare in the state of Rio Grande do Sul, and Butia capitata is endangered in the state (RIO GRANDE DO SUL 2002).

The channel may represent an important water source, since frequent water ingestion is a critical factor for the water balance of insectivorous bats (Neuweiler 2000). Since insectivorous bats use watercourses for spatial orientation (Racey & Swift 1985, Serra-Cobo et al. 2000), the channel can also be used as a flying route. Apparently, the presence of open water is important for bat activity, since the channel showed higher activity levels than the wetland, where portions of flooded grassland are completely covered by vegetation. A study carried out in temporary lakes showed that the activity of insectivorous bats decreases as the exposed water area decreases (Francl et al. 2008). Insect availability in riparian areas may be high (Racey & Swift 1985). However, the activity of bats that forage over water may be negatively correlated with the presence of floating vegetation, which hinders the detection of prey through echolocation (Ciechanowski et al. 2007, Siemers et al. 2001).

The low bat activity in the grassland was expected, since insectivorous bats, in general, use field and pasture areas less frequently (Fenton 1970, Estrada et al. 2004). The activity recorded for grasslands is probably a result of the movement of individuals between roosts and feeding areas, or between feeding areas.

In all seasons, the eucalyptus stand and the channel were the habitats with highest activity, except in the spring. During this season, strong rains caused a more intense flood in the riparian forest and the wetland, which attracted more bats in relation to other seasons of the year, probably due to an increase in insect abundance.

Our data suggest that watercourses and edges of tree stands are priority habitats for bat conservation in the region. Areas of native wetland, with and without riparian forest, were less used, but also presented expressive levels of bat activity. The association of these types of habitat with insectivorous bats indicates that alterations in aquatic habitats or removal of trees tend to negatively affect bat activity in the region.

Seasonal variations in bat activity and influence of climatic factors

As expected, bat activity varied seasonally, and activity in the winter was low. This result is consistent with the common pattern observed for insectivorous bats (Hays et al. 1992, Carmel & Safriel 1997, Hayes 1997). The winter is the most critical period for the energy balance of insectivorous bats because low temperatures reduce insect abundance and activity (Taylor 1963, Wolda 1988) and increase heat loss (Ransome 1990), making foraging less profitable. In the study area, bat activity was positively correlated with temperature, as in other temperate regions (Avery 1985, Brooks 2009, Broders et al. 2006), and the low activity levels in winter was a response of bats to low temperatures.

Like other authors (e.g., Johnson et al. 2011, Verboom & Spoelstra 1999), we did not detect a significant relationship between bat activity and humidity or wind speed. In the study area, the climate frequently varies more expressively in terms

of temperature than in terms of humidity or wind speed (Rambo 2000). In southern Brazil, rainfall is uniform throughout the year and the relative humidity of the air during our sampling period was never below 60%, which may explain the apparent lack of response of bats to variations in humidity. In addition, the wind may affect bat activity only when wind speed is high (> 4 m/s) (O'Farrell et al. 1967, O'Farrell & Bradley 1970), which was rarely observed in our sampling period. The lack of a correlation between bat activity and climatic factors may also result from the pooled analysis of the bat assemblage, since some species at an area may be influenced by specific climatic factors, such as wind speed, whereas others may not (Russo & Jones 2003, Russ et al. 2003).

We observed a tendency for high bat activity in the summer and spring and low activity in the winter in the eucalyptus stand, grassland, riparian forest and wetland. By contrast, in the channel, the spring and summer were the periods when bat activity was the lowest. This pattern is probably associated with the cultivation of rice in the area. From the beginning of spring to the end of summer, the grassland is removed, the soil is plowed and flooded, and the crop is treated with pesticides. It is likely that these actions may have negatively influenced bat activity in the channel, since bat activity is lower in areas with intensively managed crops (Walsh & Harris 1996), probably due to a reduction in insect availability due to pesticides (Carmel & Safriel 1998, Wickramasinghe et al. 2003).

Our study first demonstrated that insectivorous bats show seasonal variation in activity in the grasslands of southern Brazil (Pampa biome). Their activity patterns are markedly reduced in the winter, and are influenced by temperature.

ACKNOWLEDGMENTS

Rodrigo G. de Magalhães, Leandro V. Umann, Marcelo D. Freire, Estevão J.Comitti, Cristian M. Joenck, and Jeferson Bugoni provided us with logistic support in the study area. We thank Edison C. de Souza for the great help and friendship during fieldwork. Felipe N. Castro helped us in the statistical analysis. The company Maia Meio Ambiente LTDA granted us financial support. Coordenação de Aperfeiçoamento Pessoal de Nível Superior (CAPES) granted M.A.S. Barros a Master's scholarship.

REFERENCES

Adam, M.D.; M.J. Lacki & L.G. Shoemaker. 1994. Influence of environmental conditions on flight activity of *Plecotus townsendii virginianus* (Chiroptera: Vespertilionidae). **Brimleyana 21**: 77-85.

Almeida, M.H.; A.D. Ditchifield & R.S. Tokumaru. 2007. Atividade de morcegos e preferência por hábitat na zona urbana da Grande Vitória, ES, Brasil. **Revista Brasileira de Zoociências 9** (1): 13-18.

Avery, M.I. 1985. Winter activity by pipistrelle bats. **Journal of Animal Ecology 54** (3): 721-738.

Barclay, R.M.R. 1991. Population Structure of Temperate Zone Insectivorous Bats in Relation to Foraging Behaviour and Energy Demand. **Journal of Animal Ecology 60** (1): 165-178.

Barros, M.A.S. & A.M. Rui. 2011. Occurrence and Mortality of *Lasiurus ega* (Chiroptera, Vespertilionidae) in Monocultures of *Pinus* sp. in Rio Grande do Sul, Southern Brazil. **Chiroptera Neotropical 17** (2): 997-1002.

Bernard, E.; L.M.S. Aguiar & R.B. Machado. 2011. Discovering the Brazilian bat fauna: a task for two centuries? **Mammal Review 41** (1): 23-39. doi: 10.1111/j.1365-2907.2010.00164.x.

Boyles, J.G.; P.M. Cryan; G.F. McCracken & T.H. Kunz. 2011. Economic Importance of Bats in Agriculture. **Science 332** (6025): 41-42. doi: 10.1126/science.1201366.

Broders, H.G.; G.J. Forbes; S. Woodley & I.D. Thompson. 2006. Range extent and stand selection for roosting and foraging in forest-dwelling northern long-eared bats and little brown bats in the Greater Fundy ecosystem New Brunswick. **Journal of Wildlife Management 70** (5): 1174-1184. doi: 10.2193/0022-541X(2006)70[1174:REASSF]2.0.CO;2.

Brooks, R.T. 2009. Habitat-associated and temporal patterns of a bat activity in a diverse forest landscape of southern New England, USA. **Biodiversity and Conservation 18** (3): 529-545. doi: 10.1007/s10531-008-9518-x.

Brooks, R.T. & W.M. Ford. 2005. Bat activity in a forest landscape of Central Massachusetts. **Northeastern Naturalist 12** (4): 447-462. doi: 10.1656/1092-6194(2005)012[0447:BAIAFL]2.0.CO;2.

Carmel, Y. & U. Safriel. 1998. Habitat use by bats in a Mediterranean ecosystem in Israel–conservation implications. **Biological Conservation 84** (3): 245-250. doi: 10.1016/S0006-3207(97)00131-6.

Celuch, M. & R. Kropil. 2008. Bats in a Carpathian beech-oak forest (Central Europe): habitat use, foraging assemblages and activity patterns. **Folia Zoologica 57** (4): 358-372.

Ciechanowski, M.; T. Zając; A. Bitas & R. Dunajski. 2007. Spatiotemporal variation in activity of bat species differing in hunting tactics: effects of weather, moonlight, food abundance, and structural clutter. **Canadian Journal of Zoology 85** (12): 1249-1263. doi: 10.1139/Z07-090.

Ciechanowski, M.; T. Zając; A. Zielińska & R. Dunajski. 2010. Seasonal activity patterns of seven vespertilionid bat species in Polish lowlands. **Acta Theriologica 55** (4): 301-314. doi: 10.4098/j.at.0001-7051.093.2009.

Erickson, J.L. & S.D. West. 2003. Associations of bats with local structure and landscape features of forested stands in western Oregon and Washington. **Biological Conservation 109** (1): 95-102. doi: 10.1016/S0006-3207(02)00141-6.

Estrada, A.; C. Jiménez; A. Rivera & E. Fuentes. 2004. General bat activity measured with an ultrasound detector in a fragmented tropical landscape in Los Tuxtlas, Mexico. **Animal Biodiversity and Conservation 27** (2): 1-9.

Fenton, M.B. 1970. A technique for monitoring bat activity with results obtained from different environments in southern Ontario. **Canadian Journal of Zoology 48** (4): 847-851. doi: 10.1139/z70-148.

Fenton, M.B. 1997. Science and the Conservation of Bats. **Journal of Mammalogy 78** (1): 1-14.

Francl, K.E. 2008. Summer bat activity at woodland seasonal pools in the Northern Great Lakes Region. **Wetlands 28** (1): 117-124. doi: 10.1672/07-104.1.

Gerell, R. & K.G. Lundberg. 1993. Decline of a bat *Pipistrellus pipistrellus* population in an industrialized area in south Sweden. **Biological Conservation 65** (2): 153-157. doi: 10.1016/0006-3207(93)90444-6.

Hayes, J.P. 1997. Temporal variation in activity of bats and the design of echolocation-monitoring studies. **Journal of Mammalogy 78** (2): 514-524.

Hays, G.C.; J.R. Speakman & P.I. Webb. 1992. Why do brown long-eared bats (*Plecotus auritus*) fly in winter? **Physiological Zoology 65** (3): 554-567.

IBGE. 2004. **Mapa de Biomas do Brasil.** Ministério do Meio ambiente, Instituto Brasileiro de Geografia e Estatística. Available online at: http://www.ibge.gov.br/home/presidencia/noticias/21052004biomas.shtm [Accessed: 01/VI/2013].

Johnson, J.B.; J.E. Gates & N.P. Zegre. 2011. Monitoring seasonal bat activity on a coastal barrier island in Maryland, USA. **Environmental Monitoring and Assessment 173** (1): 685-699. doi: 10.1007/s10661-010-1415-6.

Jones, G.; D.S. Jacobs; T.H. Kunz; M.R. Willig & P.A. Racey. 2009. Carpe noctem: the importance of bats as bioindicators. **Endangered Species Research 8** (1-2): 93-115. doi: 10.3354/esr00182.

Kofoky, A.; D. Andriafidison; F. Ratrimomanarivo; H. J. Razafimanahaka; D. Rakotondravony; P.A. Racey & R.K.B. Jenkins. 2007. Habitat use, roost selection and conservation of bats in Tsingy de Bemaraha National Park, Madagascar. **Biodiversity and Conservation 16** (4): 1039-1053. doi: 10.1007/s10531-006-9059-0.

Kuenzi, A.J. & M.L. Morrison. 2003. Temporal Patterns of Bat Activity in Southern Arizona. **Journal of Wildlife Management 67** (1): 52-64.

Kunz, T.H. 1973. Resource utilization: temporal and spatial components of bat activity in central Iowa. **Journal of Mammalogy 54** (1): 14-32.

Lacki, M.J. 1984. Temperature and humidity-induced shifts in the flight activity of little brown bats. **The Ohio Journal of Science 84** (5): 264-266.

Lewis, T.S. 1970. Patterns of distribution of insects near a windbreak of tall trees. **Annals of Applied Biology 65** (2): 213-220. doi: 10.1111/j.1744-7348.1970.tb04581.x.

Limpens, H.J.G.A. & K. Kapteyn. 1991. Bats, their behaviour and linear landscape elements. **Myotis 29**: 63-71.

Lumsden, L.F. & A.F. Bennett. 2005. Scattered trees in rural landscapes: foraging habitat for insectivorous bats in southeastern Australia. **Biological Conservation 122** (2): 205-222. doi:10.1016/j.biocon.2004.07.006.

LUNDE, R.E. & A.S. HARESTAD. 1986. Activity of Little Brown Bats in Coastal Forest. **Northwest Science 60** (4): 206-209.

MORRIS, A.D.; D.A. MILLER & M.C. KALCOUNIS-RUEPPELL. 2010. Use of forest edges by bats in a managed pine forest landscape. **Journal of Wildlife Management 74** (1): 26-34. doi: 10.2193/2008-471.

NEUWEILER, G. 2000. **The biology of bats**. Oxford, Oxford University Press, VI+310p.

O'DONNELL, C.F.J. 2000. Conservation status and causes of decline of a threatened New Zealand Long-tailed bat *Chalinolobus tuberculatus* (Chiroptera: Vespertilionidae). **Mammal Review 30** (2): 89-106. doi: 10.1046/j.1365-2907.2000.00059.x.

O'FARRELL, M.J. & BRADLEY W.G. 1970. Activity Patterns of Bats over a Desert Spring. **Journal of Mammalogy 51** (1): 18-26.

O'FARRELL, M.J.; W.G. BRADLEY & G.W. JONES. 1967. Fall and winter bat activity at a desert spring in Southern Nevada. **The Southwestern Naturalist 12** (2): 163-171.

PAGLIA, A.P.; G.A.B. FONSECA & J.M.C. SILVA. 2008. A Fauna Brasileira Ameaçada de Extinção: Síntese Taxonômica e Geográfica, p. 63-70. *In*: A.B.M. MACHADO; G.M. DRUMMOND & A.P. PAGLIA (Eds). **Livro Vermelho da Fauna Brasileira Ameaçada de Extinção**. Belo Horizonte, Fundação Biodiversitas, vol. 2, 908p.

PARSONS, S. & J.M. SZEWCZAK. 2009. Detecting, Recording, and Analyzing the Vocalizations of Bats, p. 91-111. *In*: T.H. KUNZ & S. PARSONS (Eds). **Ecological and behavioral methods for the study of bats**. Baltimore, The Johns Hopkins University Press, XVII+901p.

PEEL, M.C.; B.L. FINLAYSON & T.A. MCMAHON. 2007. Updated world map of the Köppen-Geiger climate classification. **Hydrology and Earth System Sciences 11** (5): 1633-1644. doi: 10.5194/hess-11-1633-2007.

PETTIT, T.W. & K.T. WILKINS. 2012. Canopy and edge activity of bats in a quaking aspen (*Populus tremuloides*) forest. **Canadian Journal of Zoology 90** (7): 798-807. doi:10.1139/Z2012-049.

PIERSON, E.D. 1998. Tall Trees, Deep Holes, and Scarred Landscapes – Conservation Biology of North American Bats, p. 309-325. *In*: T.H. KUNZ & P.A. RACEY (Eds). **Bat biology and conservation**. Washington, D.C., Smithsonian Institution Press, XIV+365p.

PILLAR, V.P.; S.C. MÜLLER; Z.M.S. CASTILHOS & A.V.A. JACQUES. 2009. **Campos Sulinos – Conservação e Uso Sustentável da Biodiversidade**. Brasília, Ministério do Meio Ambiente, 403p.

RACEY, P.A. & S.M. SWIFT. 1985. Feeding ecology of *Pipistrellus pipistrellus* (Chiropteran: Vespertilionidae) during pregnancy and lactation – I. Foraging behaviour. **Journal of Animal Ecology 54** (1): 205-215.

RAMBO, B. 2000. **A fisionomia do Rio Grande do Sul: ensaio de monografia natural**. São Leopoldo, Editora Unisinos, 3rd ed., XXVII+473p.

RANSOME, R.D. 1990. **The Natural History of Hibernating Bats**. London, Christopher Helm, XXI+235p.

RIO GRANDE DO SUL. 2002. **Decreto Estadual Nº 42.099, de 31 de dezembro de 2002**. Declara as espécies da flora nativa ameaçadas de extinção no estado do Rio Grande do Sul e dá outras providências. Porto Alegre, Diário Oficial do estado do Rio Grande do Sul de 01/01/2003.

RUSS, J.M.; M. BRIFFA & W.I. MONTGOMERY. 2003. Seasonal patterns in activity and habitat use by bats (*Pipistrellus* spp. and *Nyctalus leisleri*) in Northern Ireland, determined using a driven transect. **Journal of Zoology 259** (3): 289-299. doi: 10.1017/S0952836902003254.

RUSSO, D. & G. JONES. 2003. Use of foraging habitats by bats in a Mediterranean area determined by acoustic surveys: conseration implications. **Ecography 26** (2): 197-209. doi: 10.1034/j.1600-0587.2003.03422.x.

SERRA-COBO, J.; M. LOPEZ-ROIG; T. MARQUES-BONET & E. LAHUERTA. 2000. Rivers as possible landmarks in the orientation flight of *Miniopterus schreibersii*. **Acta Theriologica 45** (3): 347-352.

SIEMERS, B.M.; P. STILZ & H. SCHNITZLER. 2001. The acoustic advantage of hunting at low heights above water: behavioural experiments on the European 'trawling' bats *Myotis capaccinii*, *M. dasycneme* and *M. daubentonii*. **The Journal of Experimental Biology 204** (22): 3843-3854.

SIMMONS, N.B. 2005. Order Chiroptera, p. 312-529. *In*: D.E. WILSON & D.M. REEDER (Eds). **Mammal species of the world: a taxonomic and geographic reference**. Baltimore, The Johns Hopkins University Press, vol. 1, XXXV+743p.

TAYLOR, L.R. 1963. Analysis of the Effect of Temperature on Insects in Flight. **Journal of Animal Ecology 32** (1): 99-117.

TUTTLE, M.D. 1979. Status, Causes of Decline, and Management of Endangered Gray Bats. **The Journal of Wildlife Management 43** (1): 1-17.

VAUGHAN, N.; G. JONES & S. HARRIS. 1997. Habitat use by bats (Chiroptera) assessed by means of a broad-band acoustic method. **Journal of Applied Ecology 34** (3): 716-730.

VERBOOM, B. & K. SPOELSTRA. 1999. Effects of food abundance and wind on the use of tree lines by an insectivorous bat, *Pipistrellus pipistrellus*. **Canadian Journal of Zoology 77** (9): 1393-1401. doi: 10.1139/z99-116.

WALSH, A.L. & S. HARRIS. 1996. Foraging habitat preferences of vespertilionid bats in Britain. **Journal of Applied Ecology 33** (3): 508-518.

WICKRAMASINGHE, L.P.; S. HARRIS; G. JONES & N. VAUGHAN. 2003. Bat activity and species richness on organic and conventional farms: impact of agricultural intensification. **Journal of Applied Ecology 40** (6): 984-993. doi: 10.1111/j.1365-2664.2003.00856.x.

WOLDA, H. 1988. Insect Seasonality: Why? **Annual Review of Ecology and Systematics 19**: 1-18. doi: 10.1146/annurev.es.19.110188.000245.

17

Relative growth, sexual dimorphism and morphometric maturity of *Trichodactylus fluviatilis* (Decapoda: Brachyura: Trichodactylidae) from Santa Terezinha, Bahia, Brazil

Tiago Rozário da Silva[1,3], Sérgio Schwarz da Rocha[2] & Eraldo Medeiros Costa Neto[1]

[1] *Departamento de Ciências Biológicas, Universidade Estadual de Feira de Santana. Avenida Transnordestina, Novo Horizonte, 44036-900 Feira de Santana, BA, Brazil. E-mail: apingorasilva@hotmail.com; eraldont@hotmail.com*
[2] *Laboratório de Bioecologia de Crustáceos, Centro de Ciências Agrárias, Ambientais e Biológicas, Universidade Federal do Recôncavo da Bahia. Rua Rui Barbosa 710, 44380-000 Cruz das Almas, BA, Brazil. E-mail: ssrocha@ufrb.edu.br*
[3] *Corresponding author. E-mail: apingorasilva@hotmail.com*

ABSTRACT. Freshwater crabs are important elements in the aquatic biota of brooks, rivers, lakes and ponds, from both ecological and the socio-economic aspects. Trichodactylidae comprises 51 endemic species from the Neotropical region. Among all the species of this family, *Trichodactylus fluviatilis* Latreille, 1828 has the widest geographic distribution throughout Brazil. Despite that, there are few published contributions on the biology of this species. The present study investigated the following aspects of *T. fluviatilis*: relative growth, mean size at onset of morphometric maturity, sexual dimorphism, laterality and heterochely. Specimens were collected monthly from September 2010 through August 2011, from the Velha Eugênia Brook, municipality of Santa Teresinha, State of Bahia. Carapace width (CW), carapace length (CL), major cheliped length (MaCL) and minor cheliped length (MiCL), major cheliped height (MaCH) and minor cheliped height (MiCH), and width of the fifth abdominal segment (5AB) were measured to evaluate the presence of sexual dimorphism; the major difference between the sexes was in the CW vs. 5AB ratio. Heterochely was observed in males and females,with the right cheliped larger than the left in 89% of males and 81% of females. Crab size at the onset of morphometric maturity (= puberty molt) was estimated based on the ratio between CL, cheliped dimensions, 5AB and CW (independent variable). Females were larger than males when they reached morphometric maturity in all studied relations. We recommend the use of chelipeds and abdominal width relationships to estimate the size at the morphometric maturity in males and females, respectively.

KEY WORDS. Allometry; chelipeds; crabs; maturity size; Serra da Jibóia.

Trichodactylidae comprises 51 endemic species from the Neotropical region. All members of the family are considered "true freshwater" crabs (sensu YEO *et al.* 2008), because they have direct development and complete their entire life cycle independently of the marine environment (MÜLLER, 1892, YEO *et al.* 2008). *Trichodactylus* Latreille, 1828 includes 11 valid species, nine of which have been reported from Brazil (MAGALHÃES 2003). Most trichodactylid crabs that occur in Brazil inhabit coastal-plain rivers at altitudes up to 300 m (MAGALHÃES 2003), with some specimens collected at altitudes up to 960 m (GOMIDES *et al.* 2009, MOSSOLIN & MANTELATTO 2008, ROCHA & BUENO 2004, 2011).

Trichodactylus fluviatilis Latreille, 1828 has a wide geographic distribution along the Brazilian coast, from the state of Pernambuco to the state of Rio Grande do Sul (MAGALHÃES 2003). Despite this wide range, little is known about the biology of the species (MANSUR *et al.* 2005). ALARCON *et al.* (2002) investigated the structure of a population from Ubatuba, São

Paulo, in which sex ratio was 1:1, and females were larger than males. According to these authors, ovigerous females of *T. fluviatilis* have cryptic behavior, and were therefore absent from their sample. Later, COSTA-NETO (2007) studied some etnocarcinological aspects of the species in the state of Bahia. In addition, studies have examined the association of *T. fluviatilis* with *Temnocephala* Blanchard, 1849 flatworms (AMATO *et al.* 2005, 2006) and also the use of this species as a bioindicator of some metals, including aluminum, manganese, cadmium and lead (CHAGAS *et al.* 2009, FRANCHI *et al.* 2011).

Recently, LIMA *et al.* (2012) reported on the allometric growth and average size of *T. fluviatilis* crabs from southeastern Brazil at the onset of morphometric maturity. Published studies on relative growth in brachyuran crabs often find a relationship between allometric growth and morphometric maturity, since the puberty molt is usually marked by a significant change in the allometric growth rate of some body structures such as the carapace, chela and abdomen (HARTNOLL 1978).

Considering the ecological importance of freshwater crabs and the continuous degradation of their habitats, it is important to study the biological aspects of these crustaceans, aiming the conservation and sustainable use of this natural resource. This study investigated the relative growth and average size of *T. fluviatilis* crabs at the onset of morphometric maturity, in order to provide information that could contribute to efforts to protect this species in the state of Bahia, Brazil.

MATERIAL AND METHODS

Specimens were sampled monthly, from September 2010 through August 2011, in a 100-m long section of Velha Eugênia Brook (12°50'42.9"S, 39°29'46.4"W), Pedra Branca Village, Santa Terezinha Municipality, Bahia, Brazil (Fig. 1). The sampling site had a sandy substrate, with slow water flow, and the water depth rarely exceeded 50 cm. In backwaters in some areas of the brook, large amounts of organic matter (mainly leaf litter) accumulated. Some partially submerged vegetation was present along the stream banks, and impacts caused by humans and farm animals were evident.

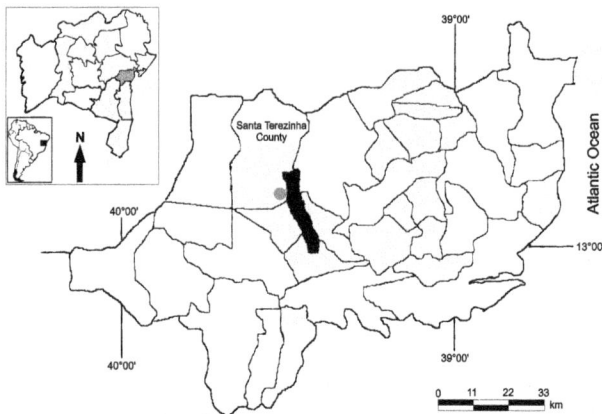

Figure 1. Map of southeast Bahia. Black area, Serra da Jibóia; gray spot, sampling site at Pedra Branca Village.

Trichodactylids were collected during daytime, with the aid of sieves (diameter 50 cm, mesh 0.5 mm). Crabs were sexed according to the location of the genital pores, external morphology of the abdomen, and the presence of developed pleopods (females) or two pairs of gonopods (males).

Body dimensions measured were: carapace width (CW), carapace length (CL), major claw propodus length (MaCL), major claw height (MaCH), minor claw propodus length (MiCL), minor claw height (MiCH), and fifth abdomen segment width (5AB). All measurements were taken to the nearest 0.01 mm with the aid of a digital caliper, and were used in

the relative growth analysis and to estimate the average size at the onset of morphometric maturity.

Except for a few animals fixed as voucher material (INPA 1398), all other specimens caught were returned alive to the sampling site after their measurements were taken.

Data from all measurements taken were converted to the linear form by means of natural logarithm transformation. Log-transformed values of carapace width (lnCW) were used as the independent variable, and all other body dimensions [Ln (claw dimension); ln(CL); and ln(5AB)] were considered as dependent variables. Data were subjected to successive linear regression analysis, and data with corresponding absolute values of the standardized residuals higher than 2.57 ($p < 0.01$) were considered outliers and excluded from the analysis. Then, each data group from the regression analysis was subjected to non-hierarchical k-means clustering, to separate the data set into juveniles and adults. Discriminant analysis was used to re-allocate any misclassified data. All statistical analyses were performed with the PAST – Paleontological Statistics Software (version 2.14) computer program (HAMMER et al. 2001). Finally, the equations obtained from each linear regression (lny = lna + bx) were tested statistically using covariance analysis (ANOVA), and all of them showed high significance levels (ZAR 1996).

The state of allometry of each body dimension was determined by testing the slope and elevation of each linear equation obtained from regression analysis, using Student's t test (ZAR 1996). Whenever the "b" value was statistically equal to 1 (H_0) growth was considered isometric; when "b" was different from 1 (H') growth was considered positively ($b > 1$) or negatively allometric ($b < 1$) (HARTNOLL 1978, 1982, MARTINEZ-MAYÉN et al. 2000, BUENO & SHIMIZU 2009). Student's t test was also applied to compare the parameters of allometric equations of juveniles and adults, in both sexes.

Mean size at the onset of morphometric maturity in each sex was calculated only for dimensions in which significant differences were detected by Student's t test when comparing the allometric equations from data groups for juveniles and adults. As performed by BUENO & SHIMIZU (2009), the size of morphometric maturity (L50) was calculated from the function $y = 1/1 + e^{(a-bx)}$ (y = proportion of adults; x = carapace width size class), with the software CurveExpert, version 1.3 (HYAMS 2001).

Handedness was verified in adult males and females separately with the Yates corrected goodness-of-fit chi-square test. The heterochely of the claws was tested with the non-parametric Wilcoxon test for paired samples (ZAR 1996).

Sexual dimorphism was determined based on body (CW and 5AB) and claw dimensions of adult specimens (MOSSOLIN & BUENO 2003, BUENO & SHIMIZU 2009). Carapace width of males and females was compared with the non-parametric Mann-Whitney test because data for both males ($W = 0.88$, $p < 0.001$) and females ($W = 0.89$, $p < 0.01$) showed a non-normal distribution (ZAR 1996). Fifth abdominal segment and claw dimensions were compared with Student's *t* test (ZAR 1996), used to compare regression parameters between adult specimens.

RESULTS

Of the total of 623 specimens of *T. fluviatilis* sampled, 212 were juveniles, 220 males and 191 females. Carapace width ranged from 5.39 to 30.42 mm (Mean ± S.D. = 11.46 ± 4.09 mm) in males and from 5.50 to 28.97 mm (Mean ± S.D. = 13.10 ± 5.04 mm) in females. Although males attained the largest carapace width recorded, the mean size of females exceeded males by 1.64 mm and the Student's *t* test detected a significant difference in the mean size of the sexes (t = 3.55; p < 0.001).

Relative Growth

All linear regression parameters obtained for each data set are shown in Table I. Juveniles of both sexes showed the same state of allometry in the linear regression equations for carapace and claw dimensions. However, the growth of the fifth abdominal segment width was negatively allometric in juvenile males and positively allometric in juvenile females.

Adult males showed positive allometry for all body dimensions, except for the fifth abdominal segment, which showed isometric growth. Adult females showed positive allometry in most of the claw dimensions, except for the minor cheliped height, which showed isometric growth. Unlike males, females showed positive allometry for the fifth abdominal segment and negative allometry for the carapace length.

Comparison between juveniles and adults indicated significant differences in all linear equations obtained from regression analyses, except for the MiCH (juvenile males vs. adult males) and MiCL (juvenile females vs. adult females) growth relationships (Table II). Therefore, different growth patterns were found between the two life phases that mark the transition from immature to adult.

Sexual dimorphism

Females were larger than males (Mann-Whitney test: U = 16820, p = 0.0005). Except for the MiCH relationship, all other dimensions were strongly sexually dimorphic (Table III). The most pronounced sexual dimorphism was detected in the fifth abdominal segment (Fig. 2 and Table I).

Table I. *Trichodactylus fluviatilis*. Linear regression parameters and equations, and the state of allometry corresponding to each regression.

Category	Relationship[1]	n	Linear regression equation and coeficient of determinance (r²)	Student's t test for allometry (H₀: b = 1)	State of allometry[2]
Young Males	CL vs. CW	203	y = 0.9694x + 0.0162; r² = 0.96	t = 2.31; p < 0.05	(−)
	MaCL vs. CW	119	y = 0.9812x − 0.6415; r² = 0.81	t = 0.43; p > 0.50	(0)
	MaCH vs. CW	117	y = 1.1529x − 1.6983; r² = 0.94	t = 5.55; p < 0.001	(+)
	MiCL vs. CW	120	y = 0.9777x − 0.7432; r² = 0.78	t = 0.46; p > 0.50	(0)
	MiCH vs. CW	114	y = 1.0407x − 1.6321; r² = 0.88	t = 1.10; p > 0.20	(0)
	5AB vs. CW	121	y = 0.865x − 0.9985; r² = 0.66	t = 109.67; p < 0.001	(−)
Adult Males	CL vs. CW	179	y = 1.0482x − 0.0225; r² = 0.99	t = 7.47; p < 0.001	(+)
	MaCL vs. CW	162	y = 1.2889x − 1.2718; r² = 0.90	t = 8.54; p < 0.001	(+)
	MaCH vs. CW	145	y = 1.4655x − 2.3886; r² = 0.97	t = 21.57; p < 0.001	(+)
	MiCL vs. CW	163	y = 1.0747x − 0.8824; r² = 0.91	t = 2.86; p < 0.005	(+)
	MiCH vs. CW	159	y = 1.0484x − 1.6369; r² = 0.95	t = 2.63; p < 0.01	(+)
	5AB vs. CW	69	y = 1.0335x − 1.2606; r² = 0.90	t = 0.79; 0.20 < p < 0.50	(0)
Young Females	CL vs. CW	199	y = 0.9403x + 0.0537; r² = 0.99	t = 7.42; p < 0.001	(−)
	MaCL vs. CW	139	y = 1.0068x − 0.6821; r² = 0.89	t = 0.22; p > 0.50	(0)
	MaCH vs. CW	129	y = 1.1103x − 1.6292; r² = 0.95	t = 4.85; p < 0.001	(+)
	MiCL vs. CW	127	y = 1.0051x − 0.7822; r² = 0.82	t = 0.12; p > 0.50	(0)
	MiCH vs. CW	118	y = 0.9941x − 1.5538; r² = 0.90	t − 0.19; p > 0.50	(0)
	5AB vs. CW	117	y = 1.3485x − 1.9851; r² = 0.89	t = 7.91; p < 0.001	(+)
Adult Females	CL vs. CW	172	y = 0.9722x − 0.0261; r² = 0.99	t = 5.82; p < 0.001	(−)
	MaCL vs. CW	116	y = 1.2184x − 1.1534; r² = 0.94	t = 7.35; p < 0.001	(+)
	MaCH vs. CW	109	y = 1.2704x − 1.9811; r² = 0.97	t = 11.66; p < 0.001	(+)
	MiCL vs. CW	124	y = 1.0497x − 0.8521; r² = 0.94	t = 2.03; p < 0.05	(+)
	MiCH vs. CW	117	y = 1.0121x − 1.54; r² = 0.9764	t = 0.83; p < 0.50	(0)
	5AB vs. CW	57	y = 1.8019x − 3.1106; r² = 0.94	t = 12.76; p < 0.001	(+)

[1] CW, carapace width; CL, carapace length; 5AB, width of the fifth abdominal segment; MaCL, major cheliped propodus lenght; MaCH, major cheliped height; MiCL, minor cheliped propodus length; MiCH, minor cheliped height.
[2] State of allometry: (−) negatively allometric; (+) positively allometric; (0) isometric.

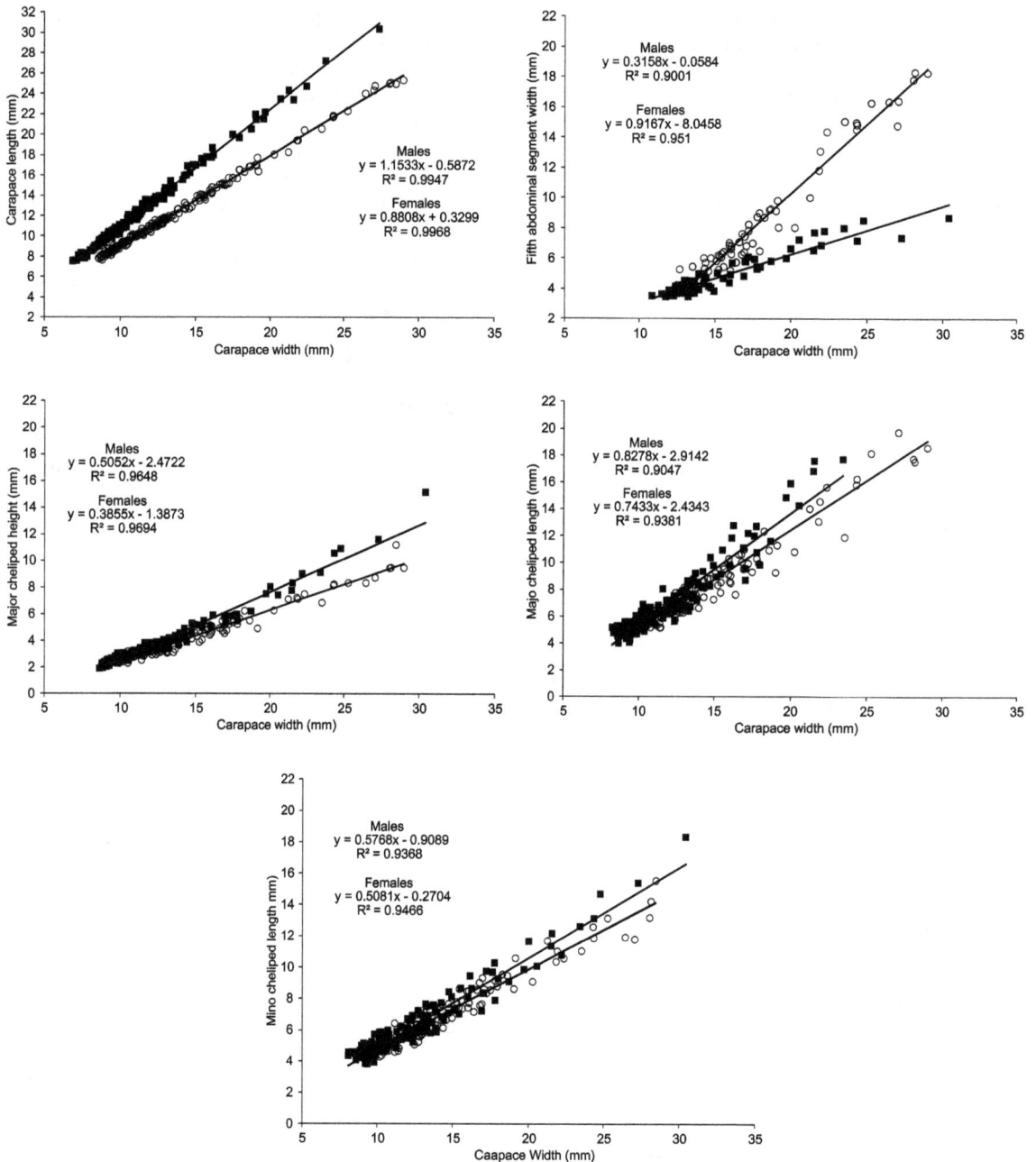

Figure 2. *Trichodactylus fluviatilis*. Sexual dimorphism in biometric relationships (non-linearized data). Solid quadrat, adult males; empty circle, adult female.

Laterality and Heterochely

Laterality analysis showed that 89% of males ($\chi^2 = 110.45$, $p < 0.001$) and 81% of females ($\chi^2 = 60.31$, $p < 0.001$) showed right-handedness. Males and females of *T. fluviatilis* are heterochelous, since comparisons of all chela dimensions gave significant results (Wilcoxon test, $p < 0.001$).

Morphometric maturity

The mean sizes of carapace width at the onset of morphometric maturity estimated for each variable are shown in Table IV. For males and females, the maturation size (= puberty molt) could not be estimated from the minor cheliped height and the minor cheliped length relationships, respectively, be-

Table II. *Trichodactylus fluviatilis*. Student's t comparisons of linear regression parameters between juvenile and adult life stages. v, degrees of freedom.

Life stages	Relationship[1]	Comparison of slopes	Comparison of elevations
Young males vs. Adult males	CL vs. CW	t = 5.24; v = 378; p < 0.001	
	MaCL vs. CW	t = 5.56; v = 277; p < 0.001	
	MaCH vs. CW	t = 8.80; v = 258; p < 0.001	
	MiCL vs. CW	t = 1.88; v = 279; p > 0.05	t = 2.79; v = 280; p < 0.01
	MiCH vs. CW	t = 0.20; v = 269; p > 0.50	t = 0.72; v = 270; p > 0.20
	5AB vs. CW	t = 2.29; v = 186; p < 0.05	
Young females vs. Adult females	CL vs. CW	t = 3.42; v = 367; p < 0.001	
	MaCL vs. CW	t = 4.92; v = 251; p < 0.001	
	MaCH vs. CW	t = 4.93; v = 234; p < 0.001	
	MiCL vs. CW	t = 0.94; v = 247; p > 0.20	t = 1.23; v = 248; p > 0.10
	MiCH vs. CW	t = 0.55; v = 231; p > 0.20	t = 3.53; v = 232; p < 0.001
	5AB vs. CW	t = 5.90; v = 170; p < 0.001	

[1]Legend as in Table I.

Table III. *Trichodactylus fluviatilis*. Student's t comparisons of linear regression parameters between males and females, to test for sexual dimorphism. v, degrees of freedom.

Relationship[1]	n (males)	n (females)	Comparison of slopes	Comparison of elevations	Sexual dimorphism
CL vs. CW	179	172	t = 9.53; v = 347; p < 0.001		Yes
MaCL vs. CW	162	116	t = 1.56; v = 274; p > 0.10	t = 5.05; v = 275; p < 0.001	Yes
MaCH vs. CW	145	109	t = 6.18; v = 250; p < 0.001		Yes
MiCL vs. CW	163	124	t = 0.69; v = 283; p > 0.20	t = 3.30; v = 284; p < 0.02	Yes
MiCH vs. CW	159	117	t = 1.51; v = 272; p > 0.10	t = 0.71; v = 273; p > 0.20	No
5AB vs. CW	69	57	t = 10.41; v = 122; p < 0.001		Yes

[1]Legend as in Table I.

cause no significant differences in regression slope and elevation were detected for these data when comparing juveniles and adults (Table II).

The lowest value of maturity size for both sexes (males = 8.35 mm CW; females = 9.03 mm CW) was obtained from the data set for carapace length. While the largest sizes at the onset of maturity (males = 12.63 mm CW; females = 14.22 mm CW) were obtained from the data set for the fifth abdominal segment (Table IV).

All claw dimensions for which the puberty molt was detected, provided very close estimates of maturation sizes in both sexes. For males, these estimates ranged from 8.99 to 9.12 mm CW, whereas for females they ranged from 9.69 to 10.15 mm CW (Table IV). As proposed by Bueno & Shimizu (2009), we suggest the adoption of the mean of the estimated values obtained from the claw dimensions vs. CW relationships, which would represent the size at the onset of morphometric maturity based on these data sets. Thereby, the puberty molt (obtained with the claw dimensions) would occur when specimens attain CW = 9.05 mm (males) and CW = 9.97 mm (females).

Table IV. *Trichodactylus fluviatilis*. Average carapace width size (mm) at onset of maturity, regarding the biometric relationships for which significant differences were found between juveniles and adults.

Relationship[1]	Male	Female
MaCL vs. CW	8.99	10.15
MaCH vs. CW	9.12	10.08
MiCL vs. CW	9.06	–
MiCH vs. CW	–	9.69
5AB vs. CW	12.63	14.22
CL vs. CW	8.35	9.03
Mean of the estimated values obtained with the claws dimensions	9.05 (± 0.06)	9.97 (± 0.25)

[1]Legend as inTtable I.

DISCUSSION

Alarcon *et al*. (2002) found a population of *T. fluviatilis* in Ubatuba, São Paulo in which the mean carapace width was greater in females than males. On the other hand, Zimmermann *et al.* (2009) found that males of *Trichodactylus panoplus* von Martens, 1869 were larger than females, whereas VENÂNCIO & LEME (2010) and LIMA *et al.* (2012) found similar sizes for males and females in populations of *Trichodactylus petropolitanus* (Goeldi, 1886) from Caçapava, São Paulo, and *T. fluviatilis* from Ubatuba, respectively.

Within the Brachyura, males are usually larger than females. This pattern is frequently related to the investment in reproduction over growth by females (HARTNOLL 1982, 1985, TADDEI & HERRERA 2010). Therefore, most crab species are sexually dimorphic in size, with males larger and with larger structures than females (NG *et al.* 2008).

Males of *T. fluviatilis* crabs from Velha Eugênia brook (present study) grew proportionally more in carapace length than in carapace width (positive allometry), while females showed an opposite growth pattern (negative allometry) (Fig. 2, Table I). With this growth pattern, females from the Velha Eugênia Brook attain a larger CW than males because males grow proportionally more in carapace length. Figure 2 shows that males started investing more energy than females in growing the chelipeds (over CW) when they attained approximately 20 mm CW. This growth pattern could also affect the mean value of CW calculated for males, since they would direct more energy to chelae growth than to carapace width.

Chelipeds are often used by males in agonistic behaviors and/or courtship and to protect females during copulation and spawning (CRANE 1975, GHERARDI & MICHELI 1989, BRANCO 1993, MASUNARI & DISSENHA 2005). Therefore, males with more robust chelipeds are better equipped to win the inter-male competition, and are more likely to be selected by their female counterparts (MARIAPPAN *et al.* 2000, QURESHI & SAHER 2011). On the other hand, brachyuran females with direct development, for instance *T. fluviatilis*, invest a significant amount of energy in abdominal growth, since this body part is used as an incubating chamber for eggs and newly hatched young (see HARTNOLL 1974, VOGT 2013) (Fig. 2). In males, there is no need for the abdomen to grow faster than the carapace, since it only has the reproductive function of protecting the gonopods (DANIELS 2001, CASTIGLIONI & NEGREIROS-FRANZOSO 2004).

In the present study, the biometric relationships in juveniles of both sexes were mostly isometric, whereas in adults they were mostly positively allometric (Table I). Similar results were found by LIMA *et al.* (2012) for the same species studied at Ubatuba, São Paulo. According to HARTNOLL (1982), brachyurans are usually strongly allometric in adulthood. Among all variables investigated, the length and height of the major cheliped in adult males and the fifth abdominal segment in adult females showed the highest coefficient of allometric growth (Table I). These results are consistent with the relative growth pattern observed in brachyurans (HARTNOLL 1974), and reveal the differences in allocation of resources (energy) between males and females. Females divert energy to abdominal growth, while males prioritize cheliped growth, as discussed above.

According to SCALICI & GHERARDI (2008), in brachyurans, right-handedness is usually the most common condition. This condition was also recorded for males and females of *T. fluviatilis* from Brazilian populations (LIMA *et al.* 2012, present study). Similar patterns have been recorded for freshwater crabs of the families Potamidae and Potamonautidae (GHERARDI & MICHELI 1989, DANIELS 2001) and some marine taxa (see table II in MARIAPPAN *et al.* 2000 for review).

In many decapod crustaceans, the right and left chelae are differentiated morphologically into crusher and cutter, making them heterochelous and sexually dimorphic (MARIAPPAN *et al.* 2000). Generally, the major cheliped is associated with defense, reproduction and crushing of shells, whereas the minor cheliped is used for feeding and cleaning parts of the body (LEVINTON *et al.* 1995, DANIELS 2001).

Like most freshwater crabs, trichodactylid species are omnivorous (MAGALHÃES 2003, YEO *et al.* 2008, CUMBERLIDGE *et al.* 2009, ZIMMERMANN *et al.* 2009). According to DANIELS (2001), in freshwater crabs such as *T. fluvialitis*, the major claw may play a less significant role during feeding, since they are either detritivores, or feeding generalists. Therefore, sexual selection is probably the reason why males of freshwater crabs develop a major cheliped that can be used in courtship, agonistic behaviors, and defense of home territory (STEIN 1976, MARIAPPAN *et al.* 2000, DANIELS 2001). On the other hand, the major cheliped of females may act as a signal of sexual vitality and reproductive vigor to males, allowing females to be more successful at mating and have a better chance of defending the developing young (since they show direct development) against predators (DANIELS 2001).

In crustaceans, some aspects such as gonadal development, presence of ovigerous females, and morphometric data are used to calculate the mean size at the onset of maturity. In the present study, the mean size at the onset of morphometric maturity calculated from the fifth abdominal segment data set (male: 12.63 mm; female: 14.22 mm) were quite different from those calculated from the cheliped data set (male: 9.05 mm; female: 9.97 mm). Considering how important cheliped size is for the reproductive success of males, and abdomen width for females, as discussed above, we recommend the use of chelipeds and abdominal width relationships to estimate the size at morphometric maturity in males and females, respectively. In the case of *T. fluviatilis*, females were larger than males when they attained morphometric maturity (Table IV).

The size at first maturity was also greater in females than in males, in studies conducted by COBO (2006), COBO & ALVES (2009), and HARTNOLL (2009). Delayed maturity could allow females of *T. fluviatilis* to live longer, grow larger, and therefore have higher fecundity (RAMIREZ-LLODRA 2002). Furthermore, the size at onset of maturity depends on exogenous factors, such as temperature and/or food availability. Therefore, it is not a fixed character and may vary in different populations of the same species (HARTNOLL 1978).

Considering that specimens of *T. fluviatilis* are commonly collected by the residents of Pedra Branca Village for medicinal purposes and food (Costa Neto 2007), the results presented here are important to establish a minimum catch size that would contribute to the sustainable use and conservation of this species in the state of Bahia.

ACKNOWLEDGMENTS

The authors express their sincere gratitude to CAPES (Coordenação de Aperfeiçoamento de Pessoal de Nível Superior) for providing financial support (PROAP) and a research grant (740766) to one of us (TRS), and to the State University of Feira de Santana for providing all laboratory facilities. Finally, thank you to Janet W. Reid for her help in the English revision.

REFERENCES

Alarcon, D.T.; M.H. Leme & V.J. Cobo. 2002. Population structure of the freshwater crab *Trichodactylus fluviatilis* Latreille, 1828 (Decapoda, Trichodactylidae) in Ubatuba, Northern Coast of Sao Paulo, p. 179-182. *In*: F. Escobar-Briones (Orgs.). **Modern approaches to the study of Crustacea.** New York, Kluwer Academic, Plenum Publishers, 376p.

Amato, J.F.R.; S.B. Amato & S.A. Seixas. 2005. *Temnocephala lutzi* ectosymbiont on two species of *Trichodactylus* from southern Brazil. **Revista Brasileira de Zoologia 22** (4): 1085-1094. doi: 10.1590/S0101-81752005000400038.

Amato, J.F.R.; S.B. Amato & S.A. Seixas. 2006. A new species of *Temnocephala* ectosymbiont on *Trichodactylus fluviatilis* Latreille (Crustacea, Decapoda, Trichodactylidae) from southern Brazil. **Revista Brasileira de Zoologia 23** (3): 796-806. doi: 10.1590/S0101-81752007000400022.

Branco, J.O. 1993. Aspectos bioecológicos do caranguejo *Ucides cordatus* (Linnaeus 1763) (Crustacea, Decapoda) do manguezal do Itacorubi, Santa Satarina, BR. **Arquivos de Biologia e Tecnologia 36** (1): 133-148.

Bueno, S.L.S & R.M. Shimizu. 2009. Allometric growth, sexual maturity and adult male chelae dimorphism in *Aegla franca* (Decapoda: Anomura: Aeglidae). **Journal of Crustacean Biology 29** (3): 317-328. doi: 10.1651/07-2973.1.

Castiglioni, D.S.E & M.L. Negreiros-Fransozo. 2004. Comparative analysis of the relative growth of *Uca rupax* (Smith) (Crustacea, Ocypodidae) from two mangroves in São Paulo, Brazil. **Revista Brasileira de Zoologia 21** (1): 137-144. doi: 10.1590/S0101-81752008000400004.

Chagas, G.C.; A.L. Brossi-Garcia; A.A. Menegário; A.C.S. Pião & J.S.Govone. 2009. Use of the freshwater crab *Trichodactylus fluviatilis* to biomonitoring Al and Mn contamination in river water. **Holos Environment 9** (2): 289-300.

Cobo, V.J. 2006. Population biology of the spider crab, *Mithraculus forceps* (A. Milne-Edwards, 1875) (Majidae, Mithraci-nae) on the southeastern Brazilian coast. **Crustaceana 78** (9): 1079-1087.

Cobo, V.J & D.R.F. Alves. 2009. Relative growth and sexual maturity of the spider crab, *Mithrax tortugae* Rathbun, 1920 (Brachyura, Mithracidae) on a continental island off the southeastern Brazilian coast. **Crustaceana 82** (10): 1265-1273. doi: 10.1163/00112 1609X12481627024490.

Costa-Neto, E.M. 2007. Caranguejo de água-doce, *Trichodactylus fluviatilis* (Latreille, 1828) (Crustacea, Decapoda, Trichodactylidae), na concepção dos moradores do povoado de Pedra Branca, Bahia, Brazil. **Biotemas 20** (1): 59-68.

Crane, J. 1975. **Fiddler crabs of the world.** Princeton, Princeton University Press, 725p.

Cumberlidge, N.; P.K.L. Ng; D.C.J. Yeo; C. Magalhães; M.R. Campos; F. Alvarez; T. Naruse; S.R. Daniels; L.J. Esser; F.Y.K. Attipoe; F.L. Clotilde-Ba; W. Darwall; A. McIvor; J.E.M. Baillie; B. Collen & M. Ram. 2009. Freshwater crabs and the biodiversity crisis: Importance, threats, status, and conservation challenges. **Biological Conservation 142**: 1665-1673. doi:10.1016/j.biocon.2009.02.038

Daniels, S.R. 2001. Allometric growth, handedness, and morphological variation in *Potamonautes warreni* (Calman, 1918) (Decapoda, Brachyura, Potamonautidae) with a redescription of the species. **Crustaceana 74** (3): 237-253. doi: 10.1163/156854001505488.

Franchi, M.; A.A. Menegário; A.L. Brossi-Garcia; G.C. Chagas; M.V. Silva; A.C.S. Pião & J.S. Govone. 2011. Bioconcentration of Cd and Pb by the river crab *Trichodactylus fluviatilis* (Crustacea: Decapoda). **Journal of the Brazilian Chemical Society 22** (2): 230-238. doi: 10.1590/S0103-50532011000200007.

Gherardi, F. & F. Micheli. 1989. Relative growth and population structure of the freshwater crabs, *Potamon potamios palestinensis*, in the Dead Sea area. **Israel Journal of Zoology 36**: 133-145.

Gomides, S.C.; I.A. Novelli; A. O. Santos; S.S.S. Brugiolo & B.M. Sousa. 2009. Novo registro altitudinal de *Trichodactylus fluviatilis* (Latreille, 1828) (Decapoda, Trichodactylidae) no Brazil. **Acta Scientiarum. Biological Sciences 31** (3): 327-330. doi: 10.4025/actascibiolsci.v31i3.785.

Hammer, O.; D.A.T. Harper & P.D. Ryan. 2001. Past: Palaeontological Statistics Software Package for Education and Data Analysis. **Palaeontological Electronica 4** (1): 9p. Available online at: http://palaeo-electronica.org/2001_1/past/past.pdf [Accessed: 03/IV/2012].

Hartnoll, R.G. 1974. Variation in growth pattern between some secondary sexual characters in crabs (Decapoda, Brachyura). **Crustaceana 27** (2): 151-156.

Hartnoll, R.G. 1978. The determination of relative growth in Crustacea. **Crustaceana 34**: 281-293.

Hartnoll, R.G. 1982. Growth. *In*: D.E. Bliss (Ed.), p. 111-196. **The Biology of Crustacea: Embryology, Morphology and Genetics.** New York, Academic Press.

Hartnoll, R.G. 1985. Growth, sexual maturity and reproductive output, p.101-128. *In*: A.M. Wenner (Ed.). **Crustacean Issues 3: Factors in adult growth.** Boston, Balkema.

HARTNOLL, R.G. 2009. Sexual maturity and reproductive strategy of the rock crab *Grapsus adscensionis* (Osbeck, 1765) (Brachyuara, Grapsidae) on Ascension Island. **Crustaceana 82** (3): 275-291 doi: 10.1163/156854009X409090.

HYAMS, D. 2001. CurveExpert 1.3: a comprehensive curve fitting package for Windows. Available online at: www.ebicom.net/ ,DHYAMS/CFTP.HTML [Accessed: 03/IV/2012].

LEVINTON, J.S.; M.L. JUDGE & J.P. KURDZIEL. 1995. Functional differences between the major and minor claws of fiddler crabs (*Uca*, family Ocypodidae, Order Decapoda, Subphylum Crustacea): A result of selection or developmental constraint? **Journal of Experimental Marine Biology and Ecology 193:** 147-160. doi: 10.1016/0022-0981(95)00115-8.

LIMA, D.J.M.; V.J. COBO; D.F.R. ALVES; S.P. BARROS-ALVES & V. FRANSOZO. 2012. Onset of sexual maturity and relative growth of the freshwater crab *Trichodactylus fluviatilis* (Trichodactyloidea) in southeastern Brazil. **Invertebrate Reproduction & Development 57** (2): 105-112. doi: 10.1080/07924259.2012.689263.

MAGALHÃES, C. 2003. Brachyura: Famílias Pseudothelphusidae e Trichodactylidae, p. 143-197. *In*: G.A.S. MELO (Ed.). **Manual de identificação dos crustáceos decápodos de água doce Brasileiros.** São Paulo, Loyola.

MANSUR, C.B.; N.J. HEBLING & J.A. SOUZA. 2005. Crescimento relativo de *Dilocarcinus pagei* Stimpson, 1861 e *Sylviocarcinus austalis* Magalhães e Türkay (Crustacea,Decapoda, Trichodactylidae) no Pantanal do Rio Paraguai, Porto Murtinho, Mato Grosso do Sul. **Boletim do Instituto de Pesca 31** (2): 103-107.

MARIAPPAN, P.C.; C. BALASUNDARAM & B. SCHMITZ. 2000. Decapod crustacean chelipeds: an overview. **Journal of Bioscience 25** (3): 301-313.

MARTÍNEZ-MAYÉN, M.; R. ROMÁN-CONTRERAS; A. ROCHA-RAMÍREZ & S. CHAZARO-OLVERA. 2000. Relative growth of *Atya margaritacea* A. Milne-Edwards, 1864 (Decapoda, Atyidae) from the southern Pacific coast of Mexico. **Crustaceana 73** (5): 525-534.

MASUNARI, S. & N. DISSENHA. 2005. Alometria no crescimento de *Uca mordax* (Smith) (Crustacea, Decapoda, Ocypodidae) na Baía de Guaratuba, Paraná, Brazil. **Revista Brasileira de Zoologia 22** (4): 984-990. doi: 10.1590/S0101-81752005000400026.

MOSSOLIN, E.C. & S.L.S. BUENO. 2003. Relative growth of the second pereiopod in *Macrobrachium olfersi* (Wiegmann, 1836) (Decapoda, Palaemonidae). **Crustaceana 76** (3): 363-376.

MOSSOLIN, E.C. & F.L. MANTELATTO. 2008. Taxonomic and distributional results of a freshwater crab fauna survey (Family Trichodactylidae) on São Sebastião Island (Ilhabela), South Atlantic, Brazil. **Acta Limnologica Brasiliensia 20** (2): 125-129.

MULLER, F. 1892. *Trichodactylus*, siri de água doce sem metamorfose. **Arquivo do Museu Nacional do Rio de Janeiro 8:** 125-133.

NG, P.K.L.; D. GUINOT & P.J.F. DAVIE. 2008. Systema Brachyurorum: Part 1. An annotated checklist of extant Brachyuran crabs of the world. **Raffles Bulletin of Zoology 17:** 1-286.

QURESHI, N.A. & N.U. SAHER. 2011. Relative growth and morphological sexual maturity of *Macrophthalmus (venitus) dentipes* Lucas, in Guérin-Méneville, 1836 from two mangrove areas of Karachi Coast. **Biharean Biologist 5** (1): 56-62.

RAMIREZ-LLODRA, E. 2002. Fecundity and life-history strategies in marine invertebrates. **Advances in Marine Biology 43:** 87-170.

ROCHA, S.S. & S.L.S. BUENO. 2004. Crustáceos decápodes de água doce com ocorrência no Vale do Ribeira de Iguape e rios costeiros adjacentes, São Paulo, Brazil. **Revista Brasileira de Zoologia 21** (4): 1001-1010. doi: 10.1590/S0101-817520040004 00038.

ROCHA, S.S. & S.L.S. BUENO. 2011. Extension of the known distribution of *Aegla strinatii* Türkay, 1972 and a checklist of decapod crustaceans (Aeglidae, Palaemonidae and Trichodactylidae) from the Jacupiranga State Park, South of São Paulo State, Brazil. **Nauplius 19** (2): 163-167.

SCALICI, M. & F. GHERARDI. 2008. Heterochely and handedness in the river crab, *Potamon potamios* (Oliver, 1804) (Decapoda, Brachyura). **Crustaceana 81:** 507-511. doi:10.1163/ 156854008783797525.

STEIN, R.A. 1976. Sexual dimorphism in crayfish chelae: functional significance linked to reproductive activities. **Canadian Journal of Zoology 54:** 220-227.

TADDEI, F.G. & D.R. HERRERA. 2010. Crescimento do caranguejo *Dilocarcinus pagei* na represa Barra Mansa, Mendonça, SP. **Boletim do Instituto de Pesca 35** (2): 99-110.

VENÂNCIO, F.A. & M.H.A. LEME. 2010. The freshwater crab *Trichodactylus petropolitanus* associated with roots of *Hedychium coronarium*. **Pan-American Journal of Aquatic Sciences 5** (4): 501-507.

VOGT, G. 2013. Abbreviation of larval development and extension of brood care as key features of the evolution of freshwater Decapoda. **Biological Reviews of the Cambridge Philosophical Society 88** (1): 81-116.

YEO, D.C.J.; P.K.L. NG; N. CUMBERLIDGE; C. MAGALHÃES; S.R. DANIELS & M.R. CAMPOS. 2008. Global diversity of crabs (Crustacea: Decapoda: Brachyura). In: freshwater. **Hydrobiologia 595:** 275-286. doi: 10.1007/978-1-4020-8259-7_30

ZAR, J.H. 1996. **Biostatistical Analysis.** New Jersey, Prentice Hall, 662p.

ZIMMERMANN, B.L; A.W. AUED; S. MACHADO; D. MANFIO; L.P. SCARTON & S. SANTOS. 2009. Behavioral repertory of *Trichodactylus panoplus* (Crustacea: Trichodactylidae) under laboratory conditions. **Zoologia 26** (1): 5-11. doi: 10.1590/S1984-46702009000100002.

Seasonal habitat selection of the red deer (*Cervus elaphus alxaicus*) in the Helan Mountains, China

Mingming Zhang[1], Zhensheng Liu[1,2] & Liwei Teng[1,2,3]

[1] *College of Wildlife Resources, Northeast Forestry University, No.26 Hexing Road, Xiangfang District, Harbin 150040, P.R. China.*
[2] *Key Laboratory of Conservation Biology, State Forestry Administration, No.26 Hexing Road, Xiangfang District, Harbin 150040, P.R. China.*
[3] *Corresponding author. E-mail: tenglw@gmail.com*

ABSTRACT. We studied the seasonal habitat selection of the red deer, *Cervus elaphus alxaicus* Bobrinskii & Flerov, 1935, in the Helan Mountains, China, from December 2007 to December 2008. Habitat selection varied widely by season. Seasonal movements between high and low elevations were attributed to changes in forage availability, alpine topography, the arid climate of the Helan Mountains, and potential competition with blue sheep, *Pseudois nayaur* (Hodgson, 1833). The use of vegetation types varied seasonally according to food availability and ambient temperature. Red deer used montane coniferous forest and alpine shrub and meadow zones distributed above 2,000 m and 3,000 m in summer, alpine shrub and meadows above 3,000 m in autumn, being restricted to lower elevation habitats in spring and winter. The winter habitat of *C. elaphus alxaicus* was dominated by *Ulmus glaucescens* Franch. and *Juglans regia* Linnaeus, deciduous trees, and differed from the habitats selected by other subspecies of red deer. *Cervus elaphus alxaicus* preferred habitats with abundant vegetation coverage to open habitats in winter, but the reverse pattern was observed in summer and autumn. Red deer preferred gentle slopes (<10°) but the use of slope gradient categories varied seasonally. Red deer avoidance of human disturbance in the Helan Mountains varied significantly by season. Information on red deer habitat selection can help understand the factors affecting seasonal movements and also support decision making in the management and conservation of red deer and their habitats.

KEY WORDS. Habitat use; migration; seasonal movement; ungulates.

The red deer, *Cervus elaphus alxaicus* Bobrinskii & Flerov, 1935 is an endemic subspecies in China, distributed in the central region of the Helan Mountains, which range from the border between Alxa League of the Inner Mongolia Autonomous Region to the Ningxia Hui Autonomous Region in west-central China. The Helan Mountains support the only known population of this subspecies (Wang *et al.* 1999). Its population is small and is isolated from other subspecies of red deer (Zhang *et al.* 1999). Historically, the survival of *C. e. alxaicus* has been threatened by the deterioration of habitats on the Alxa Plateau, and its reduced population has since been isolated in the relatively poor ecological environment of the Helan Mountains for a long period. Population estimates for *C. e. alxaicus* in the Helan Mountains ranged from 850 to 1,060 in 1983, increasing to 1,705 ± 523 in 2005, after the creation of the Ningxia Helan Mountain Nature Reserve in 1982 (Zhang *et al.* 2006). While *C. e. alxaicus* is a threatened subspecies, *C. elaphus* is categorized as a species of Least Concern in the World Conservation Union's Red List of Threatened Animals (Lovari *et al.* 2008).

Habitat selection by animals is considered as an optimization process that involves factors such as food supply, conspecific population density, body size, competitors, predators, and landforms (Morrison *et al.* 1998). Information on which resources are preferred or avoided by organisms improves our understanding of how they meet their requirements for survival and reproduction (Manly *et al.* 2002). The distribution and availability of trophic resources are important factors that affect habitat selection. In most temperate habitats, food is scarce during the winter months and abundant in spring and early summer (Moen 1976, Schmitz 1991). This forces animals in temperate regions to adapt to seasonal changes in food supply. Ungulates that inhabit temperate and boreal regions often exhibit cyclical seasonal movements between summer and winter ranges in response to environment factors (e.g., snow conditions, food availability), social constraints and predation risk (Fryxell & Sinclair 1988, Pépin *et al.* 2008). Regular, round-trip movements between seasonal home ranges (White & Garrott 1990) have evolved to enable animals to avoid undesirable conditions at a particular time of the year (VAUGHAN *et al.* 2000). Generally, seasonal movement patterns of ungulates include short-distance movements, dispersal, and migration (GROVENBURG *et al.* 2009). Seasonal migration of cervids involves dispersal to areas of lower elevation, particularly in winter, when the environment is less hospitable at higher elevations (ALBON & LANGVATN 1992, IGOTA *et al.* 2004, PÉPIN *et al.* 2008). Mixed strategies of migration have, however, been found among cervids that inhabit temperate and boreal mountainous regions (IGOTA *et al.* 2004, BRINKMAN *et al.* 2005, GROVENBURG *et al.* 2009).

Despite numerous studies of other subspecies of red deer, little is known about the seasonal habitat use or movement patterns of *C. e. alxaicus* during different seasons. The objective of this study was to: 1) compare habitats used by *C. e. alxaicus* during different seasons to document differences in habitat selection; and 2) examine environmental variables that affect the seasonal movement of *C. e. alxaicus*.

MATERIAL AND METHODS

This study was conducted over four seasons from December 2007 to December 2008 in the Helan Mountain region, which is located between the eastern Yinchuan plain in Ningxia Hui Autonomous Region and the western Alxa Plateau in Inner Mongolia Autonomous Region (105°44'-106°42'E, 38°21'-39°22'N) (Fig. 1). The Helan Mountain region, located in northwestern China, is on the transitional zone between steppe and desert regions of central Asia (TAKHTAJAN 1986). It generally lies at 2,000-3,000 masl, with a maximum elevation of 3,556 masl. It covers an area of 2,740 km² [including Ningxia Helan Mountain National Nature Reserve (2,063 km²) and Inner Mongolia Helan Mountain National Nature Reserve (677 km², Fig. 1)], with a north-south length of about 250 km and an east-west width of about 20-40 km (Z.S. Liu unpublished data).

The region has a typical continental climate, characterized by cool and dry conditions, with annual mean temperature of -0.9°C and mean annual rainfall of 420 mm. The local climate is influenced by the topography of the Helan Mountain, the low temperature center of the northern Ningxia. The maximum monthly mean temperature is 11.9°C, in July, and the minimum is -14.2°C, in January, 8.8 ~9.8°C lower than in Yinchuan, the capital city of Ningxia Hui Autonomous Region. Precipitation varies seasonally, with 62% falling as rain in summer. There is little precipitation in winter, about 10.1% of the annual total. Snow cover is limited in the Helan Mountains (GENG & YANG 1990). The vegetation distribution is strongly influenced by moisture conditions. The elevation differential between the Helan Mountains and the plains to the south and east is about 2100 m, which creates an elavational climatic gradient that results in the formation of four elevational vegetation zones. The mountain steppe zone (MS) occurs at 1,400-1,600 masl and covers an area of 1,241 km² dominated by *Stipa breviflora* Griseb., *Ajania fruticulosa* (Ledeb.) Poljak,

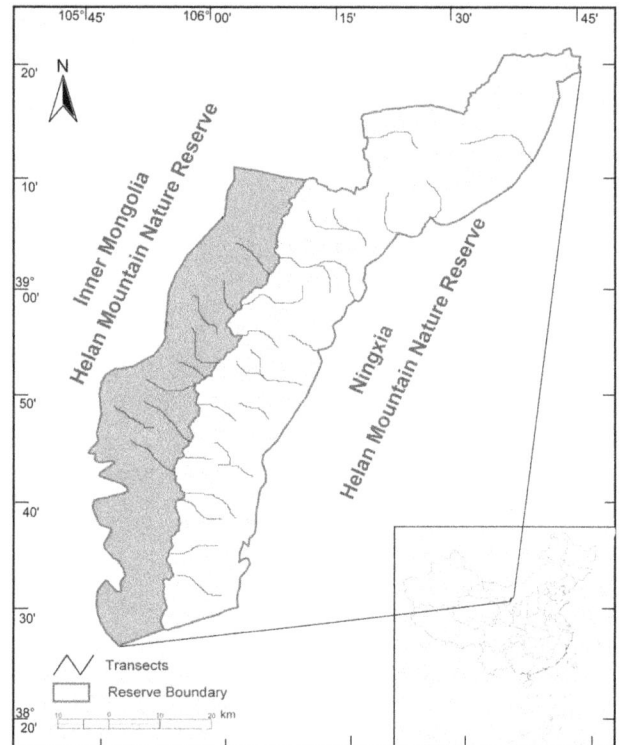

Figure 1. Location and distribution of the study area and transects in the Helan Mountain region, China. Grey shading represents Helan Mountain Nature Reserve in Inner Mongolia Autonomous Region, while the unshaded area represents the reserve in Ningxia Hui Autonomous Region.

Ptilagrostis pelliotii (Danguy) Grubov.), *Oxytropis aciphylla* Ledeb.), *Convolvulus gortschakovii* Schrenk ex Fisch. & C.A. Mey., and *Salsola laricifolia* Turcz. ex Litv. The open mountain forest and steppe zone (MOFS) (1,600-2,000 masl, 1,155 km²) is dominated by *Ulmus glaucescens* Franch., *Prunus mongolica* Maxim., *Stipa grandis* P.A. Smirn., and *S. bungeana* Trin. The mountain coniferous forest zone (MCF) (1,900-3,000 masl, 319 km²) is dominated by *Picea crassifolia* Kom., *Pinus tabulaeformis* Carrière, *Juniperus rigida* Siebold & Zucc., and *Potentilla parvifolia* Fisch. ex Lehm. The alpine shrub and meadow zone (ABM) (3,000-3,556 masl, 23 km²) is dominated by *Salix cupularis* var. *lasilogyne* Rehd., *Caragana jubata* (Pall.) Poir., *Kobresia* spp., *Polygonum viviparum* Linnaeus and *Arenaria* spp. (JIANG *et al.* 2000, DI 1986). Mammals found in the area include insectivores: Daurian hedgehog, *Mesechinus dauuricus* (Sundeyall, 1842); carnivores: red fox, *Vulpes vulpes* (Linnaeus, 1758); artiodactyls: blue sheep, alpine musk deer: *Moschus chrysogaster* (Hodgson, 1839); lagomorphs: Daurian pika, *Ochotona dauurica* (Pallas, 1776); and chiropterans and rodents. Blue sheep and *C. e. alxaicus* are the two dominant ungulate species in the Helan Mountain region (LIU 2009).

From December 2007 to December 2008, we carried out four surveys, one per season, to determine the distribution of *C. e. alxaicus* throughout the Helan Mountain area. Each seasonal survey sampled 32 line transects established along the valleys and each survey took about one month to sample all topographic types. Differences in topographic relief prevented walking along the transects at the same velocity and length, therefore, transects ranged in length from 4.5 to 8.5 km, for a total of 350 km traversing the whole study area and covering all four elevational vegetation zones. The distance between any two transects was at least 2 km, to ensure the independence of each transect.

Because of the rarity of *C. e. alxaicus* in the area and their high sensitivity to anthropogenic disturbance (JEPPESEN 1987), we documented habitat use mainly by recording fresh signs, such as evidence of bedding or the presence of feces. We also observed deer in order to record their activities. It was easy to differentiate the signs left by *C. e. alxaicus* from those of other ungulates (e.g., *P. nayaur*, *M. chrysogaster*) in the Helan Mountains, because *C. e. alxaicus* is larger than *P. nayaur* and *M. chrysogaster*, and the size and shape of their feces are different (CHANG & XIAO 1988). To ensure the accu-

racy of data, we recorded fresh feces only (3-5 days old as estimated by the color and water content). When deer were observed during line transect surveys, we first used telescopes to observe their feeding or bedding behavior, without disturbing the animals. After deer had departed the area, we carried out detailed sampling.

We recorded terrestrial coordinates according to a Global Positioning System (GPS) after a feeding or bedding habitat was identified. We then established a plot with five sample quadrants (Fig. 2) to collect data of 18 topographic and biological variables (Table I) using the methods described by LIU *et al.* (2002). The distance between any two plots was at least 500 m to ensure the independence of each plot (Fig. 2).

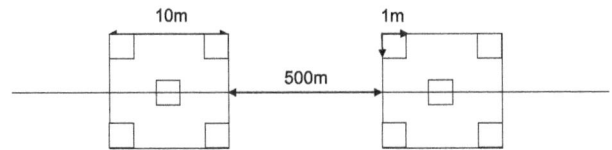

Figure 2. Diagramatic presentation of the survey transects, plots and sample quadrats used in this study.

Table I. Variables collected in feeding and bedding habitat plots used by *Cervus elaphus alxaicus* in the Helan Mountain region, China.

Variables	Categorization and Criterion	Abbreviation
Altitude (m)	The altitude of the plot accorded to GPS	AL
Vegetation types	Mountain steppe zone (MS); Mountain open forest and steppe zone (MOFS); Mountain coniferous forest zone (MCF); Alpine bush and meadow zone (ABM)	VT
Topography	Categorized by the slope and fault of a hillside, divided into 5 levels: Smooth undulating slope; Moderately broken slope; Distinctly broken slope; Scree/landslide; Cliff	TO
Dominant tree	The tree covers 70% of the density in the 10×10 m plot. It usually was *Ulmus glaucescens*, *Ziziphus jujube*, *Salix* spp., *Juniperus rigida*, *Pinus tabulaeformis*, *Picea crassifolia* Mixture or Open land with no tree	DT
Tree density (trees/100 m²)	The total number of trees in the 10×10 m plot	TD
Tree height (m)	The mean height of trees in the 10×10 m plot	TH
Distance to the nearest tree (m)	Distance from the center of the 10×10 m plot to the nearest tree	DtT
Shrub density (trees/100 m²)	The number of shrubs in the 10×10 m plot	SD
Shrub height (m)	The mean height of shrubs in the 10×10 m plot	SH
Distance to the nearest shrub (m)	Distance from the center of the 10×10 m plot to the nearest shrub	DtS
Herb coverage (%)	The mean herb coverage of the 5 sample quadrats in the 10×10 m plot	HC
Slope gradient (°)	Slope gradient of the hillside where the spot located measured with military compass	SG
Slope location	A visual assessment of the site location relative to the macroslope which is usually from valley bottom to ridge top, classed as: lower slope (includes valley bottom and flat), middle slope and upper slope (includes ridge top)	SL
Slope aspect	Aspect was surveyed to eight compass points, translated as 0°, 45°, 90°, 135°, 180°, 225°, 270° and 315° from North, as 0° is equivalent to 360°. And the slope aspect was grouped into 3 main directions: sunny slope (135°~225°), partial shade slope (45°~135° and 225°~315°) or shady slope (315°~45°)	SA
Distance to water resource (m)	The distance from the spot to the nearest water resource	DtW
Distance to human disturbance (m)	The distance from the spot to the nearest place of human activity such as highway, road and shelter forest station, etc	DtH
Distance to bare rock (m)	The distance from the spot to the nearest bare rock	DtR
Hiding cover (%)	The coverage of the hiding conditions. Percent hiding cover was determined by visually estimating the percent of a deer or a substitute (a 1 m stick) obscured at 30 m in the four cardinal directions (KUNKEL & PLETSCHER 2001)	HiC

Generally, research on resource selection requires a comparison of the habitats used by *C. e. alxaicus* (observed plots) with those that are available (expected plots, assuming no differential selection by deer). To provide comparison plots for the analysis of habitat selection, 617 randomly located plots were surveyed. The random plots were established along the survey transects, in areas with no obvious evidence of *C. e. alxaicus* use. The distance between random plots and occupied plots or any two random plots was at least 500 m to reduce the possibility of overlap between used and unused plots. Comparison plots were surveyed in each vegetation zone according to the proportion of used plots in each zone. Data were recorded for comparison plots using the same methods used in the occupied plots.

Data were analyzed to quantify habitat selection by *C. e. alxaicus* by season. To assess seasonal differences between the 18 factors recorded at used plots and comparison plots, we used a chi-square goodness-of-fit test within classified categories for each variable (MARCUM & LOFTSGAARDEN 1980). P-values less than 0.05 were considered statistically significant. Bonferroni confidence intervals were calculated by the following formula to identify variables that indicate preference or avoidance, following the method developed by NEU *et al.* (1974) and BYERS *et al.* (1984). $(p_i - r_i) \pm Z_{1-\alpha/2k} \times \sqrt{p_i(1-p_i)/n_i + r_i(1-r_i)/m_i}$; where, n_i is the number of comparison plots in category *i*, and p_i is the proportion of the comparison plots that fall in category *i*; m_i is the total number of plots used by *C. e. alxaicus*, r_i is the proportion of plots used by *C. e. alxaicus* in category *i*; $Z_{1-a/2k}$ is the upper standard normal table value corresponding to a probability tail area of 1-*a*/2k; *a* is the level of significance; and *k* is the number of categories tested. The confidence intervals indicated that *C. e. alxaicus* showed avoidance (marked "-") of category *i* when $(p_i - r_i) - Z_{1\alpha/2k} \times \sqrt{p_i(1-p_i)/n_i + r_i(1-r_i)/m_i} > 0$; whereas *C. e. alxaicus* showed preference (marked "+") for category *i* when $(p_i - r_i) + Z_{1\alpha/2k} \times \sqrt{p_i(1-p_i)/n_i + r_i(1-r_i)/m_i} < 0$; and *C. e. alxaicus* showed no obvious selection (marked " = ") for category *i* when $(p_i - r_i) - Z_{1\alpha/2k} \times \sqrt{p_i(1-p_i)/n_i + r_i(1-r_i)/m_i} < 0$ and $(p_i - r_i) + Z_{1\alpha/2k} \times \sqrt{p_i(1-p_i)/n_i + r_i(1-r_i)/m_i} > 0$.

Data for non-numeric ecological factors (VT, TO, DT, SL and SA; see Table I) were examined with chi-square tests. Data for the remaining numeric ecological variables were initially analyzed with one-sample Kolmogorov-Smirnov tests to determine if they were normally distributed. The normally distributed data were analyzed with independent-samples t-tests, while the non-normally distributed data were analyzed with Kruskal-Wallis H tests.

RESULTS

A total of 602 plots used by *C. e. alxaicus* (observed plots) were recorded and compared among the four vegetation zones (Fig. 3). Across the whole study period, 209 used plots in the coniferous forest mountain zone were measured, and this type was the most common vegetation type, followed by open mountain forest and steppe zone of 169 plots, mountain steppe zone of 108 plots and alpine shrub and meadow zone of 106 plots (Fig. 3).

Across the entire study area in spring, we sampled 181 plots used by deer (Fig. 3) and 181 comparison plots (Appendix 1). Deer selected habitats characterized by gentle (<10°), undulating, sunny slopes in spring. Deer preferred habitats with 4-6 m high trees, near dense shrubs (> 10 trees/100 m²) taller than 1.3 m, with high herb coverage of more than 50%, good hiding conditions (hiding cover lower than 50%) (Fig. 4), and distant from bare rock. However, no significant preference was shown during spring with respect to dominant tree species (x^2 = 9.65, df = 10, p > 0.05), distance to the nearest tree (x^2 = 4.02, df = 2, p > 0.05), altitude (x^2 = 0.59, df = 3, p > 0.05), distance to water resource (x^2 = 0.59, df = 2, p > 0.05), or distance to human disturbance (x^2 = 0.00, df = 2, p > 0.05) (Appendix 1).

In summer, 146 plots were used by deer (Fig. 3) and 167 comparison plots (Appendix 1) were surveyed. Deer preferred habitats with gentle, undulating slopes (<20°) on the south side of the MCF zone above 2,000 m and the ABM zone above 3,000 m, respectively, during summer. Habitats used in summer were on lower slopes near dense stands of trees (> 4 tree/100 m²) with mixed tree species of 4-6 m height, near dense shrub stands (> 10 tree/100 m²) taller than 1.3 m, with high herb coverage (> 80%), good hiding condition (hiding coverage lower than 50%) (Fig. 4) and far from bare rock (> 50 m). Distance to human disturbance did not affect habitat use by deer in summer (Appendix 1).

A total of 144 plots used by deer (Fig. 3) and 138 comparison plots (Appendix 1) were surveyed in autumn. Autumn habitat use was similar to that during summer: *C. e. alxaicus* preferred habitats with gentle (<10°), undulating and partially shaded slopes in the ABM zone, at elevations above 3,000 m in autumn. They also preferred habitats with more gentle slopes, high tree density (> 4 trees/100 m²), mixed tree species of 4-6 m height, dense shrubs (> 10 trees/100 m²), high herb coverage (> 80%) good hiding condition (hiding coverage lower than 50%) (Fig. 4), and distant from bare rock (> 100 m). There was no significant difference between the used plots and comparison plots in autumn with respect to distance to the nearest tree (x^2 = 0.36, df = 2, p > 0.05), shrub height (x^2 = 4.00, df = 2, p > 0.05), distance to water resource (x^2 = 2.95, df = 2, p > 0.05), or distance to human disturbance (x^2 = 5.41, df = 2, p > 0.05) (Appendix 1).

131 plots used by *C. e. alxaicus* (Fig. 3) and 131 comparison plots (Appendix 1) were surveyed in winter. Deer preferred winter habitats with gentle (<10°), undulating, sunny slopes, lower slopes with high herb coverage (> 50%), and average hiding condition (hiding coverage lower than 75%) (Fig. 4). However, deer maintained a distance of more than 1.5 m from shrubs

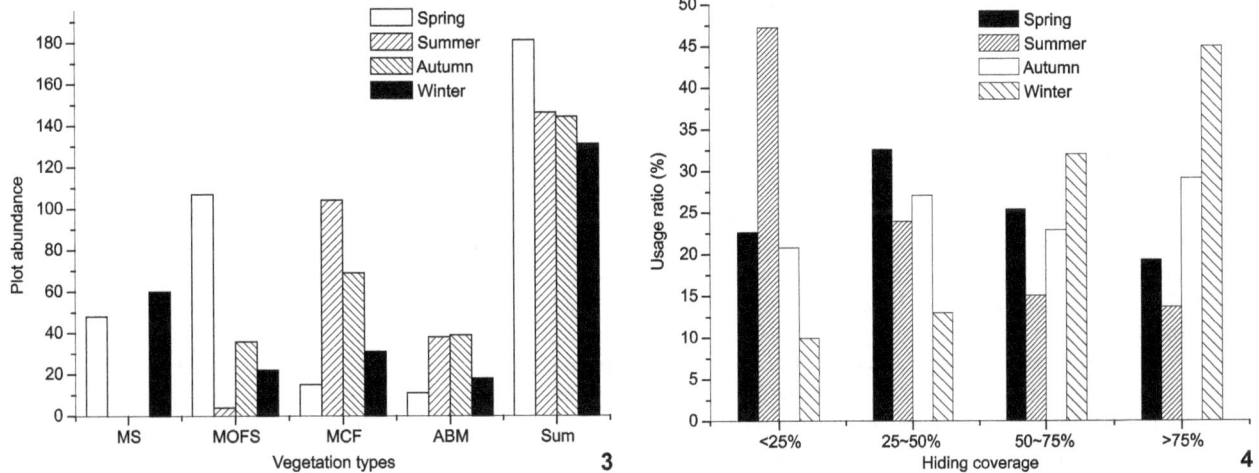

Figures 3-4. (3) Abundance and proportion of plots used by *Cervus elaphus alxaicus* in different vegetation type zones among four seasons, in the Helan Mountain region, China. (MS) Mountain steppe zone, (MOFS) Mountain open forest and steppe zone, (MCF) Mountain coniferous forest zone, (ABM) Alpine bush and meadow zone. (4) Seasonal changes in the usage ratio of hiding coverage on *C. e. alxaicus* in the Helan Mountain region, China.

and 50-100 m from bare rock. Deer showed no significant preference with respect to vegetation type ($\chi^2 = 0.90$, df = 3, p > 0.05), shrub height ($\chi^2 = 2.40$, df = 2, p > 0.05), altitude ($\chi^2 = 1.59$, df = 3, p > 0.05), distance to water source ($\chi^2 = 0.14$, df = 2, p > 0.05), or distance to human disturbance ($\chi^2 = 2.13$, df = 2, p > 0.05) (Appendix 1).

There were statistically significant differences in the use of five non-numeric ecological factors by season. *Cervus elaphus alxaicus* selected habitats by season based on vegetation type ($\chi^2 = 58.611$, df = 3, p < 0.001), topography ($\chi^2 = 969.841$, df = 3, p < 0.001), dominant tree species ($\chi^2 = 820.947$, df = 10, p < 0.001), slope aspect ($\chi^2 = 74.355$, df = 2, p < 0.001) and slope location ($\chi^2 = 143.096$, df = 2, p < 0.001). There were also statistically significant differences in the use of 13 numeric ecological factors by season (Table II).

DISCUSSION

Cervus elaphus alxaicus in the Helan Mountains displayed a pattern of seasonal elevational migration similar to that of other red deer subspecies in mountainous areas (ALBON & LANGVATN 1992, JARNEMO 2008, PÉPIN et al. 2008). However, we observed differences between the habitat selection of *C. e. alxaicus* and that of other subspecies. HUTTO (1985) described the mechanisms determining habitat selection as: geographic restrictions, genetic evolution, influence of experience, and settlement decisions following exploration.

Vegetation type determines the composition and distribution of deer forage, and is determined by soil type, climate, sunlight, topography, landform and many microhabitat factors. The heterogeneous distribution of biotic and abiotic fac-

tors in environments leads to spatial heterogeneity in vegetation types. Physiological and ecological requirements of deer are met to varying degrees by different vegetation types. Thus the geographic and seasonal variation in vegetation types affect red deer habitat selection. In the Helan Mountains, *C. e. alxaicus* range annually from the mountain steppe zone below 1,600 m to alpine shrub and meadow zone above 3,000 m (LIU 2009). Deer preferred habitats in the montane coniferous forest zone (> 2,000 masl) and alpine shrub and meadow zone (> 3,000 masl), in summer, and those in the alpine shrub and meadow zone (> 3,000 m) in autumn. Deer showed no preference for any vegetation type in spring or winter (Appendix 1), using habitats in proportion to their availability during those seasons. IGOTA et al. (2004) reported that deer rarely change their summer home ranges for breeding and nursing of offspring. We predicted that some *C. e. alxaicus* might migrate down from the alpine shrub and meadow zone during winter and spring, while some might stay at high altitude. Further study with GPS collars on *C. e. alxaicus* has been conducted to test the prediction.

CUI et al. (2007) and CHANG et al. (2010) reported that *U. glaucescens, Populus davidiana* Dode, *P. monglica, Potentilla* spp., Graminoids (*Stipa* spp., *Poa* spp.), *Caragana* spp. were important in the winter diet of *C. e. alxaicus*. In summer, *C. e. alxaicus* ate 18 plant species of 11 families, which were mostly *Salix microstachya* var. *bordensis* (Nakai) C.F. Fang, *P. davidiana, U. glaucescens* and *Agropyron cristatum* (L.) Gaertn. which were mainly distributed in the mountainous open forest and steppe zone, ranging from 1,600 m to 2,000 m altitude, and *C. e. alxaicus* of migratory group migrated down from alpine shrub and meadow zone during winter and spring (Table II) as the amount of food resources declined. Food availability is clearly critical for the nutrition of ungulates, especially for fawns and females (Pettorelli et al. 2005). Adaptation to varying feeding conditions throughout seasons or years was also confirmed by Groot & Hazebroek (1995).

Habitat selection by deer is determined by the presence of both food and hiding cover (Borkowski & Ukalska 2008). The role of hiding cover in habitat use may be especially important in winter, when cervids reduce their food intake and survive, to a large extent, using their fat reserves (Putman 1988). Peek et al. (1982) studied the role of cover in habitat selection. Two cover types were recognized: thermal cover, in which a forest overstory protects against weather and sun, and hiding (security) cover used to escape and avoid predators and humans. C. e. alxaicus in the Helan Mountains preferred habitats near cover to open habitats in winter, but the reverse was found in summer and autumn (Fig. 4). Habitats dominated by U. glaucescens and Juglans regia Linnaeus were frequently used in winter (Appendix 1). C. e. alxaicus are likely to be under low predation risk except from poaching in the Helan Mountains. Therefore, selection of habitat by deer mainly reflects their food requirements and need for protection against severe weather. The habitat selection process presumably results from demands to maximize their energy efficiency while minimizing their movements when searching for food, water and cover (Qiao et al. 2006). It differed from other subspecies of red deer, such as Cervus elaphus xanthopygus Milne-Edwards, 1867 in northeast China and Cervus elaphus nelsoni (Bailey, 1935) in the Rocky Mountains, which preferred coniferous forest and sapling trees and shrubs, feeding on shrub and epicormics in winter (Unsworth et al. 1998). C. e. alxaicus preferred montane coniferous forest and alpine shrub and meadow zones in summer and autumn, but preferred mountain steppe and mountain open forest and steppe zones in winter and spring, similar to Cervus elaphus scoticus Lönnberg, 1906 (Welch et al. 1990).

The special alpine topography and arid climate in the Helan Mountain region were the two important factors explaining the habitat selection of C. e. alxaicus. Given that precipitation is low in the Helan Mountain region, especially in winter, and the snow coverage is also very low (Geng & Yang 1990), the snow depth and coverage are not important factors initiating the habitat selection and movement of C. e. alxaicus. The local population of blue sheep is widespread and numbers more than 10,000 (Liu 2009, Liu et al. 2007b). Blue sheep inhabit the mountain open forest and steppe zone, and probably compete with deer for forage, especially in winter and early spring, when food is scarce (Liu et al. 2007a). We recorded migration of C. e. alxaicus during winter and early spring to the mountain open forest and steppe zone from higher altitude areas (Appendix 1). A reasonable assumption is that blue sheep density, to a certain degree, has a negative influence on C. e. alxaicus density, which might also explain deer movements to higher el-

evations in summer. Further study on the niche overlap between blue sheep and C. e. alxaicus is underway. Ager et al. (2003) reported sympatric populations of Rocky Mountain elk (C. e. nelsoni) and mule deer, Odocoileus hemionus (Rafinesque, 1817), in northeastern Oregon, where elk exhibited strong daily and seasonal patterns of movements and habitat use under competition from mule deer.

Topography and slope gradient were important factors affecting habitat selection by C. e. alxaicus. Generally, deer preferred gentle (<10°), undulating slopes (Appendix 1), but this varied seasonally according to food availability and temperature (Fig. 5). Deer selection of lower slopes was also caused by the predominantly steep topography of the Helan range, which includes only a small proportion of the total area as gentle slopes (Di 1986). C. e. alxaicus preferred sunny slopes with direct solar radiation, especially during the cold winter and spring (Fig. 5).

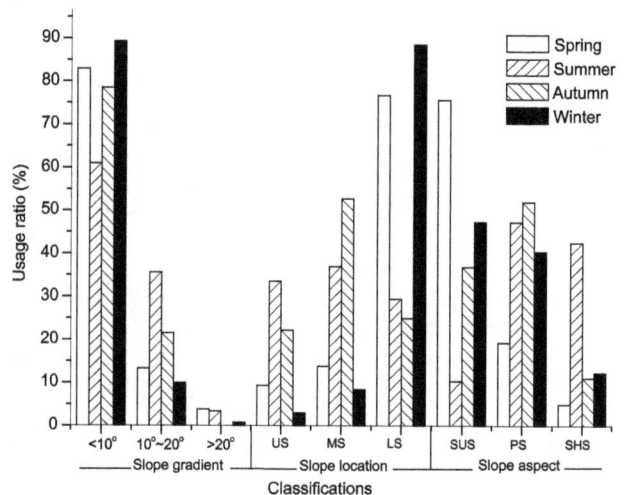

Figure 5. Seasonal changes in the effect of slope characteristics on Cervus elaphus alxaicus. (US) Upper Slope, (MS) Middle Slope, (LS) Lower Slope, (SUS) Sunny Slope, (PS) Partial shade slope, (SHS) Shady slope.

Pressure from predators and human disturbance are often considered to be important factors that influence ungulate behavior and habitat selection (Bonenfant et al. 2004, Borkowski 2004). However, historically, the main predators of C. e. alxaicus, the snow leopard, Uncia uncia (Schreber,1775); the gray wolf, Canis lupus Linnaeus, 1758; and lynx, Lynx lynx (Linnaeus, 1758), became extinct during the 1980s in the Helan mountain region (Wang & Schaller 1996). Numerous studies have demonstrated that deer behavior and habitat use are mostly influenced by humans (Borkowski & Ukalska 2008), including hunting, other human activities (Licoppe 2006), travel routes and road traffic (Yost & Wright 2001). Jiang *et al.* (2007) reported that *C. e. xanthopygus* in northeastern China showed some behavioral plasticity in response to human influences, and the effect of habitat loss or fragmentation caused by human activities are expected to be great (Jaeger *et al.* 2005). However, in the Helan Mountain region, we found that the deer neither preferred nor avoided human disturbance (p > 0.05) (Appendix 1), and they profited from the programs of Forest-Grass Conservation Project and Shelterbelt Forestry Project and Returning Husbandry to Forestry (Grass) launched in 1996 (Zhao *et al.* 2000) and 1999 (Zhao *et al.* 2004), respectively. Wild animal hunting and livestock grazing have been forbidden in most of the Helan Mountain region and habitats are generally well-preserved. Therefore, *C. e. alxaicus* has not suffered much human disturbance, apart from regular ranger, activities whose impact is probably negligible. However, in our data, there was a significant amount of seasonal variation in deer avoidance of human disturbance in the Helan Mountains (Table II). The migration of *C. e. alxaicus* individuals to lower altitudes for foraging in winter and spring has brought deer closer to areas that are impacted by humans. Deer preferred habitats far from human disturbance at higher elevations during summer and autumn. During our survey, human-related *C. e. alxaicus* fatalities were low compared to those of other studies (Licoppe 2006, Pépin *et al.* 2008); therefore, limited persecution and food availability may have facilitated and encouraged *C. e. alxaicus* individuals to use human-impacted areas in winter and spring.

Many studies on ungulate seasonal movement and habitat selection are based on data provided by GPS-collars or radio-collars, which are very precise. However, because of the difficulties involved in capturing *C. e. alxaicus* and the rugged topographic conditions in the Helan Mountains, we conducted our surveys by comparing use versus availability of habitats to quantify the seasonal habitat selection of *C. e. alxaicus*. This study was based on several years of observation and one year of data collection, but the biodiversity and natural conditions in the Helan Mountains have remained stable for years due to its continental climate. Understanding habitat selection by the red deer and its strategies of seasonal movement can be used for decision making on appropriate management and conservation measures (Guisan & Thuiller 2005) for *C. e. alxaicus* and its habitat.

ACKNOWLEDGMENTS

Our project was financially supported by the National Nature Science Foundation of China (#30970371), Program for New Century Excellent Talents in University (#NCET-08-0753), the Fundamental Research Funds for the Central Universities (#DL09CA03, #DL13EA01-01), and the Optional Project of Ningxia Hui Autonomous Region (2011-017). We are grateful for the support of Ningxia and Inner Mongolia the Helan Mountain National Nature Reserve. We would like to thank Wang Zhaoding for his great efforts and expertise in conducting field research. We also thank two anonymous reviewers and Tom Dahmer for valuable review of this manuscript.

REFERENCES

Ager, A.A.; B.K. Johnson; J.W. Kern & J.G. Kie. 2003. Daily and seasonal movements and habitat use by female rocky mountain elk and mule deer. **Journal of Mammalogy 84** (3): 1076-1088.

Albon, S.D. & R. Langvatn. 1992. Plant phenology and the benefits of migration in a temperate ungulate. **Oikos 65**: 502-513.

Table II. Characteristics of 13 ecological factors in the habitats used by *Cervus elaphus alxaicus* during the year in the Helan Mountain region, China. Significant P-values: * $p \leq 0.001$.

Ecological factors	Spring (mean ± SE)	Summer (mean ± SE)	Autumn (mean ± SE)	Winter (mean ± SE)	Kruskal-Wallis H Test χ^2
TD (individuals/100 m²)	2.06 ± 5.065	8.89 ± 10.258	3.13 ± 5.634	2.90 ± 4.642	58.051*
TH (m)	1.91 ± 2.050	3.09 ± 2.561	1.66 ± 2.423	1.90 ± 1.879	3.319*
DtT (m)	13.28 ± 11.243	9.83 ± 11.437	17.01 ± 10.845	12.81 ± 11.129	60.433*
SD (individuals/100 m²)	12.08 ± 10.954	19.20 ± 13.947	12.29 ± 11.127	6.70 ± 10.808	76.191*
SH (m)	0.95 ± 1.012	0.90 ± 0.567	0.64 ± 0.611	0.64 ± 3.499	71.366*
DtS (m)	3.24 ± 3.782	1.51 ± 2.143	3.29 ± 3.795	6.30 ± 4.252	78.617*
HC (%)	67.84 ± 23.194	82.40 ± 23.873	88.56 ± 17.755	66.98 ± 24.664	126.156*
SG (°)	5.55 ± 6.151	10.21 ± 10.977	6.40 ± 3.350	4.61 ± 6.081	75.136*
AL (m)	1,903.62 ± 427.717	2,512.13 ± 375.169	2,512.46 ± 493.990	1,890.61 ± 570.243	203.163*
DtW (m)	1,060.92 ± 1351.187	1,176.24 ± 863.845	1,510.63 ± 1221.745	1,327.50 ± 1431.247	26.920*
DtH (m)	2,933.31 ± 1907.579	5,380.26 ± 3124.869	7,255.49 ± 3343.345	3,134.27 ± 1913.725	158.822*
DtR (m)	120.04 ± 185.693	186.05 ± 168.370	247.47 ± 222.321	75.61 ± 217.001	178.969*
HiC (%)	47.36 ± 27.710	35.80 ± 28.239	52.68 ± 27.888	67.25 ± 24.434	86.234*

BONENFANT, C.; E.L. LEIF; A. MYSTERUD; R. LANGVATN; N.C. STENSETH; J.M. GAILLARD & F. KLEIN. 2004. Multiple causes of sexual segregation in European red deer: enlightenments from varying breeding phenology at high and low latitude. **Proceedings of the Royal Society B Biological Sciences 271:** 883-892.

BORKOWSKI, J. & J. UKALSKA. 2008. Winter habitat use by red and roe deer in pine-dominated forest. **Forest Ecology and Management 255:** 468-475.

BORKOWSKI, J. 2004. Distribution and habitat use by red and roe deer following a large forest fire in South-western Poland. **Forest Ecology and Management 201:** 287-293.

BRINKMAN, T.J.; C.S. DEPERNO; J.A. JENKS; B.S. HAROLDSON & R.G. OSBORN. 2005. Movement of female white-tailed deer: effects of climate and intensive row-crop agriculture. **The Journal of Wildlife Management 69** (3): 1099-1111.

BYERS, C.R.; K. STEINHORST & P.R. KRAUSMAN. 1984. Clarification of a technique for analysis of utilization-availability data. **The Journal of Wildlife Management 48:** 1050-1053.

CHANG, H. & Q.Z. XIAO. 1988. Selection of winter habitat of Red deer in Dailing region. **Acta Theriologica Sinica 8** (2): 81-88.

CHANG, Y.; M.M. ZHANG; Z.S. LIU; T.H. HU & Z.G. LI. 2010. Summer Dies of Sympatric Blue Sheep (*Pseudois nayaur*) and Red Deer (*Cervus elaphus alxaicus*) in the Helan Mountains, China. **Acta Ecologica Sinica 30** (6): 1486-1493.

CUI, D.Y.; Z.S. LIU; X.M. WANG; H. ZHAI; T.H. HU & Z.G. LI. 2007. Winter food-habits of red deer in the Helan Mountains, China. **Zoological Research 28** (4): 383-388.

DI, W.Z. 1986. **Plantae vasculares the Helan Mountain.** Xi'an, Northwestern University Press.

FRYXELL, J.M. & A.R.E. SINCLAIR. 1988. Causes and consequences of migration by large herbivores. **Trends in Ecology & Evolution 3** (9): 237-241.

GENG, K. & Z.R. YANG. 1990. Climatic characteristics and climatic landforms in Helan Mountain. **Yantai Teacher's College Journal (Natural Science Edition) 6** (2): 49-56.

GROOT, B. & E. HAZEBROEK. 1995. Ingestion and diet composition of red deer (*Cervus elaphus L.*) in the Netherlands from 1954 till 1992. **Mammalia 59:** 187-195.

GROVENBURG, T.W.; J.A. JENKS; R.W. KLAVER; C.C. SWANSON; C.N. JACQUES & D. TODRY. 2009. Seasonal movements and home ranges of white-tailed deer in north-central South Dakota. **Canadian Journal of Zoology 87** (10): 876-885.

GUISAN, A. & W. THUILLER. 2005. Predicting species distribution: offering more than simple habitat models. You have full text access to this content. **Ecology Letters 8** (9): 993-1009.

HUTTO, R.L. 1985. **Habitat selection in birds.** New York, Academic Press, 558p.

IGOTA, H.; M. SAKURAGI; H. UNO; K. KAJI; M. KANEKO; R. AKAMATSU & K. MAEKAWA. 2004. Seasonal migration patterns of female sika deer in eastern Hokkaido, Japan. **Ecological Research 19** (2): 169-178.

JAEGER, J.A.G; J. BOWMAN; J. BRENNAN; L. FAHRIG; D. BERT; J. BOUCHARD; N. CHARBONNEAU; K. FRANK; B. GRUBER & K.T. VON TOSCHANOWITZ.

2005. Predicting when animal populations are at risk from roads: an interactive model of road avoidance behavior. **Ecological Modelling 185** (2-4): 329-348.

JARNEMO, A. 2008. Seasonal migration of male red deer (*Cervus elaphus*) in southern Sweden and consequences for management. **European Journal of Wildlife Research 54** (2): 327-333.

JEPPESEN, J.L. 1987. Impact of human disturbance on home range, movements and activity of red deer (*Cervus elaphus*) in a Danish environment. **Danish Review of Game Biology 13** (2): 35-38.

JIANG, G.S.; J.Z. MA & M.H. ZHANG. 2007. Effects of human disturbance on movement, foraging and bed selection in red deer *Cervus elaphus xanthopygus* from the Wandashan Mountains, northeastern China. **Acta Theriologica 52** (4): 435-446.

JIANG, Y.; M. KANG; S. LIU; L.S. TIAN & M.D. LEI. 2000. A study on the vegetation in the east side of Helan Mountain. **Plant Ecology 149** (2): 119-130.

KUNKEL, K. & D.H. PLETSCHER. 2001. Winter hunting patterns of wolves in and Near Glacier National Park, Montana. **The Journal of Wildlife Management 65** (3): 520-530.

LICOPPE, A.M. 2006. The diurnal habitat used by red deer (*Cervus elaphus L.*) in the Haute Ardenne. **European Journal of Wildlife Research 52** (3): 164-170.

LIU, Z.S. 2009. **Notes of vertebrates in the Helan Mountain.** Yinchuan, Ningxia People's Publishing House.

LIU, Z.S.; X.M. WANG; Z.G. LI; D.Y. CUI & X.Q. LI. 2007a. Feeding habitats of blue sheep (*Pseudois nayaur*) during winter and spring in the Helan Mountains, China. **Frontiers of Biology in China 2** (1): 100-107.

LIU, Z.S.; X.M. WANG; Z.G. LI; H. ZHAI & T.H. HU. 2007b. Distribution and abundance of blue sheep in the Helan Mountains, China. **Chinese Journal of Zoology 42** (3): 1-8.

LIU, Z.S.; J.P. WU & L.W. TENG. 2002. Time budget and behavior pattern of semi free *Cervus nippon* in spring. **Chinese Journal of Ecology 21** (6): 29-32.

LOVARI, S.; J. HERRERO; J. CONROY; T. MARAN; G. GIANNATOS; M. STUBBE; S. AULAGNIER; T. JDEIDI; M. MASSETI; I. NADER; K. DE SMET & F. CUZIN. 2008. *Cervus elaphus. In:* IUCN 2010 (Ed). **IUCN Red List of Threatened Species. Version 2010.4.** Available online at: http://www.iucnredlist.org/apps/redlist/details/41785/0 [Accessed: 12/X/2011].

MANLY, B.F.J.; L.L. MCDONALD; D.L. THOMAS; T.L. MCDONALD & W.P. ERICKSON. 2002. **Resource selection by animals: statistical design and analysis for field studies.** London, Chapman & Hall.

MARCUM, C.L. & D.O. LOFTSGAARDEN. 1980. A nonmapping technique for studying habitat preferences. **The Journal of Wildlife Management 44** (4): 963-968.

MOEN, A.N. 1976. Energy conservation by white-tailed deer in the winter. **Ecology 57:** 192-198.

MORRISON, M.L.; B.G. MARCOT & R.W. MANNAN. 1998. **Wildlife-habitat relationships: concepts and applications.** Madison, The University of Wisconsin Press.

Neu, C.W.; C.R. Byers & J.M. Peek. 1974. A technique for analysis of utilization-availability data. **The Journal of Wildlife Management 38** (3): 541-545.

Peek, J.M.; M.D. Scott; L.J. Nelson & D.J. Pierce. 1982. Role of cover in habitat management for big game in northwestern United States. **Transactions of North American Wildlife and Natural Resources Conference 47**: 363-373.

Pépin, D.; C. Adrados; G. Janeau; J. Joachim & C. Mann. 2008. Individual variation in migratory and exploratory movements and habitat use by adult red deer (*Cervus elaphus*) in a mountainous temperate forest. **Ecological Research 23** (2): 1005-1013.

Pettorelli, N.; A. Mysterud; N.G. Yoccoz; R. Langvatn & N.C. Stenseth. 2005. Importance of climatological downscaling and plant phenology for red deer in heterogeneous landscapes. **Proceedings of the Royal Society Biological Sciences 272**: 2357-2364.

Putman, R. 1988. **The Natural History of Deer.** London, Christopher Helm.

Qiao, J.F.; W.K. Yang & X.Y. Gao. 2006. Natural diet and food habitat use of the Tarim red deer, *Cervus elaphus yarkandensis.* **Chinese Science Bulletin 51** (Supp. I): 147-152.

Schmitz, O.J. 1991. Thermal constrains and optimization of winter feeding and habitat choice in white-tailed deer. **Ecography 14**: 104-111.

Takhtajan, A. 1986. **Floristic regions of the world.** Berkeley, University of California Press, 62p.

Unsworth, J.W.; L. Kuck; E.O. Garton & B.R. Butterfield. 1998. Elk habitat selection on Clearwater national forest, Idaho. **The Journal of Wildlife Management 62** (4): 1255-1263.

Vaughan, T.A.; J.M. Ryan & N.J. Czaplewski. 2000. **Mammalogy.** Orlando, Saunders College Publishing, 4th ed.

Wang, S. 1998. **China red data book of endangered animals: mammals.** Beijing, Science Press.

Wang, X.M.; M. Li; S.Y. Tang; Z.X. Liu; Y.G. Li & H.L. Sheng. 1999. The Study of resource and conservation of artiodactyls in the Helan Mountain. **Chinese Journal of Zoology 34** (5): 26-29.

Wang, X.M. & G.B. Schaller. 1996. Status of large mammals in Inner Mongolia, China. **Journal of East China Normal University 6** (Special Issue): 94-104.

Welch, D.; B.W. Staines; D.C. Catt & D. Scott. 1990. Habitat usage by red (*Cervus elaphus*) and roe deer (*Capreolus capreolus*) in a Scottish Sitka Apruce plantation. **Journal of Zoology 221** (3): 453-476.

White, C.G. & R.A. Garrott. 1990. **Analysis of Wildlife Radio-Tracking Data.** San Diego, Academic Press.

Yost, A.C. & R.G. Wright. 2001. Moose, Caribou,and Grizzly Bear distribution in relation to road traffic in Denali National Park, Alaska. **Arctic 54** (1): 41-48.

Zhang, X.L.; Z.G. Li; Z. Li; Y.X. Ma; T.S. Zhang & H. Zhai. 2006. Studies on population quantity and dynamics of red deer in spring for Helanshan Mountain of Ningxia. **Journal of Ningxia University (Natural Science Edition) 27** (3): 263-265.

Zhang, X.L.; Z.G. Li; H.J. Lü & H.L. Guo. 1999. Studies on ecological habits and population dynamics of Ningxia red deer. **Ningxia Journal of Agriculture and Forestry Science and Technology** (Supp. I): 22-27.

Zhao, C.L.; Z.G. Li; H.J. Lü; T. Li; T.H. Hu; H. Zhai; H.L. Wang; Y.X. Li; Z.C. Wang; Z.L. Chang; R.F. Jiao; H.Y. Shi & H.P. Xu. 2000. Vegetation cover degree monitoring in the Helanshan Mountain project area of Sino-Germany cooperation Ningxia shelter-forest project. **Ningxia Journal of Agriculture and Forestry Science and Technology** (Supp. I): 6-14.

Zhao, Y.S.; P. Sun; X.Q. Zhou & D.J. Cui. 2004. The role of closing hills for reforestation on eco-environment in the Helanshan Mountains. **Inner Mongolia Forestry Investigation and Design 27** (4): 7-9.

19

Culicidae (Diptera: Culicomorpha) from the central Brazilian Amazon: Nhamundá and Abacaxis Rivers

Rosa Sá Gomes Hutchings[1], Roger William Hutchings[1,3] Honegger & Maria Anice Mureb Sallum[2]

[1] Coordenação de Biodiversidade, Instituto Nacional de Pesquisas da Amazônia. Caixa Postal 2223 AC Andre Araujo, 69080-971 Manaus, AM, Brazil.
[2] Laboratório de Sistemática e Ecologia de Culicidae, Faculdade de Saúde Pública, Universidade de São Paulo. 01246-904 São Paulo, SP, Brazil.
[3] Corresponding author. E-mail: rwhutch@inpa.gov.br

ABSTRACT. Mosquito fauna (Culicidae) from remote areas along the geographical limits of the State of Amazonas were assessed by employing CDC, Shannon, Malaise and Suspended traps, together with net sweeping and immature collections. Two hundred and six collections were performed in seven localities along the Nhamundá and Abacaxis Rivers, State of Amazonas, Brazil, during May and June 2008. The northernmost locality was 120 km from Nhamundá, whereas the southernmost locality was 150 km from the mouth of the Abacaxis River. The 5,290 mosquitoes collected are distributed in 16 genera, representing 109 different species, of which eight are new distributional records for the State of Amazonas. Furthermore, there are nine morphospecies which may represent undescribed new taxa, five of which are also new records for the State of Amazonas. Culex presented the highest number of species and the largest number of individuals. Anopheles, which represents 3% of the total sample, had the second highest number of species, followed by Wyeomyia. Psorophora and Aedes, represent the third and fourth largest number of individuals. The most abundant species was Cx. (Mel.) vaxus Dyar, 1920 followed by Cx. (Mel.) eknomios Forattini & Sallum, 1992, Cx. (Cux.) mollis Dyar & Knab, 1906, Cx. (Mel.) theobaldi Lutz, 1904, and Cx. (Cux.) declarator Dyar & Knab, 1906. The epidemiological and ecological implications of mosquito species found are discussed and are compared with other mosquito inventories from the Amazon region. The results presented represent the largest standardized inventory of mosquitoes of the Nhamundá and Abacaxis rivers, with the identification of 118 species level taxa distributed in seven localities, within four municipalities (Nhamundá, Maués, Borba, Nova Olinda do Norte), of which we have only few or no records in the published literature.

KEY WORDS. Amazonia; distribution; mosquitoes.

There are about 3,523 species of mosquitoes (Culicidae) described throughout the world (HARBACH 2012). Mosquitoes have a worldwide distribution with at least 553 species present in the Neotropical region, of which 468 are recorded from Brazil (GAFFIGAN et al. 2012). Records of geographical distribution are essential to improve our knowledge of the systematics of mosquitoes, as well as the need for the correct identification of species in studies of biodiversity, ecology and vector incrimination. In general, the knowledge of the biodiversity of Culicidae is of public health interest, since it enables a better understanding of the dynamics of transmission of infectious agents and the role of mosquito species as vectors, facilitating the adoption of control measures.

Because of its extensive and complex geographical structure, the Amazon region has many remote areas, such as the basins of the Nhamundá and Abacaxis rivers, located north and south of the Amazon River along the eastern border of the Brazilian State of Amazonas, where the Culicidae fauna is un-

known. Unfortunately, very little is known about the geographic distribution of mosquitoes in the State of Amazonas. CERQUEIRA (1961), in a pioneering work, using information gathered from the collections of the defunct Serviço Nacional de Febre Amarela (National Yellow Fever Service) and material collected by the Instituto Nacional de Pesquisas da Amazônia (National Institute of Amazonian Research), reported the presence of 148 species in 24 locations within the State of Amazonas. Later, several papers were published on the geographical distribution of Culicidae in the Amazon, using information gathered from bibliographical references and material from the Entomology Museum of the Centro de Pesquisas René Rachou (FIOCRUZ) (René Rachou Research Center), adding new locality records for the state, where the number of known species increased to 175 in 114 locations representing 61% of the state's municipalities (XAVIER & MATTOS 1976). Unfortunately, after XAVIER & MATTOS (1976), there has not been any new publication compiling and updating the distribution records of species which can be found in more recent publications. Most of these new records of distribution are found in publications resulting from inventories (Barbosa *et al.* 2008, Hutchings *et al.* 2002, 2005, 2008, 2010, 2011, Suárez-Mutis *et al.* 2009) and as a result of the description of new species (Forattini & Sallum 1992, Sallum & Hutchings 2003, Sallum *et al.* 1997).

It should be noted that many of the published records are not results of collections made with the purpose of studying the entire mosquito community, but mainly had epidemiological objectives (Cerqueira 1961, Deane 1947). Therefore, any list of species prepared for a given location which is based on published records may be incomplete and/or biased. For example, after collecting 119 species in the Jau National Park, 25% (30 species) were new records for the State of Amazonas (Hutchings *et al.* 2005) and of 145 species collected north of Manaus (Hutchings *et al.* 2011), 16% (23 species) are also new records for the State of Amazonas, including seven new records for Brazil. Outside of being biased, the geographical distribution of published records is still unrepresentative given the low coverage of the municipalities. Although the coverage includes 61% of the municipalities within the State of Amazonas, the sampled area of each municipality is still very small.

It is important to consider that an increase in the knowledge of the mosquito fauna of the Amazon region will permit us to obtain basic information of the faunal diversity, distribution and variety of ecosystems where mosquitoes occur, thus providing basic knowledge for studies on the control of diseases which affect humans and animals, whose infectious agents are transmitted by mosquitoes. In this work, we present the first results of mosquito collections from remote regions located near the political boundaries of the State of Amazonas, as part of the project *"Amazonas: Diversidade de insetos ao longo de suas fronteiras"* of the Programa de Apoio a Núcleos de Excelência (FAPEAM-CNPq).

Therefore, with the objective of serving as a base inventory for future surveys of Culicidae from the Amazon, the mosquito species collected inside the riparian and *terra firme* forests along the basins of the Nhamundá and Abacaxis rivers, Amazonas, Brazil, are reported herein.

MATERIAL AND METHODS

A mosquito inventory, conducted during a river expedition in areas near the eastern border of the State of Amazonas, Brazil (Fig. 1), includes collections of mosquitoes from seven different localities: two localities along the Nhamundá River, Municipality of Nhamundá (between 01°35'S, 057°37'W and 01°53'S, 057°03'W); and five localities along the Abacaxis River, including the Municipalities of Maués, Borba and Nova Olinda do Norte (between 05°15'S, 058°41'W and 04°28'S, 058°33'W). These localities are characterized by having most of their area covered by dense upland (*terra firme*) ombrophilous forests with low plateaus, together with riparian rain forests having dense alluvial and lowland vegetation (*Floresta Ombrófila Densa Aluvial e de Terras Baixas*) along the rivers, intermixed with areas of transition including Amazonian white sand (*campinarana*) and floodplain (*varzea*) forests. The tropical rainforest climate is warm and wet, characterized by being constantly humid, with temperature and precipitation with little annual variation. Based on climatic data from Parintins and Maues (RADAMBRASIL 1976), the

Figure 1. Localities sampled along the Nhamundá and Abacaxis Rivers, State of Amazonas, Brazil (The stars indicate the collecting locations described in Table I).

region has an annual relative humidity of 86% and a mean annual temperature of 26°C. A shorter dry season occurs from July to November with the lowest monthly precipitation being less than 50 mm along the Nhamundá River and over 150 mm along the Abacaxis River. The rainy season occurs between December and June, with the maximum precipitation in April. For the Nhamundá basin the mean annual precipitation is 1,750 mm, while the Abacaxis basin is greater than 2,750 mm.

Located in the far eastern region of the State of Amazonas, there are many difficulties in accessing the collection locations because of the long distance from urban centers. The most remote locality surveyed along the Nhamundá River is 240 km from Parintins and 630 km from Manaus, while the localities along the Abacaxis River are 360 km from Maués and 530 km from Manaus. The region, with a very low demographic density, only has small settlements which occupy marginal areas along the rivers. The main means of transport is by boat.

Mosquitoes specimens were mostly collected inside the riparian forest along existing and/or newly created trails, perpendicular to the river banks, and within continuous upland *terra firme* forest using a variety of capture methods including: CDC traps with different types of lighting (incandescent lamp (CDC) or ultraviolet fluorescent tube (UV CDC)); flight intercept traps (Malaise, Shannon, Suspended); and sweeping with nets. The CDC traps were installed every 50 m along trails, placed at 1, 10, 15 or 20 m above the ground, and were activated at dusk for a period of 12 hours, between 18:00 and 06:00 h. The Shannon traps, placed within small open understory areas, using an internal light source for attraction and a portable battery powered aspirator for capturing specimens, were used between 18:00 and 21:00 h. The 6 m long Malaise flight intercept traps, also placed within small open understory areas, were used for periods of up to three days and the Suspended flight intercept traps were hung one meter above the water level, along river margins, or at tree canopy level, also for periods of up to three days. Each sweeping collection, using entomological nets, was performed during a minimum of two hours at each location. Immature mosquitoes were collected from

breeding sites found along trails, in the same areas where adults were captured. The immature mosquitoes were reared for the purpose of obtaining adult males and females, associated with larval and pupal exuviae. These reared specimens were mostly used to obtain a more accurate identification of adult female specimens captured using other methods.

Adult and immature mosquitoes were captured, preserved and mounted following techniques detailed by Belkin *et al.* (1967). Specimens were identified in the laboratories at INPA in Manaus, and were confirmed at the Laboratório de Sistemática e Ecologia de Culicidae (LASEC), of the Faculdade de Saúde Pública (FSP/USP), in São Paulo and in the Laboratório de Transmissores de Hematozoários of the Instituto Oswaldo Cruz (IOC), in Rio de Janeiro, using the identification keys in Lane (1953a, b), Forattini (1965a, b, 2002), Zavortink (1972, 1979), Arnell (1973), Valencia (1973), Berlin & Belkin (1980), Sallum & Forattini (1996), as well as the Pecor *et al.* (1992) catalog for *Culex* (*Melanoconion*). Whenever possible, anatomical characteristics of the male genitalia were examined to confirm the identifications of both females and males. The collected material will be deposited in the Coleção de Invertebrados of the Instituto Nacional de Pesquisas da Amazônia (INPA-Manaus), in the Coleção Entomológica de Referência of the Faculdade de Saúde Pública, Universidade de São Paulo (FSP/USP) and in the Coleção de Culicídeos of the Instituto Oswaldo Cruz (FIOCRUZ). The collection and specimen data was digitized, stored, archived and organized using the relational database structure provided by the Biota software version 2.04 (Colwell 2012).

RESULTS

A total of 206 collections distributed in seven locations along the Nhamundá and Abacaxis Rivers, in the State of Amazonas, from May 14 to June 6, 2008, resulted in the capture of over 5,000 mosquitoes (Table I). Each collection corresponds to the capture yield of a trap (i.e. CDC, CDC UV, Shannon, Malaise, and Suspended) or method (i.e. sweeping,

Table I. Collections of mosquitoes distributed in seven localities along the Nhamundá and Abacaxis Rivers in the State of Amazonas, Brazil.

Locality	Locality name*	Municipality	Coordinates	Number of collections	Number of specimens
ProN-001	Areia, Igarape do Areia (LM), Rio Nhamundá (RM)	Nhamundá	01°35′22″S, 57°37′06″W	55	674
ProN-002	Cuipiranga, Lago do Aburi, Rio Nhamundá (RM)	Nhamundá	01°53′42″S, 57°03′25″W	43	259
ProN-004	Picada Pirarara, Rio Abacaxis (RM)	Maues	05°15′09″S, 58°41′52″W	26	512
ProN-005	Picada Borba, Rio Abacaxis (LM)	Borba	05°13′19″S, 58°41′22″W	20	659
ProN-006	Pacamiri, Rio Abacaxis (RM)	Maues	04°35′49″S, 58°13′14″W	29	1,118
ProN-007	Paxiuba, Rio Abacaxis (LM)	Borba	04°29′00″S, 58°34′14″W	32	2,044
ProN-008	Paxiuba, Rio Abacaxis (RM)	Nova Olinda do Norte	04°28′36″S, 58°33′46″W	1	24
Total				206	5,290

* (LM) left margin of the basin; (RM) right margin of the basin.

immature rearing) (Table II). The CDC traps were used during 1,200 trap-hours and the CDC UV during 744 trap-hours. The Shannon traps were used in eight collections, totaling 28 trap-hours. The 6m Malaise flight intercept traps were used during 15 trap-days and the Suspended traps during nine trap-days. The net sweeping collections were done during 40 hours. A total of eight immature collections were performed in different habitats: Bromeliaceae leaf axils (three samples); lakes and streams (3); in a tree hole; and in a *Bertholletia* pixidium. More specific information on the sampling effort for each locality is presented in Table II.

Of the 5,290 specimens captured, 5,231 were identified and are distributed in 16 genera, representing 118 different taxa (among species and morphospecies) (Appendix 1). The morphospecies (identified as near F#) are similar to a known species, but it is believed that some may represent undescribed new taxa. Some identification could not be exact because of the absence of males, whose genitalia usually possess anatomical features that allow the specific diagnosis. These individuals were identified as morphotypes, indicating the species to which they are most similar.

Unfortunately, among the mounted, sorted and examined material, it was not possible to identify 1,815 specimens ($\approx$ 34%) to the species level for several reasons: either there are no known characters to separate female individuals of different species or the characters used to separate these species were damaged. For some of these individuals it was only possible to identify to ge-nus level because the characters which are used for identification are damaged and/or lost, and the rest of the collected material was recognized to subgeneric or informal taxonomic group (sections or groups) level (shown with the prefix "gr.", "sG." or "sec." or the suffix "sp." in Appendix 1). Most of the individuals that could not be identified to species level are females (1790 H" 98%) and belong to the genus *Culex* (91%) (Appendix 1). It is interesting to note that only 13% of the specimens collected in this inventory were males.

Culex presented the highest number of species (45 H" 42%) and the largest number of individuals (4,653 H" 89%). The genus *Anopheles*, which represents 3% of the total sample (166 specimens), had the second highest number of species (13 H" 12%), followed by *Wyeomyia* with 11 species (H" 10%), and less than 1% of individuals. *Psorophora* and *Aedes*, respectively with 9 and 8 species each (H" 7%), represent the third largest (178 H" 3%) and the fourth largest number of individuals (90 < 2%).

The most abundant species was *Culex* (*Mel.*) *vaxus* Dyar, 1920 (587 individuals collected, representing 17% of the material identified to species level) followed by *Cx.* (*Mel.*) *eknomios* Forattini & Sallum, 1992, *Cx.* (*Cux.*) *mollis* Dyar & Knab, 1906, *Cx.* (*Mel.*) *theobaldi* Lutz, 1904, and *Cx.* (*Cux.*) *declarator* Dyar & Knab, 1906 (with 481, 456, 415 and 255 individuals, respectively). The five most abundant species (<5% of the recorded species) represent 66% of specimens identified to the species level.

Table II. Method of capture, sampling effort and number of mosquitoes collected along the Nhamundá and Abacaxis Rivers in the State of Amazonas, Brazil.

Method of capture	Species/Method Total number *Exclusive number*	Number of specimens *Sampling effort*		
		Total	Nhamundá River	Abacaxis River
CDC trap	51	2,573	152	2,421
	7	*100c:1,200h*	*44c:528h*	*56c:672h*
CDC (UV) trap	62	1,777	299	1478
	12	*62c:744h*	*34c:408h*	*28c:336h*
Shannon Trap	31	406	155	251
	3	*8c:28h*	*5c:17h*	*3c:11h*
Net sweeping	51	376	219	157
	17	*20c:40h*	*9c:18h*	*11c:22h*
Malaise Trap	18	56	41	15
	4	*5c:15d*	*3c:9d*	*2c:6d*
Suspended Trap	18	69	48	21
	6	*3c:6d*	*1c:3d*	*2c:3d*
Immature collections	9	33	19	14
	4	*8c*	*2c*	*6c*
Total	118	5,290	933	4,357
	53	*206c*	*98c*	*108c*

The values in italics indicate the sampling effort for the method used: number of collections (#c); trap-hours (#h); or trap-days (#d).

Among the 110 species identified, there are eight new species distribution records for the State of Amazonas: *Psorophora (Jan.) discrucians* (Walker, 1856); *Culex (And.) luteopleurus* (Theobald, 1903); *Culex (Mel.) rooti* Rozeboom, 1935; *Culex (Mel.) trigeminatus* Clastrier, 1970; *Culex (Mcx.) aureus* Lane & Whitman, 1951; *Onirion brucei* (Del Ponte & Cerqueira, 1938); *Wyeomyia (Spi.) aningae* Motta & Lourenço, 2005; and *Wyeomyia surinamensis* Bruijning, 1959. There are also 203 specimens of at least nine morphospecies (marked as near F # in Appendix 1), of which five also represent new geographical records for the State of Amazonas. These morphospecies, which probably represent species not yet described, belong to three different genera. *Aedes (Ochlerotatus)* has a total of four specimens of two morphotypes: *Ae. (Och.)* near *pectinatus* F1 and *Ae. (Och.)* near sG Infirmatus F1. *Culex (Melanoconion)* has 196 specimens of six morphotypes: *Cx. (Mel.)* near *creole* F1, *Cx. (Mel.)* near *eastor* F1, *Cx. (Mel.)* near *silvai* F1, *Cx. (Mel.)* near *vaxus* F1, *Cx. (Mel.)* near *vaxus* F3 and *Cx. (Mel.)* near *venezuelensis* F1. *Wyeomyia (Hystatomyia)* has two specimens of one morphotype: *Wy. (Hys.)* near *baltae* F1. Among the nine morphotypes identified (*Anopheles (Nys.) goeldi/***dunhami**, *Anopheles (Nys.) konderi/oswaldoi*, *Anopheles (Ste.) nimbus/thomasi*, *Aedes (Och.)* **hastatus**/*oligopistus*, *Aedes (Och.) serratus/***nubilus**, *Culex (Ads.)* **clastrieri**/*guyanensis*, *Culex (Car.) urichii/***anduzei**, *Culex (Cux.)* **coronator**/*usquatus* and *Culex (Cux.) mollis/declarator*) there are seven species (indicated in bold above) which could potentially also increase the number of species recorded within each sampled locality.

Together, the nocturnal collecting methods (CDC, CDC-UV and Shannon Traps) were responsible for 90% of the captured mosquitoes, of which the CDC traps (with a total combined sampling effort of 162 trap-nights) were responsible for more than 83%. Net sweeping accounted for 7%, followed by the Suspended and Malaise flight intercept traps, with 1.3% and 1% of the specimens, respectively (Table II). Both types of CDC traps together were responsible for collecting 62% (73) of the species level taxa, of which the CDC-UV trap alone collected 53% of the species level taxa. The adult specimens of *Aedeomyia*, *Orthopodomyia* and *Uranotaenia* were only collected at night (CDC, CDC-UV and Shannon), while *Haemagogus* was only collected during the day and *Onirion* was only registered by rearing larvae. The methods of capture for each taxon can be seen in the final columns of Appendix 1. Net sweeping, and the CDC type traps combined, were responsible for the highest number of species which were only and exclusively collected with a specific method of capture, although every method did collect exclusive species (see details in Table II). The diurnal mosquitoes are not equally represented in this inventory, compared to the nocturnal mosquitoes because the sampling effort was greater for the nocturnal collecting methods.

Of the 118 different species level taxa identified during this inventory, 48 (41%) were collected in both river basins, while 29 species (24%) were found only along the Nhamundá River and 41 species (35%) were only found along the Abacaxis

River (Appendix 1). The results of the mosquito inventory for each separate river basin are presented below.

Nhamundá River

The inventory along the Nhamundá river basin, sampled from May 16 to 19, includes specimens from 98 collections in two localities (Table I), resulting in 933 mosquitoes from 15 genera, representing 77 different taxa identified to species level (between species and morphospecies) (Appendix 1). It was not possible to identify 256 specimens (H" 27%) to the species level for the reasons previously discussed. For this basin, the sampling effort included 528 CDC trap-hours, 408 CDC (UV) trap-hours, 17 Shannon trap-hours, 9 Malaise trap-days, 3 Suspended trap-days, 18 net sweeping hours and two immature collections (in a *Bertholletia* pixidium and a Bromeliaceae leaf axil) (Table II).

In the Nhamundá River basin, the genus *Culex* presented the highest number of species (26 H" 38%) and the largest number of individuals (712 H" 77%). The genus *Wyeomyia*, which represents 4% of the total sample (34 specimens), had the second highest number of species (11 H" 16%), followed by *Psorophora* with only seven species represents the second largest number of specimens (71 H" 8%). *Anopheles* had 6 species (H" 9%), and less than 4% of the individuals, while *Aedes* with five species had the third largest number of individuals (45 < 5%). The most abundant taxon was the morphospecies *Culex (Mel.)* near *vaxus* F3 (142 individuals collected, representing 23% of the specimens identified to species level) followed by *Cx. (Mel.) vaxus* Dyar, 1920, *Cx. (Mel.) bequaerti* Dyar & Shannon, 1925, *Psorophora (Jan.) ferox* (Humboldt, 1819), and *Cx. (Cux.) mollis* Dyar & Knab, 1906 (with 122, 64, 38 and 29 individuals, respectively). The five most abundant species (<7% of the recorded species) represent 64% of specimens identified to the species level.

Among the 70 species, collected along the Nhamundá River, there are six new species distribution records for the state of Amazonas. There are also 159 specimens of at least seven morphospecies (9% of the species level taxa), of which four represent new geographical records for the State of Amazonas. These morphospecies, which probably represent species not yet described, belong to three different genera: *Ae. (Och.)* near sG Infirmatus F1, *Cx. (Mel.)* near *creole* F1, *Cx. (Mel.)* near *silvai* F1, *Cx. (Mel.)* near *vaxus* F1, *Cx. (Mel.)* near *vaxus* F3, *Cx. (Mel.)* near *venezuelensis* F1 and *Wy. (Hys.)* near *baltae* F1.

Abacaxis River

The inventory along the Abacaxis river basin, sampled from May 26 to June 4, includes specimens from 108 collections in five localities (Table I), resulting in 4,357 mosquitoes, from 15 genera, representing 89 different taxa identified to species level (including species and morphospecies) (Appendix 1). It was not possible to identify 1558 specimens (H" 36%) to the species level for the reasons previously discussed. For this basin, the sampling effort included 672 CDC trap-hours, 336

CDC (UV) trap-hours, 11 Shannon trap-hours, six Malaise trap-days, three Suspended trap-days, 22 net sweeping hours and six immature collections (in lakes and streams (three samples), in two Bromeliaceae leaf axils, and in a tree hole (Table II).

In the Abacaxis River basin, *Culex* presented the highest number of species (37 H" 44%) and the largest number of individuals (3,941 H" 92%). *Anopheles*, which represents 3% of the total sample (136 specimens), had the second highest number of species (13 H" 15%), followed by *Psorophora* and *Aedes* with six species (H" 7%) each, representing the third largest (107 H" 3%) and the fourth largest number of individuals (45 = 1%). The most abundant species was *Cx.* (*Mel.*) *eknomios* Forattini & Sallum, 1992 (479 individuals collected, representing 18% of the material identified to species level) followed by *Culex* (*Mel.*) *vaxus* Dyar, 1920, *Cx.* (*Cux.*) *mollis* Dyar & Knab, 1906, *Cx.* (*Mel.*) *theobaldi* Lutz, 1904, and *Cx.* (*Cux.*) *declarator* Dyar & Knab, 1906 (with 465, 427, 392 and 245 individuals, respectively). The five most abundant species (<6% of the recorded species) represent 74% of specimens identified to the species level.

Among the 84 species, collected along the Abacaxis River, there are three new species distribution records for the State of Amazonas. There are also 43 specimens of at least five morphospecies (H"6%), of which two represent new geographical records for the State of Amazonas. These morphospecies, which probably represent species not yet described, belong to two different genera: *Ae.* (*Och.*) near *pectinatus* F1, *Cx.* (*Mel.*) near *eastor* F1, *Cx.* (*Mel.*) near *silvai* F1, *Cx.* (*Mel.*) near *vaxus* F1, and *Cx.* (*Mel.*) near *vaxus* F3.

DISCUSSION

Among the 118 species level taxa collected in this inventory, there are 13 (11%) new geographical distribution records for the State of Amazonas. Other mosquito surveys from upland terra firme sites have similar results: of the 145 species collected, north of Manaus, 16% (23 species) were new records for the State of Amazonas (HUTCHINGS et al. 2011); of the 119 species collected in the Jau National Park, 25% (30 species) were new records (HUTCHINGS et al. 2005); and of the 44 species recorded in Querari, 27% (12 species) were also new records for the state (HUTCHINGS et al. 2002). We found no previously published mosquito distributional records for the municipalities of Nhamundá and Nova Olinda do Norte. Therefore, the results of this inventory represent the first published report of mosquito taxa for these municipalities.

Epidemiologically, the presence of Anophelines may be important because this genus includes *Plasmodium* vector species, which cause malaria in humans. Within the *Anopheles*, it is worth noting the presence of *Anopheles* (*Nys.*) *konderi s.l.*, *An.* (*Nys.*) *oswaldoi s.l.* and *Anopheles* (*Nys.*) *triannulatus*, and absence of *An.* (*Nys.*) *darlingi* Root, 1926 and any species of the *An.* (*Nys.*) *albitarsis* complex. *Anopheles* (*Nys.*) *konderi s.l.*, *An.* (*Nys.*) *oswaldoi s.l.* and *Anopheles* (*Nys.*) *triannulatus* are consid-

ered secondary vectors, but they can take the role of local or regional primary vectors (FORATTINI 2002). Considering that *Anopheles* (*Nys.*) *konderi s.l.* and *An.* (*Nys.*) *oswaldoi s.l.* were demonstrated to be species complexes (MOTOKI et al. 2009, SALLUM et al. 2008), the vector status of each species needs to be determined in further studies conducted in areas of malaria transmission where species of these complexes are present. Additionally, the absence of *An.* (*Nys.*) *darlingi* and also of species of the *An.* (*Nys.*) *albitarsis* complex may be indicative of an undisturbed natural environment. In several studies conducted inside pristine areas of the State of Amazonas, no specimens of *An.* (*Nys.*) *albitarsis s.l.* and only a few specimens of *An.* (*Nys.*) *darlingi* were found. For example, only seven *An.* (*Nys.*) *darlingi* specimens (2% and 4% of the Anophelines, respectively) were collected in both the Jau National Park (HUTCHINGS et al. 2005) and in the Juami-Japura Ecological Station (HUTCHINGS et al. 2010), while only one *An.* (*Nys.*) *darlingi* specimen (<0.4%) was found in areas north of Manaus (HUTCHINGS et al. 2011). In contrast, *An.* (*Nys.*) *darlingi* can be the most prevalent species inside deforested areas of the Amazon region (CASTRO et al. 2006), whereas species of *An.* (*Nys.*) *albitarsis* complex can become more frequent depending on the land use (CONN et al. 2002).

Furthermore, there are *Culex* species which are potential vectors of arboviruses. For example, *Cx. gnomatos*, the second most common *Culex* species in these samples, is highly susceptible to infection by enzootic (ID and IE) and epizootic strains (IAB and IC) of the Venezuelan Equine Encephalitis Virus (VEEV) (TURELL et al. 2000). It is worth mentioning that *Cx. pedroi*, also a common species in these collections, is considered a potential enzootic vector of the Eastern Equine Encephalitis Virus (EEEV), in Brazil, as well as of the VEEV and other arboviruses (GALINDO & SRIHONGSE 1967, GALINDO et al. 1966, SRIHONGSE & GALINDO 1967). Moreover, it is interesting to note that AITKEN (1972) observed that *Cx. portesi* may be involved in the of epizootic and enzootic transmission cycles of the Mucambo virus. *Cx. spissipes* is a potential vector of the Bimiti, Caraparu, Oriboca and Itaqui viruses, of the Bunyaviridae family and of the VEEV III-B subtype (SHOPE et al. 1988, WALTON & GRAYSON 1988).

Considering the number of specimens and/or species resulting from the different methods of capture and sampling efforts of this inventory (Table II), future mosquito surveys should give priority to the use of CDC type traps and net sweeping in order to maximize collecting results, when time and field resources are limited.

This mosquito inventory is part of a larger entomological inventory of different locations within remote and sparsely populated areas near the border regions of the State of Amazonas which also resulted in the collection of a large number of other insects, including Lepidoptera (CASAGRANDE et al. 2012). The information presented here represents the largest standardized mosquito inventory ever executed, within the Nhamundá and Abacaxis river basins, with the identification of 118 taxa distributed in seven different locations within four different counties (Nhamundá, Maués, Borba, Nova Olinda do Norte), of which few or no geographical records have been previously published.

ACKNOWLEDGMENTS

We wish to thank Monique de A. Motta (FIOCRUZ-RJ) for help with the identification of the *Wyeomyia* and the technicians, Luis Aquino, Jose M. da S. Vilhena, and Isis S. Menezes, for their help collecting and processing specimens. All specimens were collected using the "Autorização para Atividades com Finalidade Científica" (IBAMA/ICMBio/SISBIO) #103281 (Roger W.H. Honegger). This research was financed with resources from the Fundação de Amparo à Pesquisa do Estado do Amazonas (FAPEAM) and the Conselho Nacional de Desenvolvimento Técnico e Científico (CNPq) through the project "Amazonas: diversidade de insetos ao longo de suas fronteiras" (Programa de Apoio a Núcleos de Excelência (PRONEX) Grant 1437/2007 coordinated by José A. Rafael) and from the Instituto Nacional de Pesquisas da Amazônia (INPA-PRJ12.10 Entomologia na Amazônia: Diversidade de insetos (2008/2012). Maria A.M. Sallum was financially supported by the Fundação de Amparo à Pesquisa do Estado de São Paulo (FAPESP) (Grant #2011/20397-7) and the CNPq (BPP, Grant #301666/2011-3).

REFERENCES

AITKEN, T.H.G. 1972. Habits of some mosquito hosts of VEE (Mucambo) virus from northeastern South America, including Trinidad, p. 254-256. *In:* Proceedings of the Workshop Symposium on Venezuelan Encephalitis Virus. Washington, D.C., Pan American Health Organization (PAHO) Scientific Publ., vol. 243.

ARNELL, J.H. 1973. Mosquito Studies (Diptera, Culicidae). 32. A revision of the genus *Haemagogus*. Contributions of the American Entomological Institute 10 (2): 1-174.

BARBOSA, M.D.G.V.; N.F. Fé; A.H.R. Marcião; A.P.T.D. Silva; W.M. Monteiro; M.V.D.F. Guerra & J.A.D.O. Guerra. 2008. Registro de Culicidae de importância epidemiológica na área rural de Manaus, Amazonas. Revista da Sociedade Brasileira de Medicina Tropical 41 (6): 658-663. doi: 10.1590/S0037-86822008000600019.

BELKIN, J.N.; C.L. Hogue; P. Galindo; T.H. Aitken; R.X. Schick & W.A. Powder. 1967. Estudios Sobre Mosquitos (Diptera, Culicidae) Ia. Un proyecto para un estudio sistematico de los mosquitos de Meso-America. IIa. Metodos para coleccionar, criar y preservar mosquitos. Contributions of the American Entomological Institute 1 (2a): 1-89.

BERLIN, O.G.W. & J.N. Belkin. 1980. Mosquito Studies (Diptera, Culicidae). 36. Subgenera *Aedinus, Tinolestes* and *Anoedioporpa* of *Culex*. Contributions of the American Entomological Institute 17 (2): 1-104.

CASAGRANDE, M.M.; O.H.H. Mielke; E. Carneiro; J.A. Rafael & R.W. Hutchings. 2012. Hesperioidea e Papilionoidea (Lepidoptera) coligidos em expedição aos Rios Nhamundá e Abacaxis, Amazonas, Brasil: novos subsídios para o conhecimento da biodiversidade da Amazônia Brasileira. Revista Brasileira de Entomologia 56 (1): 23-28. doi: 10.1590/S0085-56262012005000012.

CASTRO, M.C.D.; R.L. Monte-Mór; D.O. Sawyer & B.H. Singer. 2006. Malaria risk on the Amazon frontier. Proceedings of the National Academy of Sciences 103 (7): 2452-2457. doi: 10.1073/pnas.0510576103.

CERQUEIRA, N.L. 1961. Distribuição geográfica dos mosquitos da Amazônia (Diptera, Culicidae, Culicinae). Revista Brasileira de Entomologia 10: 111-168.

COLWELL, R.K. 2012. Biota: The biodiversity database manager. Version 2.04. Storrs, University of Connecticut [Originally Published by Sinauer Associates, Sunderland, Massachusetts]. User's Guide and application. Avalaible online at: http://viceroy.eeb.uconn.edu/Biota/ [Accessed: 11/VIII/2012]

CONN, J.E.; R.C. Wilkerson; M.N.O. Segura; R.T.L. de Souza; C.D. Schlichting; R.A. Wirtz & M.M. Póvoa. 2002. Emergence of a new Neotropical malaria vector facilitated by human migration and changes in land use. American Journal of Tropical Medicine and Hygiene 66 (1): 18-22.

DEANE, L.M. 1947. Observações sôbre a malária na Amazônia brasileira. Revista do Serviço Especial de Saúde Pública MES 1 (1): 3-59.

FORATTINI, O.P. 1965a. Entomologia Médica: Culicini – *Culex, Aedes, Psorophora*. São Paulo, Editora Universidade de São Paulo, vol. 2, 506p.

FORATTINI, O.P. 1965b. Entomologia Médica: Culicini – *Haemagogus, Mansonia, Culiseta*, Sabethini, Toxorhynchitini, Arboviroses, Filariose Bancroftiana, Genética. São Paulo, Editora Universidade de São Paulo, vol. 3, 416p.

FORATTINI, O.P. 2002. Culicidologia Médica: Identificação, Biologia, Epidemiologia. São Paulo, Editora Universidade de São Paulo, vol. 2, 860p.

FORATTINI, O.P. & M.A.M. Sallum. 1992. A new species of *Culex* (*Melanoconion*) from the Amazonian region (Diptera: Culicidae). Memórias do Instituto Oswaldo Cruz 87 (2): 265-274. doi: 10.1590/S0074-02761992000200015.

GAFFIGAN, T.V.; R.C. Wilkerson; J.E. Pecor; J.A. Stoffer & T. Anderson. 2012. Systematic Catalog of Culicidae. Suitland, Walter Reed Biosystematics Unit. Available online at: www.mosquitocatalog.org [Accessed: 02/VII/2012]

GALINDO, P. & S. Srihongse. 1967. Transmission of arboviruses to hamsters by the bite of naturally infected *Culex* (*Melanoconion*) mosquitoes. American Journal of Tropical Medicine and Hygiene 16 (4): 525-530.

GALINDO, P.; S. Srihongse; E. de Rodaniche & M.A. Grayson. 1966. An ecological survey for arboviruses in Almirante, Panama, 1959-1962. American Journal of Tropical Medicine and Hygiene 15 (3): 385-400.

HARBACH, R.E. 2012. **Family Culicidae Meigen, 1818.** *In*: Mosquito Taxonomic Inventory. Available online at: http://mosquito-taxonomic-inventory.info/family-culicidae-meigen-1818 [Accessed: 11/VIII/2012]

HUTCHINGS, R.S.G.; R.W. HUTCHINGS & M.A.M. SALLUM. 2010. Culicidae (Diptera: Culicomorpha) from the Western Brazilian Amazon: Juami-Japurá Ecological Station. **Revista Brasileira de Entomologia 54** (4): 687-691. doi: 10.1590/S0085-56262010000400022.

HUTCHINGS, R.S.G. & M.A.M. SALLUM. 2008. Two new species of *Culex* subgenus *Melanoconion* (Diptera: Culicidae) from the Amazon forest. **Zootaxa 1920**: 41-50.

HUTCHINGS, R.S.G.; M.A.M. SALLUM & R.L.M. FERREIRA. 2002. Culicidae (Diptera: Culicomorpha) da Amazônia ocidental Brasileira: Querari. **Acta Amazonica 32** (1): 109-122.

HUTCHINGS, R.S.G.; M.A.M. SALLUM; R.L.M. FERREIRA & R.W. HUTCHINGS. 2005. Mosquitoes of the Jaú National Park and their potential importance in Brazilian Amazonia. **Medical and Veterinary Entomology 19** (4): 428-441. doi: 10.1111/j.1365-2915.2005.00587.x.

HUTCHINGS, R.S.G.; M.A.M. SALLUM & R.W. HUTCHINGS. 2011. Mosquito (Diptera: Culicidae) diversity of a forest-fragment mosaic in the Amazon rainforest. **Journal of Medical Entomology 48** (2): 173-187. doi: 10.1603/ME10061.

HUTCHINGS, R.W.; R.S.G. HUTCHINGS & M.A.M. SALLUM. 2008. Distribuição de Culicidae na várzea, ao longo da calha dos Rios Solimões-Amazonas, p. 133-152. *In*: **Conservação da várzea: Identificação e caracterização de regiões biogeográficas.** A.L.K.M. ALBERNAZ (Ed.). Manaus, IBAMA, ProVárzea.

LANE, J. 1953a. **Neotropical Culicidae: Dixinae, Chaoborinae and Culicinae; Tribes Anophelini, Toxorhynchitini and Culicini (Genus *Culex* only).** São Paulo, Universidade de São Paulo, vol. 1, 548p.

LANE, J. 1953b. **Neotropical Culicidae: Tribe Culicini – *Deinocerites*, *Uranotaenia*, *Mansonia*, *Orthopodomyia*, *Aedomyia*, *Aedes*, *Psorophora*, *Haemagogus*; Tribe Sabethini – *Trichoprosopon*, *Wyeomyia*, *Phoniomyia*, *Limatus* and *Sabethes*.** São Paulo, Universidade de São Paulo, vol. 2, 558p.

MOTOKI, M.T.; C.L.S.D. SANTOS & M.A.M. SALLUM. 2009. Intraespecific variation on the aedeagus of *Anopheles oswaldoi* (Peryassú) (Diptera: Culicidae). **Neotropical Entomology 38**: 144-148. doi: 10.1590/S1519-566X2009000100017.

PECOR, J.E.; V.L. MALLAMPALLI; R.E. HARBACH & E.L. PEYTON. 1992. Catalog and illustrated review of the subgenus *Melanoconion* of *Culex* (Diptera: Culicidae). **Contributions of the American Entomological Institute 27** (2): 1-228.

RADAMBRASIL. 1976. **Folha SA. 21 Santarem; geologia, geomorfologia, pedologia, vegetação e uso potencial da terra.** Rio de Janeiro, Departamento Nacional de Produção Mineral, Projeto RADAM, 522p.

SALLUM, M.A.M. & O.P. FORATTINI. 1996. Revision of the Spissipes Section of *Culex* (*Melanoconion*) (Diptera: Culicidae). **Journal of the American Mosquito Control Association 12** (3): 517-600.

SALLUM, M.A.M. & R.S.G. HUTCHINGS. 2003. Taxonomic studies on *Culex* (*Melanoconion*) *coppenamensis* Bonne-Wepster & Bonne (Diptera: Culicidae), and description of two new species from Brazil. **Memórias do Instituto Oswaldo Cruz 98** (5): 615-622. doi: 10.1590/S0074-02762003000500006.

SALLUM, M.A.M.; R.S.G. HUTCHINGS & R.L.M. FERREIRA. 1997. *Culex gnomatos* a new species of the Spissipes section of *Culex* (*Melanoconion*) (Diptera: Culicidae) from the Amazon Region. **Memórias do Instituto Oswaldo Cruz 92** (2): 215-219. doi: 10.1590/S0074-02761997000200014.

SALLUM, M.A.M.; M.T. MARRELLI; S.S. NAGAKI; G.Z. LAPORTA & C.L.S. DOS SANTOS. 2008. Insight into *Anopheles* (*Nyssorhynchus*) (Diptera: Culicidae) species from Brazil. **Journal of Medical Entomology 45** (6): 970-981.

SHOPE, R.E.; J.P. WOODHALL & A.T.D. ROSA. 1988. The epidemiology of diseases caused by viruses in Groups C and Guama (Bunyaviridae), p. 37-52. T.P. MONATH (Ed.). *In*: **The Arboviruses: epidemiology and ecology.** Boca Raton, CRC Press, vol. 3.

SRIHONGSE, S. & P. GALINDO. 1967. The isolation of eastern equine encephalitis virus from *Culex* (*Melanoconion*) *taeniopus* Dyar and Knab in Panama. **Mosquito News 27**: 74-76.

SUÁREZ-MUTIS, M.C.; N.F. FÉ; W. ALECRIM & J.R. COURA. 2009. Night and crepuscular mosquitoes and risk of vector-borne diseases in areas of piassaba extraction in the middle Negro River basin, state of Amazonas, Brazil. **Memórias do Instituto Oswaldo Cruz 104** (1): 11-17. doi: 10.1590/S0074-02762009000100002.

TURELL, M.J.; J.W. JONES; M.R. SARDELIS; D.J. DOHM; R.E. COLEMAN; D.M. WATTS; R. FERNANDEZ; C. CALAMPA & T.A. KLEIN. 2000. Vector competence of Peruvian mosquitoes (Diptera: Culicidae) for epizootic and enzootic strains of Venezuelan equine encephalomyelitis virus. **Journal of Medical Entomology 37** (6): 835-839.

VALENCIA, J.D. 1973. Mosquito Studies (Diptera, Culicidae). 31. A revision of the Subgenus *Carrollia* of *Culex*. **Contributions of the American Entomological Institute 9** (4): 1-133.

WALTON, T.E. & M.A. GRAYSON. 1988. Venezuelan equine encephalomyelitis, p. 203-231. *In*: **The Arboviruses: epidemiology and ecology.** T.P. MONATH (Ed.). Boca Raton, CRC Press, vol. 4.

XAVIER, S.H. & S.D.S. MATTOS. 1976. Geographical distribution of Culicinae in Brazil. 4. State of Amazonas (Diptera, Culicidae). **Mosquito Systematics 8** (4): 386-412.

ZAVORTINK, T.J. 1972. Mosquito Studies (Diptera, Culicidae). 28. The New World species formerly placed in *Aedes* (*Finlaya*). **Contributions of the American Entomological Institute 8** (3): 1-206.

ZAVORTINK, T.J. 1979. Mosquito Studies (Diptera, Culicidae). 35. The new Sabethine genus *Johnbelkinia* and a preliminary reclassification of the composite genus *Trichoprosopon*. **Contributions of the American Entomological Institute 17** (1): 1-61.

20

Estimating cyclopoid copepod species richness and geographical distribution (Crustacea) across a large hydrographical basin: comparing between samples from water column (plankton) and macrophyte stands

Gilmar Perbiche-Neves[1], Carlos E.F. da Rocha[1] & Marcos G. Nogueira[2]

[1] *Departamento de Zoologia, Instituto de Biociências, Universidade de São Paulo. Rua do Matão, travessa 14, 321, 05508-900 São Paulo, SP, Brazil. Email: gilmarpneves@yahoo.com.br*
[2] *Departamento de Zoologia, Instituto de Biociências, Universidade Estadual Paulista. Distrito de Rubião Júnior, 18618-970 Botucatu, SP, Brazil.*

ABSTRACT. Species richness and geographical distribution of Cyclopoida freshwater copepods were analyzed along the "La Plata" River basin. Ninety-six samples were taken from 24 sampling sites, twelve sites for zooplankton in open waters and twelve sites for zooplankton within macrophyte stands, including reservoirs and lotic stretches. There were, on average, three species per sample in the plankton compared to five per sample in macrophytes. Six species were exclusive to the plankton, 10 to macrophyte stands, and 17 were common to both. Only one species was found in similar proportions in plankton and macrophytes, while five species were widely found in plankton, and thirteen in macrophytes. The distinction between species from open water zooplankton and macrophytes was supported by non-metric multidimensional analysis. There was no distinct pattern of endemicity within the basin, and double sampling contributes to this result. This lack of sub-regional faunal differentiation is in accordance with other studies that have shown that cyclopoids generally have wide geographical distribution in the Neotropics and that some species there are cosmopolitan. This contrasts with other freshwater copepods such as Calanoida and some Harpacticoida. We conclude that sampling plankton and macrophytes together provided a more accurate estimate of the richness and geographical distribution of these organisms than sampling in either one of those zones alone.

KEY WORDS. La Plata River basin; reservoirs; rivers, zooplankton.

Freshwater copepods are a link in the trophic web, connecting producers to consumers (PERBICHE-NEVES *et al.* 2007); they inhabit lakes, rivers, pools, caves, humid rocks, etc. (BOXSHALL & DEFAYE 2008), where it is easy to find free living copepods of the order Cyclopoida.

In large spatial scales, cyclopoids are less endemic than other copepods, for instance diaptomids. Many cyclopoid species in *Mesocyclops*, *Metacyclops*, *Eucyclops*, etc. are widely distributed in the Neotropical region and in the world (REID 1985, SILVA 2008). However, there are exceptions to this rule, for instance species of *Thermocyclops* (REID 1989), which occur in the South hemisphere, and differ between the Afrotropical and Neotropical biogeographical regions. In contrast to the patterns of distribution of cyclopoids in large bio geographical areas, there are no clear patterns in the spatial distribution of these organisms among river basins in South America, as observed for diaptomids (SUÁREZ-MORALES *et al.* 2005). The low endemism of cyclopoids can be in part explained by their efficient dispersion (e.g., by birds, fishes, humans) and their recent colonization of many parts of the world (BOXSHALL & JAUME 2000, SUÁREZ-MORÁLES *et al.* 2004). Comparing among the main

rivers of the La Plata Basin, the composition of cyclopoid species is similar (PAGGI & JOSÉ DE PAGGI 1990, LANSAC-TÔHA *et al.* 2002).

Cyclopoid copepods of inland waters are more diverse than in the littoral, which can be colonized by aquatic macrophytes. For example, two studies sampling the two types of habitats, open water and macrophyte stands have documented this trend for lakes in Brazil (LANSAC-TÔHA *et al.* 2002, MAIA-BARBOSA *et al.* 2008). The habitat complexity provided by aquatic macrophytes (GENKAI-KATO 2007, LUCENA-MOYA & DUGGAN 2011) also allow several species to be more abundant in them (GERALDES & BOAVIDA 2004).

Most of zooplankton horizontal migration between limnetic zones and macrophyte stands in lentic environments can be attributed to predation pressure by planktivorous fish (GENKAI-KATO 2007, FANTIN-CRUZ *et al.* 2008). Lower richness of cyclopoid species tends to be found in limnetic waters, where generally few abundant species dominate. Clear tendencies in some ecological attributes can be observed for copepods and other crustaceans in reservoirs (SILVA & MATSUMURA-TUNDISI 2002, NOGUEIRA *et al.* 2008).

There are no comparative studies of copepod richness between habitats in large geographical scales, only in lotic stretches or in lakes (Lansac-Tôha *et al.* 2002, Maia-Barbosa *et al.* 2008). This study is the first to compare copepod species richness in a large hydrographic basin, the fourth largest in the world. We simultaneously sampled plankton and macrophyte stands in rivers and reservoirs. Based on references, we tested the alternative hypotheses that in limnetic zones there is greater richness of species in macrophytes than in zooplankton. Additionally, we tried to pinpoint particular species in each kind of habitat, and to ascertain how sampling effort can determine the species that are found.

MATERIAL AND METHODS

The "La Plata" River basin crosses Argentina, Brazil, Bolivia, Paraguay and Uruguay. Samples were taken in the summer (from January to March 2010) and in the winter (from June to July 2010), periods that have different mean temperatures and precipitation. It was established that a minimum of two sampling trips were necessary to estimate richness. Altogether 24 sites were sampled, which included 12 reservoirs in the high Paraná and Uruguay rivers (because they are dominant in this stretch), and 12 lotic stretches (in Paraguay, middle and low Paraná and Uruguay rivers (Fig. 1).

The main rivers of the La Plata" River are the Paraná, Paraguay and Uruguay rivers. In each river, we chose sampling sites that were deemed representative of three stretches of the river: high, middle and low. In the main river of the basin, the Paraná River, we sampled the first reservoir (Ilha Solteira Reservoir) after it had been built. We also sampled one more reservoir (Itaipu Reservoir) at the end of the high stretch (680 km apart from each other), and another reservoir at the end of the Paraná River in the beginning of the middle stretch (1,000 away from the first sampling site). After this last reservoir (Yacyreta Reservoir), at the middle stretch of the river, there was a long lotic stretch, where we sampled six sites, each being approximately 250 km apart from the other, until we reached the mouth of the "La Plata" River, between Buenos Aires and Uruguay.

Beyond the three main rivers of the "La Plata" basin (Paraná, Paraguay and Uruguay), we sampled five main tributary rivers of the Paraná River (Grande, Paranaiba, Tiete, Paranapanema, and Iguaçu rivers), because collectively they amount to a large area of the basin in the high stretch. All tributaries are totally dammed, with a long cascade formed by a series of reservoirs. In each tributary river, we sampled the first and the last large reservoir. In general, the first reservoir of these tributaries has a dendritic shape, a large area (more than 200 km²), is deep and has high volume and high water retention time, functioning as a regulator for the downstream reservoir series (Agostinho *et al.* 2007, Nogueira *et al.* 2012). In the Uruguay River, many large reservoirs have been constructed, especially in the high stretch, but there is an old reservoir in the low

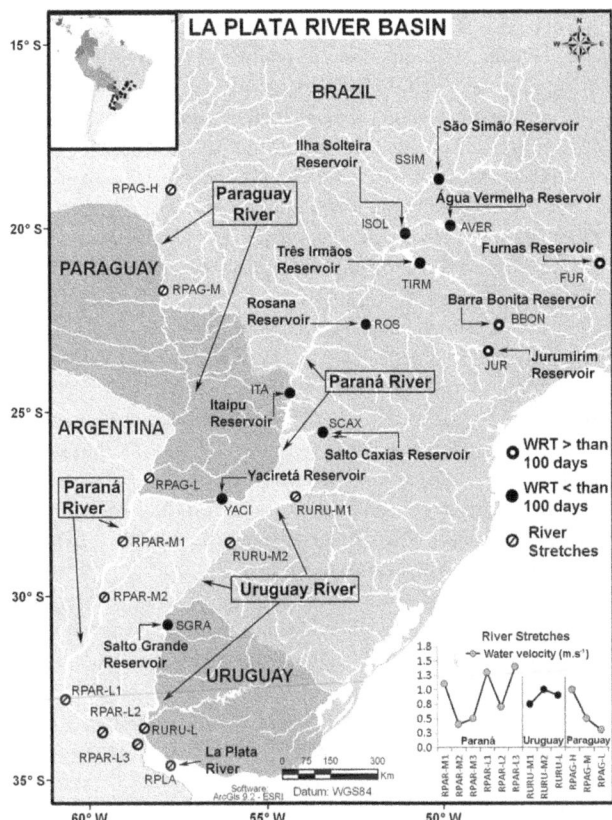

Figure 1. Sampling site map of zooplankton and macrophytes in La Plata river basin, divided in reservoirs (the first and the last in each regulated river), and free-damming lotic stretches. There are codes for abbreviation of the name of reservoirs, as also for high (H), middle (M) and low (L) stretches in lotic environments studied. Water retention time (WRT) and water flow data are given.

stretch. We sampled two reservoirs (the first and the last reservoirs in this river), and three sites in lotic stretches (two in the middle stretch, 260 km away from each other), and one in the low stretch, 5 km from the delta of Paraná River.

In the Paraguay River there are no reservoirs, only lotic stretches. Thus, the high, middle (250 km apart) and low stretches (650 km below the middle station) were sampled in the main channel. In each sampling site, we obtained zooplankton samples from open waters (limnetic region) and also zooplankton samples within macrophyte stands. In total, ninety-six samples were obtained.

As far as zooplankton are concerned, the sampling was obtained from the main river channels and from upstream zones of the reservoir, all of these with water retention time (WRT) longer than 15 days (see Agostinho *et al.* 2007, Zalocar de Domitrovic *et al.* 2007, Boltovskoy *et al.* 2013), which is considered as the lower time threshold for the development of a copepod life cycle (Rietzler *et al.* 2002). The approximate WRT of each reservoir is indicated in Fig. 1, to highlight differences between storage reservoirs and run-of-river or intermediate reservoirs. Water flow values in lotic sites are also shown in Fig. 1.

Zooplankton samples were taken by vertical hauls through water column (from close to the bottom to the surface) at each station. Values of water filtered varied between 706 L and 2,826 L, and the average value was 1,766 L. Conical plankton nets, 0.30 m mouth diameter per 0.90m side length, and 68 ìm mesh size, were used after being modified with a mouth reducing cone (TRANTER & SMITH 1979), a kind of anti-reflux bulkhead with another 0.50 m diameter circle. In deeper stations the vertical hauls were extended to a depth of 40m. In rivers, vertical hauls were taken from a drifting boat in order to ensure that the hauls were not excessively oblique. The volume of water filtered was estimated by the cylinder volume formula, using: pi * radius of the mouth net^2 * the length of the haul.

For samples obtained inside macrophyte stands, organisms were sampled with conical plankton nets of 68 ìm of mesh size, adapted with a 2 m drive cable of aluminum and with a steel screen to avoid excessive macrophyte intake. This net was passed between and alongside the macrophyte banks at standardized time of five minutes to obtain good amount of qualitative material.

Samples were fixed with 4% formalin solution. In the laboratory, they were analysed in their totality. Male and female copepods were identified to species. Copepods were examined under stereo- and compound microscopes, and identified using specialized taxonomic references (REID 1985, EINSLE 1996, ROCHA 1998, KARAYTUG 1999, ALEKSEEV 2002, UEDA & REID 2003, SILVA & MATSUMURA-TUNDISI 2005). Species richness was considered according to the number of species, in the most robust way as possible. The samples obtained in this study are deposited in the "Collection of microcrustaceans of continental waters" of "Universidade Estadual Paulista" (in Botucatu, Brazil). Vouchers are also deposited in at the Museu de Zoologia da Universidade de São Paulo MZUSP) (e.g., *Macrocyclops albidus* (Jurine, 1820) (MZUSP30601), *Eucyclops neumani* (Pesta, 1927) (MZUSP30602), and *Microcyclops ceibaensis* (Marsh, 1919) (MZUSP30603).

The non-parametric Man Whitney U test (for non-parametric data) was used to compare species richness between zooplankton and macrophytes stand. We used R Cran Project (R DEVELOPMENT CORE TEAM 2012) for this test.

A non-metric multidimensional scaling (NMDS) analysis using Bray-Curtis dissimilarity was applied for spatial ordination of the data, aiming to verify differences among sampling sites, and taking the species into consideration. We used the Vegan and Mass packages for software R Cran Project (R DEVELOPMENT CORE TEAM 2012), according to OKSANEN (2013). The iterative search was carried out using the "meta MDS" function, by several random starts, and selecting among similar solutions with the smallest stresses. For scaling we used centering, PC rotation and half change scaling.

RESULTS

We identified 32 species of cyclopoid copepods (Table I), 23 of which were found in zooplankton and 26 in macrophytes. Six species were exclusive to plankton, 10 to macrophyte stands, and 17 were common to both. Only *Mesocyclops meridianus* (Kiefer, 1926) was found in similar proportions in plankton (50% from total samples) and in macrophytes (54%), while five species (e.g., *Thermocyclops*) were widely found in plankton, and thirteen (e.g., *Microcyclops* and *Eucyclops*) in macrophytes.

Table I. List of cyclopoid species with respective abbreviations (Ab) for NMDS, and number of occurrences in sampling sites at plankton (P) of limnetic zones (from a total of 24 sites) and within macrophytes stands (M) samples (from a total of 24 sites). In bold occurrences only in plankton or macrophytes.

Cyclopoida	Ab.	P	M
Acanthocyclops robustus (Sars, 1863)	Arob	4	6
Ectocyclops herbsti Dussart, 1984	Eher	0	**11**
Ectocyclops rubescens Brady, 1904	Erub	1	2
Eucyclops elegans (Herrick, 1884)	Eele	1	0
Eucyclops ensifer Kiefer, 1936	Eens	1	0
Eucyclops serrulatus (Fischer, 1851)	Eser	0	**10**
Eucyclops solitarius Herbst, 1959	Esol	1	0
Eucyclops prionophorus Kiefer, 1931	Epri	0	**11**
Eucyclops leptacanthus Kiefer, 1956	Elep	0	**4**
Eucyclops neumani (Pesta, 1927)	Eneu	0	**6**
Homocyclops ater (Herrick, 1882)	Hat	0	**3**
Macrocyclops albidus (Jurine, 1820)	Malb	4	14
Megacyclops viridis (Jurine, 1820)	Mvir	0	**1**
Mesocyclops aspericornis (Daday, 1906)	Masp	1	2
Mesocyclops ellipticus Kiefer, 1936	Mell	1	0
Mesocyclops longisetus curvatus Dussart, 1987	Mloc	1	10
Mesocyclops longisetus longisetus (Thiébaud, 1912)	Mlol	1	0
Mesocyclops meridianus (Kiefer, 1926)	Mmer	12	13
Mesocyclops ogunnus Onabamiro, 1957	Mogu	6	3
Metacyclops laticornis (Lowndes, 1934)	Mlat	3	0
Metacyclops leptopus (Kiefer, 1927)	Mlep	0	**1**
Metacyclops mendocinus (Wierzejski, 1892)	Mmen	1	2
Microcyclops anceps anceps (Richard, 1897)	Manc	8	20
Microcyclops ceibaensis (Marsh, 1919)	Mcei	3	7
Microcyclops finitimus Dussart, 1984	Mfin	2	15
Microcyclops mediasetosus Dussart & Frutos, 1985	Mmed	8	10
Paracyclops chiltoni (Thomson, 1883)	Pchil	4	6
Thermocyclops decipiens (Kiefer, 1929)	Tdec	23	3
Thermocyclops inversus Kiefer, 1936	Tinv	**11**	0
Thermocyclops minutus (Lowndes, 1934)	Tmin	16	1
Tropocyclops prasinus meridionalis (Kiefer, 1931)	Tpram	2	1
Tropocyclops prasinus prasinus (Fischer, 1860)	Tprap	0	**1**

In general, higher median value of species richness was found in macrophyte than in plankton samples (Fig. 2), with a significant difference. For zooplankton in the summer we found a mean ± standard-deviation of 2.75 ± 1.03 species and 3.12 ± 1.74 species. For macrophytes, we found 5.08 ± 1.62 species in the summer and 5.41 ± 2.20 species per sample.

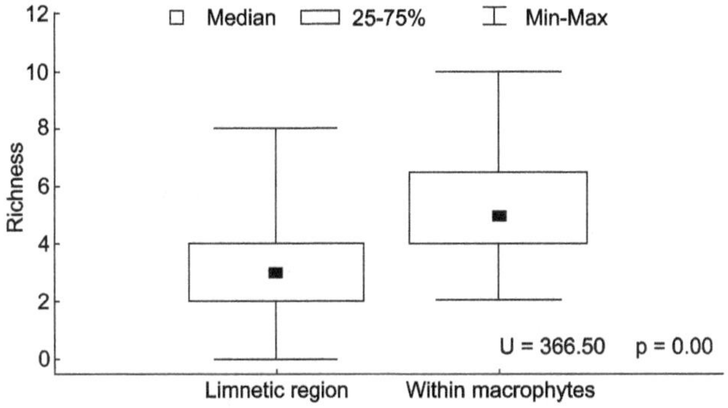

Figure 2. Median ± minimum maximum values of species richness between zooplankton and macrophytes samples.

After 20 attempts we found two convergent solutions for NMDS. This analysis (Fig. 3) classified the kind of habitat sample in basically three groups: zooplankton, macrophytes, and a median line where zooplankton and macrophyte samples were located close to each other, supporting the result shown in Table I. Geographical distribution was strongly affected by the habitat sampled. There was no clear spatial pattern of endemicity for cyclopoid species within the La Plata basin.

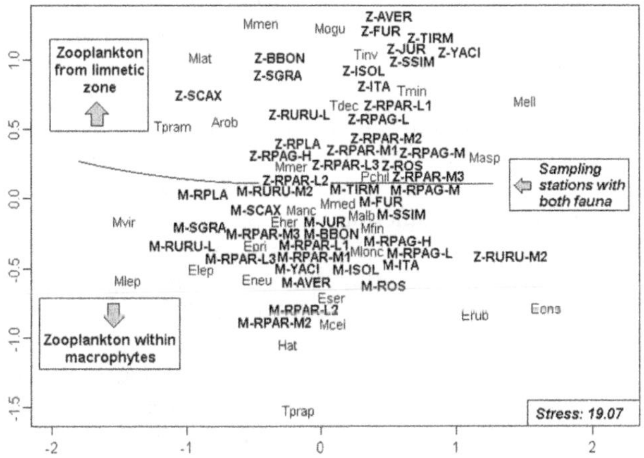

Figure 3. NMDS analysis for ordination of sampling stations and cyclopoid species in our study. See three groups of sampling sites and species: zooplankton, macrophytes and a median group where species occurred were found in zooplankton and macrophytes. For codes of species see Table I.

DISCUSSION

Our results were similar to those obtained by Maia-Barbosa et al. (2008), which found major richness of zooplanktonic species in places with aquatic macrophytes, pointing to the heterogeneity of habitats as the responsible factor. Thus, the hypothesis tested in our study was corroborated. The model of Genkai-Kato (2007) supports this statement, pointing out macrophyte areas as a refuge for zooplankton species to avoid predation of planktivorous fish.

The greatest number of exclusive species was found in macrophytes. Our results show that some genera, as *Eucyclops* and *Ectocyclops* are more frequent in macrophytes than in zooplankton, where *Thermocyclops* is common. This result agrees with Lansac-Toha et al. (2002) and Maia-Barbosa et al. (2008). Lansac-Toha et al. (2002) found 12 species of cyclopoids in the flood plain of the upper Paraná River in aquatic macrophytes. Of these species, a few were in plankton, being found more frequently and sometimes exclusively in macrophytes, similar to the findings of Maia-Barbosa et al. (2008). These authors found13 species of cyclopoid copepods in litoranean regions with aquatic macrophytes in Lake Dom Helvécio (State of Minas Gerais, Brazil), but only 7 of these were found in areas without macrophytes.

We did not find any clear pattern of geographical distribution for the cyclopoid species of our study. Many species are well distributed and some of them are cosmopolitan (Ueda & Reid 2003, Boxshall & Defaye 2008). According to Selden et al. (2010) the recent evolution of the order Cyclopoida in the Neogene (± 15 M.A.) is another possible factor for its wide distribution. The results for Cyclopoida contrasts with the results of other freshwater copepods such as diaptomids and some harpacticoids, according to the literature.

In our results, the most widely distributed species were *Ectocyclops herbsti* Dussart, 1984, *Eucyclops serrulatus* (Fischer, 1851), *Macrocyclops albidus* (Jurine, 1820), *Microcyclops anceps* (Richard, 1897), *Microcyclops finitimus* Dussart, 1984, *Microcyclops ceibaensis* (Marsh, 1919), *Thermocyclops decipiens* (Kiefer, 1929), *Thermocyclops minutus* (Lowndes, 1934), and *Thermocyclops inversus* Kiefer, 1936. It is important to highlight that some species, for instance *Mesocyclops aspericornis* (Daday, 1906) and *Mesocyclops ogunnus* Onabamiro, 1957, are invasive from Africa (Ueda & Reid 2003).The species cited above are widely distributed in the Neotropical Region (Reid 1985). The most common species, in addition to those that occur in both environments sampled, are adapted to different ecosystems. This contrasts with species that are from litoranean zones or macrophyte stands. Examples are *Homocyclops ater* (Herrick, 1882) and *E. herbsti*.

Compared to other studies in the La Plata basin (e.g., Paggi & José de Paggi 1990, Lansac-Toha et al. 2002, Silva & Matsumura-Tundisi 2002, Nogueira et al. 2008), the number of species found in this study is considerably high, and this was attributed to the combination of sampling in a large area and in two types of habitat (plankton and macrophytes), allowing the capture of exclusive species from specific habitats. Our richness of cyclopoids is also high if compared with other basins in the world (Tash 1971, Kobayashi *et al.* 1998).

The geographical distribution of cyclopoid copepods was highly influenced by the type of habitat sampled, as shown by the NMDS. This result confirms that some species are almost restricted to open waters or to macrophytes stands. Other species are common in several kinds of habitats, or can be found accidently in them, when removed by the water flow, for example. For a more accurate estimate of the diversity of these copepods it is necessary to sample at least two kinds of habitats, open waters and macrophyte stands.

ACKNOWLEDGEMENTS

We thank FAPESP (2008/02015-7; 2009/00014-6; 2011/18358-3) for financial support.

REFERENCES

Agostinho, A.A.; L.C. Gomes & F.M. Pelicice. 2007. **Ecologia e manejo de recursos pesqueiros em reservatórios do Brasil.** Maringá, Eduem, 501p.

Alekseev, V.R. 2002. Copepoda, p. 123-188. *In*: C.H. Fernando (Ed.). **A guide to tropical freshwater zooplankton: Identification, ecology and impact on fisheries.** London, Backhuys Publishers.

Boltovskoy, D.; N. Correa; F. Bordet; V. Leites & D. Cataldo. 2013. Toxic *Microcystis* (cyanobacteria) inhibit recruitment of the bloom-enhancing invasive bivalve *Limnoperna fortunei*. **Freshwater Biology 58** (9): 1968-1981. doi:10.1111/fwb.12184

Boxshall, G.A. & D. Defaye. 2008. Global diversity of Copepods (Crustacea: Copepoda) in freshwater. **Hydrobiologia 595:** 195-207.

Boxshall, G.A. & D. Jaume. 2000. Making Waves: The Repeated Colonization of Fresh Water by Copepod Crustaceans. **Advances in Ecological Research 31:** 61-79.

Einsle, U. 1996. Copepoda Cyclopoida: Genera *Cyclops, Megacyclops, Acanthocyclops*, p. 1-82. *In*: H.J.F. Dumont (Ed.). **Guides to the identification of the microinvertebrates of the continental waters of the world.** The Netherlands, Backhuys Publishers.

Fantin-Cruz, I.; K.K. Tondato; J.M.F. Penha; L.A.F. Mateus; P. Girard & R. Fantin-Cruz. 2008. Influence of fish abundance and macrophyte cover on microcrustacean density in temporary lagoons of the Northern Pantanal-Brazil. **Acta Limnologica Brasiliensia 20** (4): 339-344.

Genkai-Kato, M. 2007. Macrophyte refuges, prey behaviour and trophic interactions: consequences for lake water clarity. **Ecology Letters 10:** 105-114.

Geraldes, A.M. & M.J. Boavida. 2004. Do Littoral Macrophytes Influence Crustacean Zooplankton distribution? **Limnetica 23** (1-2): 57-64.

Karaytug, S. 1999. Genera *Paracyclops, Ochridacyclops* and key to the Eucyclopinae. *In*: H.J.F. Dumont (Ed.). **Guides to the identification of the microinvertebrates of the continental waters of the world.** The Netherlands, Backhuys Publishers.

Kobayashi, T.; R.J. Shiel; P. Gibbs & P.I. Dixon. 1998. Freshwater zooplankton in the Hawkesbury-Nepean River: comparison of community structure with other rivers. **Hydrobiologia 377:** 133-145.

Lansac-Tôha, F.A.; L.F.M. Velho; J. Higuti & E.M. Takahashi. 2002. Cyclopidae (Crustacea, Copepoda) from the upper Paraná River floodplain, Brazil. **Brazilian Journal of Biology 62** (1): 125-133.

Lucena-Moya, P. & I.C. Duggan. 2011. Macrophyte architecture affects the abundance and diversity of littoral microfauna. **Aquatic Ecology 45:** 279-287.

Maia-Barbosa, P. M.; R.S. Peixoto & A.S. Guimarães. 2008. Zooplankton in littoral Waters of a tropical lake: a revisited biodiversity. **Brazilian Journal of Biology 68** (4): 1061-1067.

Nogueira, M.G.; P.C. Reis-Oliveira & Y.T. Britto. 2008. Zooplankton assemblages (Copepoda and Cladocera) in a cascade of reservoirs of a large tropical river (SE Brazil). **Limnetica 27** (1): 151-170.

Nogueira, M.G.; G. Perbiche-Neves & D.A.O. Naliato. 2012. Limnology of two contrasting hydroelectric reservoirs (storage and run-of-river) in southeast Brazil, p. 167-184. *In*: H. Samadi-Boroujeni (Ed.). **Hydropower.** Croatia, Intech.

Oksanen, J. 2013. **Multivariate Analysis of Ecological Communities in R: vegan tutorial.** Available online at: http://cc.oulu.fi/~jarioksa/opetus/metodi/vegantutor.pdf [Accessed: 14/VII/2013].

Paggi, J.C. & S. José de Paggi. 1990. Zooplankton de ambientes lóticos e lênticas do rio Paraná médio. **Acta Limnologica Brasiliensia 3:** 685-719.

Perbiche-Neves, G.; M. Serafim-Júnior; A.R. Ghidini & L. Brito. 2007. Spatial and temporal distribution of Copepoda (Cyclopoida and Calanoida) of an eutrophic reservoir in the basin of upper Iguaçu River, Paraná, Brazil. **Acta Limnologica Brasiliensia 19** (4): 393-406.

R Development Core Team. 2012. **A language and environment for statistical computing.** Vienna, Austria, R Foundation for Statistical Computing ISBN 3-900051-07-0, URL. Available online at: http://www.R-project.org [Accessed10.XII.2012].

Reid, J.W. 1985. Chave de identificação e lista de referências bibliográficas para as species continentais sulamericanas de vida livre da Ordem Cyclopoida (Crustacea, Copepoda). **Boletim de Zoologia, Universidade de São Paulo 9:** 17-143.

Reid, J.W. 1989. The distribution of species of the genus *Thermocyclops* (Copepoda, Cyclopoida) in the western hemisphere, with description of *T. parvus*, new species. **Hydrobiologia 175:** 149-174.

Rietzler, A.C.; T. Matsumura-Tundisi & J.G. Tundisi. 2002. Life cycle, feeding and adaptive strategy implications on the co-occurence of *Argyrodiaptomus furcatus* and *Notodiaptomus iheringi* in Lobo-Broa Reservoir (SP, Brazil). **Brazilian Journal of Biology 62**: 93-105.

Rocha, C.E.F. 1998. New morphological characters useful for the taxonomy of genus *Microcyclops* (Copepoda, Cyclopoida). **Journal of Marine Systems 15**: 425-431.

Selden, P.A.; R. Huys; M.H. Stephenson; A.P. Heward & P.N. Taylor. 2010. Crustaceans from bitumen clast in Carboniferous glacial diamictite extend fossil record of copepods. **Nature Communications 1** (50): 1-6.

Silva, W.M. & T. Matsumura-Tundisi. 2002. Distribution and abundance of Cyclopoida populations in a cascade of reservoir of the Tietê River (São Paulo State, Brazil).**Verhandlungen Internationale Vereinigung für Theoretische und Angewandte Limnologie 28**: 667-670.

Silva, W.M. & T. Matsumura-Tundisi. 2005. Taxonomy, ecology, and geographical distribution of the species of the genus *Thermocyclops* Kiefer, 1927 (Copepoda, Cyclopoida) in São Paulo State, Brazil, with description of a new species. **Brazilian Journal of Biology 65** (3): 521-31.

Silva, W.M. 2008. Diversity and distribution of the free-living freshwater Cyclopoida (Copepoda:Crustacea) in the Neotropics. **Brazilian Journal of Biology 68** (4): 1099-1106.

Suárez-Morales, E.; J.W. Reid; F. Fiers & T.M. Iliffe. 2004. Historical biogeography and distribution of the freshwater cyclopine copepods (Copepoda, Cyclopoida, Cyclopinae) of the Yucatan Peninsula, Mexico. **Journal of Biogeography 31** (7): 1051-1063.

Suárez-Morales, E.; J.W. Reid & M. Elías-Gutiérrez. 2005. Diversity and Distributional Patterns of Neotropical Freshwater Copepods (Calanoida: Diaptomidae). **International Review of Hydrobiology 90** (1): 71-83.

Tash, J.C. 1971. Some Cladocera and Copepoda from the Upper Klamath River Basin. **Northwest Science 45** (4): 239-243.

Tranter, D.J. & P.E. Smith. 1979. Filtration performance, p. 27-56. *In*: D.J. Tranter (Ed.). **Zooplankton sampling.** Paris, Imprimerie Rolland, 3rd ed.

Ueda, H. & J.W. Reid. 2003. Copepoda: Cyclopoida – Genera *Mesocyclops* and *Thermocyclops*, p. 1-316. *In*: H.J.F. Dumont (Ed.). **Guides to the identification of the microinvertebrates of the continental waters of the world.** The Netherlands, Backhuys Publishers.

Zalocar de Domitrovic, Y.; A.S.G. Poi de Neiff & S.L. Casco. 2007. Abundance and diversity of phytoplankton in the Paraná River (Argentina) 220 km downstream of the Yacyretá Reservoir. **Brazilian Journal of Biology 67** (1): 53-63.

21

Geographic variation in *Caluromys derbianus* and *Caluromys lanatus* (Didelphimorphia: Didelphidae)

Raul Fonseca[1,2] & Diego Astúa[1,3]

[1]*Departamento de Zoologia, Universidade Federal de Pernambuco. Avenida Professor Moraes Rêgo, s/n, Cidade Universitária, 50670-901 Recife, PE, Brazil.*
[2]*Departamento de Zoologia, Universidade do Estado do Rio de Janeiro. 20550-013 Rio de Janeiro, RJ, Brazil.*
[3]*Corresponding author. E-mail: diegoastua@ufpe.br*

ABSTRACT. We analyzed the geographic variations in the shape and size of the cranium and mandible of two woolly opossums, *Caluromys derbianus* and *Caluromys lanatus*. Using geometric morphometrics we analyzed 202 specimens of *C. derbianus* and 123 specimens of *C. lanatus*, grouped in 7 and 9 populations, respectively. We found sexual dimorphism in shape variables only in the dorsal view of the cranium of *Caluromys derbianus*, which is not associated with geographical origin. We detected geographic variation in the size of the mandible in two populations (Nicaragua and Northern Panama), but no geographic variation in shape. The size of the cranium of *C. lanatus* varies significantly, with clinal variation in peri-Amazon populations, with a break between two populations, Bolivia and Paraguay. Shape analyses also revealed some separation between the Paraná population and all other populations. Our results suggest that the available name, *Caluromys derbianus*, should be maintained for all individuals throughout the geographic range of the species. The same is true for *Caluromys lanatus*, which can be separated into two distinct morphologic units, *Caluromys lanatus ochropus*, from the Amazon and Cerrado, and *Caluromys lanatus lanatus*, from the Atlantic forest.

KEY WORDS. Caluromyinae; geometric morphometrics; marsupial; Neotropics; skull; size and shape analysis.

The morphology and/or physiology of organisms usually vary across their distribution range. This is particularly true for species that are distributed over different biomes or biogeographic provinces (THORPE 1987). Such variation in intraspecific characters throughout a species' range is known as geographic variation (MAYR 1977). The study of geographic variation is key for understanding speciation and the role that ecological and geographical features may play in shaping biodiversity (HAFFER 1969, GOULD 1972, EMMONS 1984). Furthermore, geographic variation has been a central theme in evolutionary biology, from the works of Darwin to modern analyses based on molecular approaches (HALLGRÍMSSON & HALL 2005).

Morphological variation across geographical and environmental discontinuities occur in different small mammal groups, such as rodents (e.g., MACÊDO & MARES 1987, LESSA et al. 2005) and marsupials (e.g., LÓPEZ-FUSTER et al. 2000, HIMES et al. 2008). In the latter, variation can be found in external and cranial morphology and morphometric data (LEMOS & CERQUEIRA 2002, LÓPEZ-FUSTER et al. 2002, LÓSS et al. 2011), as well as in genetic characters (COSTA 2003, STEINER & CATZEFLIS 2004, BRAUN et al. 2005).

Woolly opossums of the genus *Caluromys* Allen, 1900 are part of a basal lineage within the living New World Didelphidae opossums (VOSS & JANSA 2009). *Caluromys* currently includes three species, *Caluromys derbianus* (Waterhouse, 1841), *Caluromys lanatus* (Olfers, 1818) and *Caluromys philander*

Linnaeus, 1758 which are widely distributed in forest areas of Central and South America (GARDNER 2008). Variation in external morphological traits has been found in *Caluromys lanatus* (Thomas, 1913) throughout its geographic range. Venezuelan populations of *Caluromys* species (LÓPEZ-FUSTER et al. 2008) also present morphometric variation. This phenotypic diversity lead to the recognition of a number of morphologically distinct groups: eight subspecies of *C. derbianus* (BUCHER & HOFFMANN 1980, GARDNER 2008), four of *C. philander* (CABRERA 1958, GARDNER 2005) and six of *C. lanatus* (CÁCERES & CARMIGNOTTO 2006, GARDNER 2008).

The purpose of this study was to evaluate and to quantify the morphological variation in the size and shape of the cranium and mandible of *Caluromys derbianus* and *Caluromys lanatus* throughout their geographic range. We used geometric morphometric tools to evaluate whether the variation supports the taxonomic status of each species and their currently recognized subspecies.

MATERIAL AND METHODS

We obtained 2D images of the crania in three views (dorsal, ventral and lateral), and lateral images of the mandibles. Only complete adult specimens, i.e., specimens with all three premolars and four molars fully erupted and functional (TRIBE

1990, Astúa & Leiner 2008) were photographed. Specimens analyzed were from the following institutions: Museu Nacional – Universidade Federal do Rio de Janeiro (MN), Museu de Zoologia da Universidade de São Paulo (MZUSP), Museu Paraense Emílio Goeldi (MPEG), Coleção de Mamíferos do Departamento de Zoologia da Universidade Federal de Minas Gerais (UFMG), Museu de História Natural Capão da Imbúia (MHNCI), Museo Argentino de Ciencias Naturales "Bernardino Rivadavia" (MACN), Museo de Historia Natural de la Universidad Nacional Mayor de San Marcos (MUSM), American Museum of Natural History (AMNH), Field Museum of Natural History (FMNH), Louisiana State University, Museum of Natural Science (LSUMZ), Museum of Southwestern Biology (MSB), Museum of Vertebrate Zoology (MVZ), Kansas University, Museum of Natural History (KU) and National Museum of Natural History (USNM).

We digitized a total of 92 landmarks – 28 in dorsal, 28 in ventral, 22 in lateral views of the cranium, and 14 landmarks on the mandible – using TPS Dig (Rohlf 2006) (Fig. 1, Appendix 1). All landmarks were tested for repeatability (Falconer & Mackay 1996), which was set at 85% for inclusion in subsequent analyses.

We applied a a Generalized Procrustes Analysis (GPA) to all landmark configurations (Rohlf & Slice 1990), to remove the effects of isometric size, orientation and position. Conse-

quently, only shape information was retained (Adams et al. 2004, 2013). We obtained two formally independent set of variables, used in the subsequent analyses. One set includes centroid size for all specimens. Centroid size is the univariate size variable resulting from the squared-root of sums of the squared distances between each landmark and the centroid of its configuration. This set was used in the analyses of geographic variation in the size of the studied structures (for more details see Zelditch et al. 2012). GPA also yields the partial warps and uniform components, a set of variables that retain all the information on the shape of the landmark configuration of the studied structures that were used in the analyses of geographic variation in shape. Further detail on the geometric morphometric procedures can be found in Zelditch et al. (2012).

We obtained the geographic coordinates of the collecting localities of each specimen from their skin tags. When coordinates were not in the tags, we used standard ornithological gazetteers (Paynter 1982, 1989, 1992, 1993, 1995, 1997) to recover them. Specimens from different localities were grouped into populations based on the features of the ecoregions (Olson et al. 2001) found in the distribution of both species (specimens from geographically close localities in the same ecoregion were pooled into populations). Next, to increase the sample size of populations resulting from the classification using ecoregions,

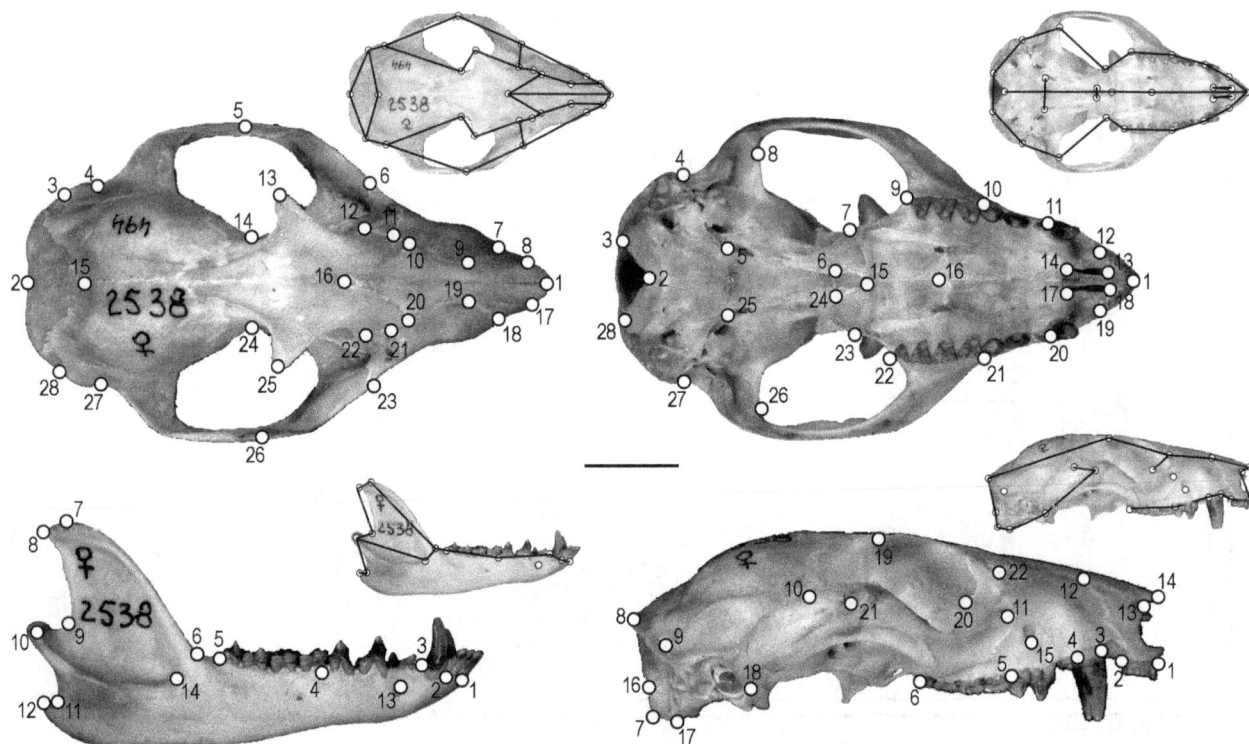

Figure 1. Landmarks used in the cranium and mandible. Smaller versions of each view include landmarks with links, as used deformation grids in subsequent figures. See Appendix 1 for detailed description of landmark locations. Scale bar: 1 cm.

we decided to pool the populations that were geographically closer to each other and which lacked morphometric divergence.

We examined a total of 202 specimens of *Caluromys derbianus* (the number of specimens analyzed in each view may vary because missing structures in one view may preclude the use of a photograph, while the photographs of the same specimen from other views can be used). The specimens were divided into seven populations: Colombian, Ecuadorian and Peruvian individuals (n = 9), Panama-Colombia (n = 16), Southern Panama (n = 22), Northern Panama (n = 51), Nicaragua (n = 54), Honduras (n = 37) and Mexico (n = 15) (Fig. 2). Likewise, we examined a total of 123 specimens of *Caluromys lanatus*, which were divided into 9 populations: Northern Venezuela (n = 11), Southern Venezuela (n = 8), Colombia (n = 17), Northern Peru (n = 22), Iquitos (n = 22), Peru-Bolivia (n = 15), Paraná (n = 8), Trombetas (n = 13), and Cerrado (n = 7) (Fig. 3). The list of all examined specimens with localities is presented in Appendix 2.

Literature information on the absence of sexual dimorphism in both species (Astúa 2010) was obtained from a smaller and geographically restricted dataset. With this in mind we re-evaluated the existence of sexual size dimorphism through a t-test on centroid size, and the existence of sexual shape dimorphism through a Hotteling T² test on shape variables. Since

several populations were represented by only a few specimens,we pooled all males into one group and all females into another regardless of their geographic origin, in order to increase sample size and to avoid a type I error. To evaluate geographic variation in size, we compared populations with ANOVAs on centroid sizes, followed by Tukey *a posteriori* tests. To evaluate geographic variation in shape we compared shape variables between populations using Canonical Variates Analyses (CVA), following Webster & Sheets (2010), given that our total sample size was much larger than [(2k − 4) + (G − 1)], where k is the number of variables and G is the amount of groups analyzed. For each view, this parameter ranged from 30 to 58 for *Caluromys derbianus*, and 32 to 60 for *C. lanatus*, indicating that running a CVA is appropriate. Because all analyses were repeated on four views of both species, we employed Bonferroni correction again, using a significant p-value of 0.0125 (0.05/4).

RESULTS

Sexual dimorphism

Neither species presented sexual dimorphism in size. Sexual dimorphism was observed only in the shape of the dorsal portion of the cranium of *Caluromys derbianus* (Hotteling

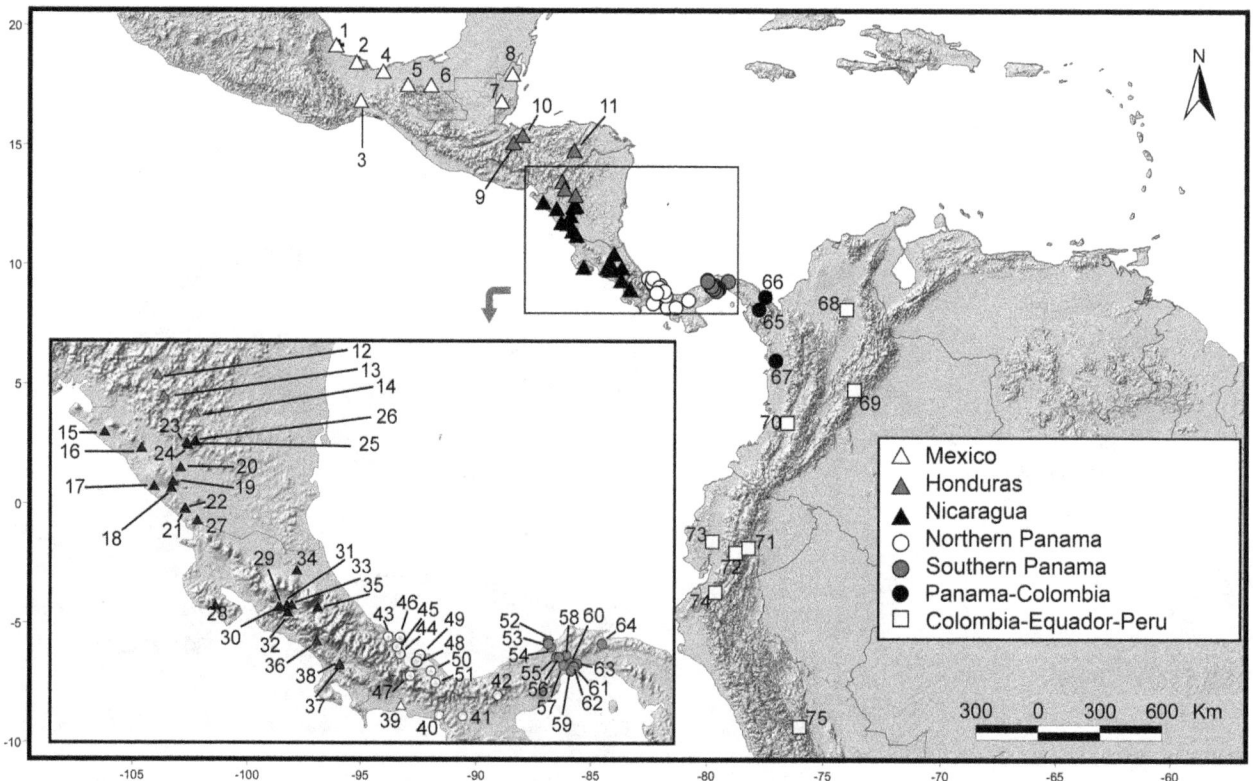

Figure 2. Distribution of the localities of *Caluromys derbianus* with specimens included in this study. Localities were grouped in populations for subsequent analyses, and are labelled accordingly. Numbers indicate localities as listed in Appendix 2.

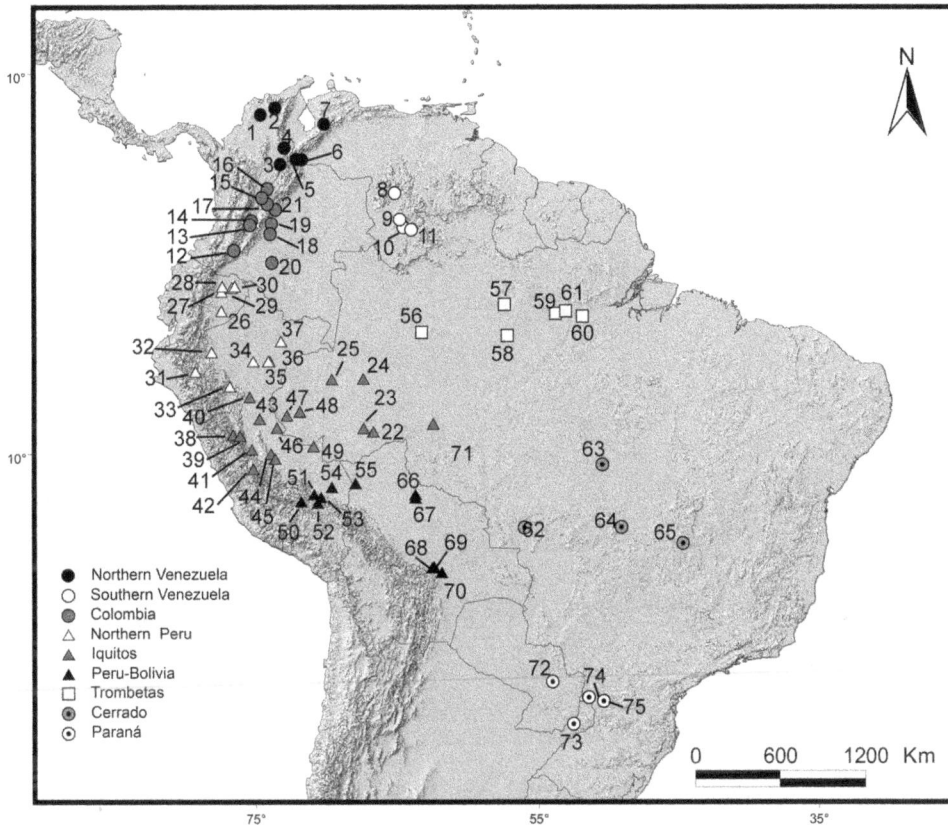

Figure 3. Distribution of the localities of *Caluromys lanatus* with specimens included in this study. Localities were grouped in populations for subsequent analyses, and are labelled accordingly. Numbers indicate localities as listed in Appendix 2.

$T^2 = 0.858$, $F = 1.82$; d.f. $= 56$, $p < 0.01$, 85 males, 90 females). In view of the absence sexual dimorphism in size and shape variables among individuals in all other views of both species, we decided to pool the sexes together within populations for subsequent analyses. This allowed us to include in the analyses specimens for which the sex was unknown.

Geographic variation in *Caluromys derbianus*

When analyzing size variation, we only found a statistically significant difference in mandible size, between the Nicaragua and Northern Panama populations (ANOVA $F = 2.89$, $p < 0.01031$, $p < 0.002$, *post-hoc* Tukey test). As for shape variation, the CVA scores overlapped considerably, indicating little morphometric divergence in size (Fig. 4). Given that the variation within each population was equal to or larger than the variation between populations, we concluded that the variation is not geographically structured and that the populations cannot be considered morphologically distinct.

Geographic variation in *Caluromys lanatus*

Under all views, size varied geographically (ANOVA, Dorsal: $F = 11.02$, $p < 0.0001$; Lateral: $F = 3.66$, $p < 0.001$; Ventral:

$F = 9.62$, $p < 0.0001$; Mandible: $F = 10.91$, $p < 0.0001$), but no clear grouping was observed among populations. However, a north-south clinal variation in skull size can be inferred, with specimens increasing in size from Colombia (smallest) to Bolivia (largest), with Ecuadorian and Peruvian specimens presenting intermediate sizes. This trend is then interrupted in southern Bolivia, with specimens from southeastern Brazil, Paraguay and Argentina being smaller than their Bolivian counterparts (Fig. 5).

Caluromys lanatus has a conserved skull shape throughout its geographic range. CVA scores show a partial separation of the Paraná population from all others, due to a variation in the morphology of the occipital and posterior roots of the squamosal, which are larger in Paraná specimens than in other individuals. The morphology of the rostrum also varies, with short and narrow nasals and basicranium with short frontals and longitudinal elongation of parietals (visualized through displacement of landmarks at the postorbital constriction) in the dorsal view of the cranium (Fig. 6). Additionally, an increase in occipital width, a more horizontally aligned molar tooth row, and shorter and narrow rostrum are found in Paraná individuals, in lateral view of the cranium (Fig. 7).

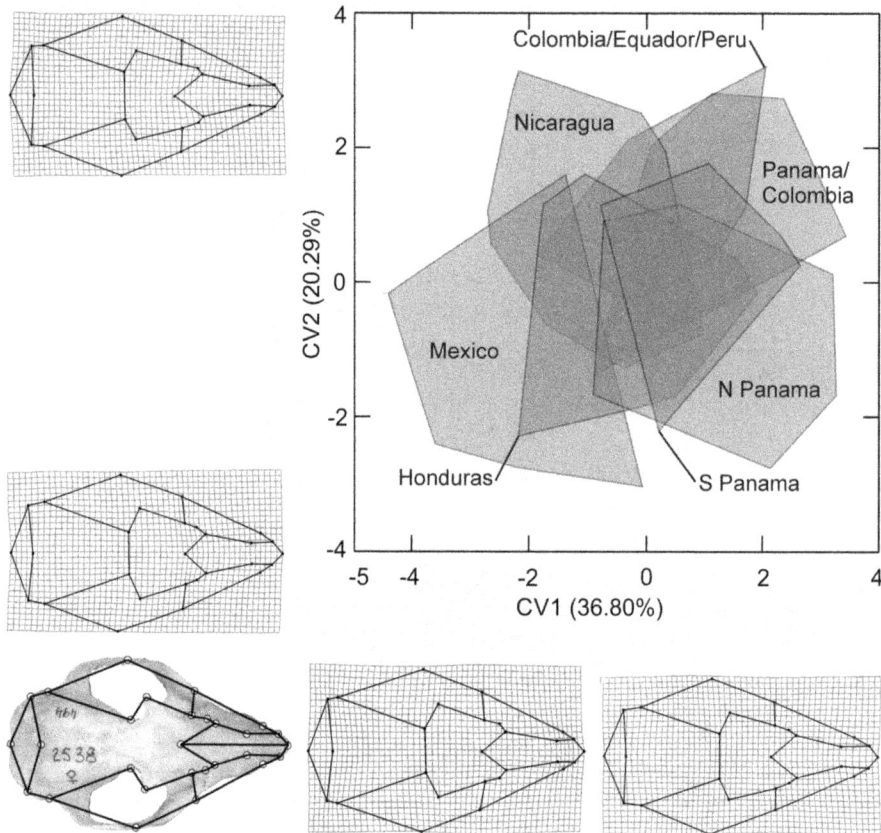

Figure 4. Canonical Variates Analysis on shape variables (partial warps and uniform components) of the skull in dorsal view of *Caluromys derbianus*, using localities as grouping factors, and percentage of variance explained by the first two CVs. Only the convex hulls for each population are shown. Grids indicate deformation associated with the extremes of each CV, from a multivariate regression of shape variables onto CV scores. Overlap for all other views are very similar, therefore only the dorsal view of the cranium is shown.

DISCUSSION

Structured geographic variation in cranial size and shape was not detected in *Caluromys derbianus*. However, it was observed in *Caluromys lanatus* populations. Despite the fact that subspecies have been recognized for *C. derbianus*, its populations belong to a single morphologic unit, which is spread throughout the geographic distribution of the species. Our results also corroborate that *Caluromys lanatus* is one species, but with two distinct morphological groups, one in the Amazon-Cerrado and the other in the Atlantic forest.

The absence of sexual size dimorphism in the skull of these species was already discussed (ASTÚA 2010), although that analysis, unlike ours, detected significant sexual shape dimorphism in both species.

Geographic variation in *Caluromys derbianus*

We did not find any evidence of structured geographical variation in the size of the skull of *Caluromys derbianus*, despite

its occurrence in the congeneric species *C. lanatus* (this study) and *C. philander* (OLIFIERS et al. 2004). We were also unable to detect a pattern in the geographic variation of the shape and size of the skull that would match the current taxonomic structures proposed for this species at the subspecific level. BUCHER & HOFFMANN (1980) and GARDNER (2008) divided *Caluromys derbianus* into seven subspecies and one trans-Andean "unspecified" population. These populations were based on morphological differences such as fur color. As we used only cranial quantitative data, it is possible that other characters, particularly in the external morphology, may be the reason for the high number of subspecies. In particular, pelage color, which was not assessed in this study, is well known to vary geographically in this and other marsupial genera (THOMAS 1913, GOODWIN 1942) and might explain the discrepancy between the existing classification and the one that results from our quantitative results from skull morphology, which failed to support a separation.

The distribution of *Caluromys derbianus* represents a continuum of populations on a N-S stripe, most of which are in

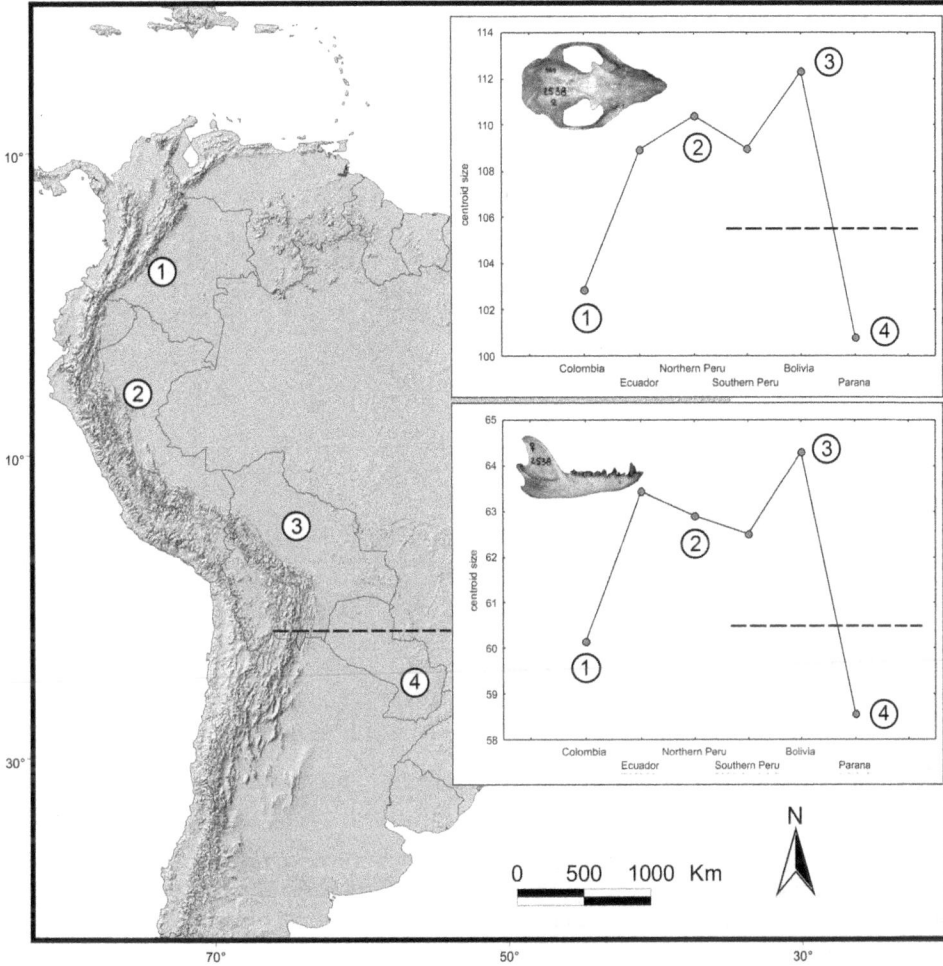

Figure 5. Clinal variation in size of the skull and mandible of populations of *Caluromys lanatus* along the east of the Andes, from Colombia to Bolivia, with a break between Bolivian and Northern Paraguay/Southern Brazil populations, indicated by the dashed line. Numbers in the map refer to the same points in the two graphs.

Central America and the remaining populations in the Andes in South America. The absence of geographic variation in the size of the cranium and mandible shape of this species is noteworthy, since several geographic and ecological discontinuities found throughout its distribution range are believed to cause variation among populations of other taxa (Savage 1987, Pérez-Emán 2005, Castoe et al. 2009).

Geographic variation in *Caluromys lanatus*

Clinal variation occurs throughout the range of many mammals (Storz et al. 2001, Cardini et al. 2007). We found clinal variation in the size of the skull of *Caluromys lanatus* from Andean populations, to the Bolivian-Paraguayan border, coinciding with those populations that overlap less in shape analyses. Even though we have not analyzed molecular data, we believe that the large overlap of CVA scores among all

Amazon populations can be associated with reduced genetic divergence in this species. The latter has been already noted for populations distributed in this area (Patton et al. 2000, Patton & Costa 2003).

The divergence among the Paraná population and the others may correspond to the geographic differences between the Amazon and the Atlantic Rainforest. A similar variation pattern has also been observed to occur in *Didelphis*, *Marmosa*, *Caluromys philander* and *Metachirus nudicaudatus* (Costa 2003, Patton & Costa 2003). Both morphological and genetic divergence were observed in these species. Similar results were also recorded for rodent genera such as *Rhipidomys*, *Oecomys*, *Hylaeamys* and *Euryoryzomys* (Costa 2003). Populations from Paraná are ecologically separated from others by the Chaco – xerophytic plant cover, located in Argentina and Paraguay (Marco & Páezw 2002, Boletta et

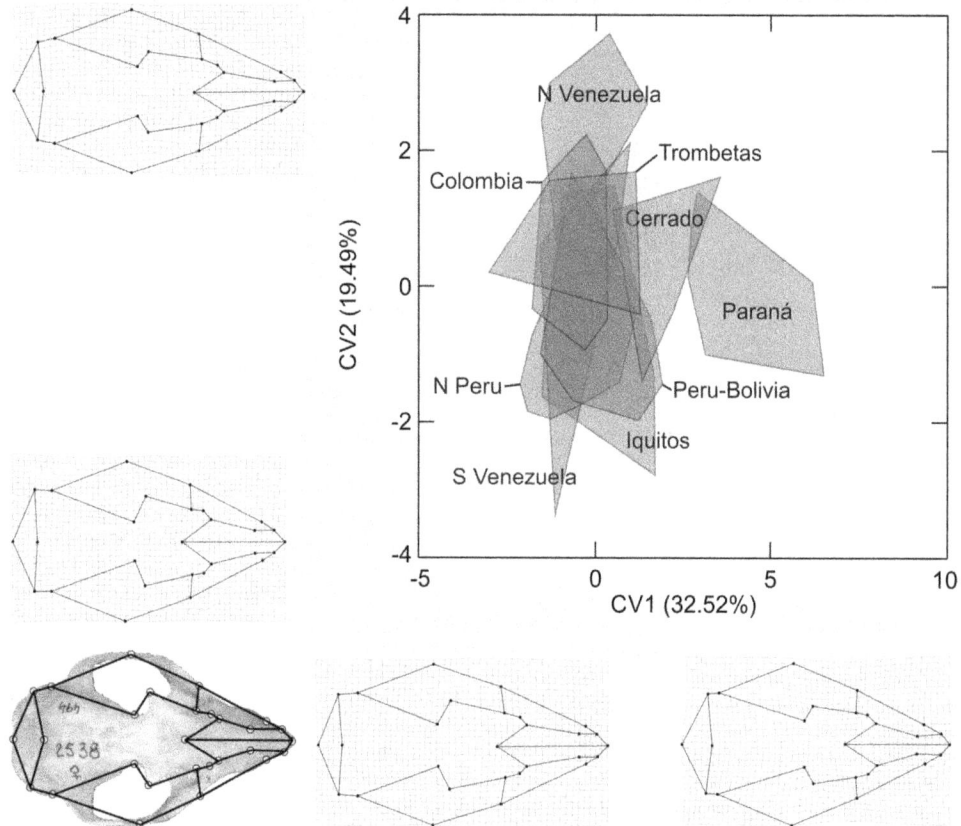

Figure 6. Canonical Variates Analysis on shape variables (partial warps and uniform components) of the cranium in dorsal view of *Caluromys lanatus*, using localities as grouping factors, and percentage of variance explained by the first two CVs. Only the convex hulls for each population are shown. Grids indicate deformation associated with the extremes of each CV, from a multivariate regression of shape variables onto CV scores.

al. 2006), which is characterized by medium and large trees such as Bignoniaceae, Leguminosae and grass fields (Pennington et al. 2000). The increase in the Araucaria cover in the early Holocene (Ledru 1993, Salgado-Labouriau et al. 1998) over open areas may have served as a bridge between the forested areas of the Atlantic forest and the Amazon (Ab'Saber 2000). This plant cover probably allowed the dispersion of *Caluromys lanatus* from the Amazon and Cerrado to the southern Atlantic Rainforest (Costa 2003, Patton & Costa 2003), where these new populations were later isolated by open lands that arose between these areas (Ledru et al. 1998, Van der Hammen & Hooghiemstra 2000, Behling 2002). This contact and subsequent isolation hypothesis is particularly likely for *Caluromys lanatus*, since this species is strictly arboreal. Environmental discontinuities that incur in canopy fragmentation may hinder population movements (Pires et al. 2002), thus providing an effective ecological barrier like the one that has been associated with speciation of the congeneric *Caluromys philander* (Lira et al. 2007). Morphologi-

cal similarities between populations from Central Brazil and the Amazon may be explained by the fact that Cerrado vegetation may not be uniformly affected by climatic changes (Salgado-Labouriau et al. 1997). At higher altitudes the plant composition was less altered even in the dry periods of the Pleistocene and may have extended to lower areas during cold periods (Bush et al. 2004). Grassland vegetation may have replaced only low-altitude forests (Salgado-Labouriau et al. 1997, 1998). Due to climatic and pluviometric oscillations, eventual expansions of gallery forests may have created ecological corridors that allowed faunal and floristic population flow among Cerrado, Llanos, Amazonia and even Gran-Sabana (Cerqueira 1982, Ledru 2002, Oliveira-Filho & Ratter 1995 apud De Oliveira et al. 2005). Gallery forests house twice as many forest-related species than the entire Cerrado *latu sensu* (Johnson et al. 1999). These forested areas may not have been totally affected by climatic changes and may have been used as a corridor that kept Amazonian and Cerrado populations in contact (Cardoso & Bates 2002).

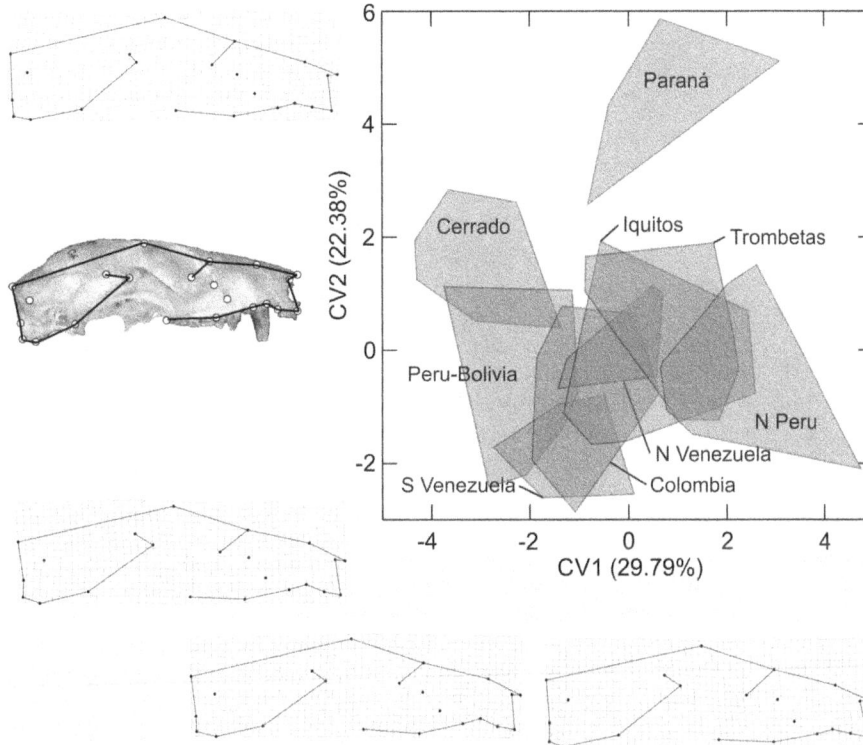

Figure 7. Canonical Variates Analysis on shape variables (partial warps and uniform components) of the cranium in lateral view of *Caluromys lanatus*, using localities as grouping factors, and percentage of variance explained by the first two CVs. Only the convex hulls for each population are shown. Grids indicate deformation associated with the extremes of each CV, from a multivariate regression of shape variables onto CV scores.

Potential implications for the taxonomic classification of *Caluromys derbianus* and *Caluromys lanatus*

The similar skull morphology shared by all populations of *Caluromys derbianus* suggest that the seven subspecies – *C. derbianus aztecus*, *C. d. canutus*, *C. d. centralis*, *C. d. derbianus*, *C. d. fervidus*, *C. d. nauticus* and *C. d. parvidus* – may be considered a unique species on morphometric grounds. Likewise, the lack of geographic variation among the Amazon and Cerrado populations of *Caluromys lanatus* suggest that three of the four subspecies recognized by Cabrera (1958) – *C. lanatus cicur*, *C. lanatus ornatus* and *C. lanatus ochropus* – and four of the six suggested by Gardner (2008) – *C. lanatus cicur*, *C. lanatus ornatus*, *C. lanatus ochropus* and *C. lanatus vitalinus* can be lumped based on morphometrical data. The geographic variation found in skull morphometric data of individuals from the southern Atlantic Forest also suggest that two subspecies proposed by Gardner (2008) – *C. lanatus lanatus* and *C. lanatus vitalinus* from southern Brazil can also be lumped.

All these subspecies were described based on external morphological characters, such as body, facial, dorsal, caudal or feet color, characters that usually present geographic variation (Thomas 1899, 1913, Allen 1904, Hollister 1914, Goodwin

1942). *Caluromys* species were first described based on morphological characters of a single or a few individuals; subspecies were generally described after comparing individual variation with the holotype. For this reason, it cannot ruled out that these subspecies were based on individual variation. In all cases, pending a proper extensive review of coat color or other morphological variation in *Caluromys*, our extensive and quantitative results do not support separation of these taxa.

However, because phenotype is mainly the expression of the underlying genotype, morphological divergence is often interpreted as evidence of specific status. In didelphids, for example, morphological evidence has been used to support splitting of the black-eared and the white-eared opossums (Cerqueira & Lemos 2000, Lemos & Cerqueira 2002) of the genus *Didelphis*, and Bolivian species of *Marmosops* (Voss et al. 2004). As such, it it possible that the morphologic groups found here may represent distinct species (see, however Lóss et al. 2011, for a situation where morphologic differentiation does not coincide with species limits). The recognition of southern and southeastern populations of South American didelphids as distinct species appears to be a recurrent pattern that emerges after a deeper analysis of the existing variation, such as in *Phi-*

lander (Patton & da Silva 1997) and *Marmosa* (Patton & Costa 2003). Such changes are actually the reflection of our still incomplete knowledge on the taxonomy and systematics of didelphids.

A proper and definite appraisal of the taxonomic status of both woolly opossums would require an integrative approach (including other phenotypical and genetic characters) to unveil their actual status. Especially among *Caluromys lanatus* populations, a molecular approach may be useful to assess if these divergent groups constitute distinct evolutionary lineages that would ultimately validate their status as distinct species. Pending this, we suggest that the available name *Caluromys derbianus* (Waterhouse, 1841) is maintained for all individuals across the geographic distribution of its populations. The name *Caluromys lanatus* (Olfers, 1818) should also be considered valid, with at least two distinct morphometric units, namely *Caluromys lanatus ochropus*, representing Amazon and Cerrado populations, and *Caluromys lanatus lanatus*, encompassing Atlantic forest individuals.

ACKNOWLEDGEMENTS

We are grateful to the following curators and/or collection managers for support and access to the specimens under their care: R. Voss (AMNH), J.L. Patton and C. Conroy (MVZ), J. Salazar-Bravo and W. Gannon (MSB), R. Timm (KU), B. Patterson and M. Schulenberg (FMNH), A. Gardner, L. Gordon and C. Ludwig (USNM), L. Costa, Y. Leite and B. Andrade (UFMG), J.A. Oliveira, L.F. Oliveira, L. Salles and S. Franco (MN), M. de Vivo and J. Barros (MZUSP), S. Aguiar (MPEG), V. Pacheco and E. Vivar-Pinares (MUSM), G. Tebet e T.C. Margarido (MHNCI), and M. Hafner (LSUMZ). RF was supported with a M.Sc. fellowship from CNPq, and funds for travel and equipment to DA were available by FAPESP, FACEPE and an American Society of Mammalogists Grants-in-Aid. DA is currently supported by a CNPq fellowship (306647/2013-3). We are grateful to M. de Vivo, L.P. Costa, M. Montes, I. Bandeira, P. Pilatti, R. Carvalho, A. Cardini, N. Caceres and one anonymous reviewer for their helpful suggestions in the numerous versions of this manuscript.

REFERENCES

Adams DC, Rohlf FJ, Slice DE (2004) Geometric morphometrics: ten years of progress following the 'revolution'. **Italian Journal of Zoology 71**(1): 5-16. doi: 10.1080/11250000409356545

Adams DC, Rohlf FJ, Slice DE (2013) A field comes of age: geometric morphometrics in the 21st century. **Hystrix, the Italian Jornal of Mammalogy 24**(1): 7-14. doi: 10.4404/hystrix-24.1-6283

Ab'saber AN (2000) Spaces occupied by the expansion of dry climates in South America during the Quaternary ice ages. **Revista do Instituto Geológico 21**(1/2): 71-78.

Allen JA (1904) Mammals from southern Mexico and Central and South America. **Bulletin American Museum of Natural History 20**(4): 29-80.

Astúa D (2010) Cranial sexual dimorphism in New World marsupials and a test of Rensch's rule in Didelphidae. **Journal of Mammalogy 91**(4): 1011-1024. doi: 10.1644/09-MAMM-A-018.1

Astúa D, Leiner NO (2008) Tooth eruption sequence and replacement pattern in woolly opossums, genus *Caluromys* (Didelphimorphia: Didelphidae). **Journal of Mammalogy 89**(1): 244-251. doi: 10.1644/06-MAMM-A-434.1

Behling H (2002) South and southeast Brazilian grasslands during Late Quaternary times: a synthesis. **Palaeogeography, Palaeoclimatology, Palaeoecology 177**: 19-27. doi: 10.1016/S0031-0182(01)00349-2

Boletta PE, Ravello AC, Planchuelo AM, Grilli M (2006) Assessing deforestation in the Argentine Chaco. **Forest Ecology and Management 228**: 108-114. doi: 10.1016/j.foreco.2006.02.045

Braun JK, Van den Bussche RA, Morton PK, Mares MA (2005) Phylogenetic and biogeographic relationships of mouse opossums *Thylamys* (Didelphimorphia, Didelphidae) in southern South America. **Journal of Mammalogy 86**(1): 147-159. Available online at: http://jmammal.oxfordjournals.org/content/86/1/147 [Accessed: 20 April 2015]

Bucher JE, Hoffmann RS (1980) *Caluromys derbianus*. **Mammalian Species 140**: 1-4.

Bush MB, De Oliveira PE, Colinvaux PA, Miller MC, Moreno JE (2004) Amazonian paleoecological histories: one hill, three watersheds. **Palaeogeography, Palaeoclimatology, Palaeoecology 214**(4): 359-393. doi: 10.1016/j.palaeo.2004.07.031

Cabrera A (1958) Catalogo de los mamiferos de America del Sur. **Revista del Museo Argentino de Ciencias Naturales "Bernardino Rivadavia", Ciencias Zoológicas 4**(1): 1-46.

Cáceres NC, Carmignotto AP (2006) *Caluromys lanatus*. **Mammalian Species 803**: 1-6. doi: 10.1644/803.1

Cardini A, Jansson A, Elton S (2007) A geometric morphometric approach to the study of ecogeographical variation in vervet monkeys. **Journal of Biogeography 34**: 663-678. doi: 10.1111/j.1365-2699.2007.01731.x

Cardoso JMD, Bates JM (2002) Biogeographic patterns and conservation in the South American Cerrado: A tropical Savanna spot. **BioScience 52**(3): 225-233. doi http://dx.doi.org/10.1641/0006-3568(2002)052[0225:BPACIT]2.0.CO;2

Castoe TA, Daza JM, Smith ER, Sasa MM, Kuch U, Campbell JA, Chippindale PT, Parkinson CL (2009) Comparative phylogeography of pitvipers suggests a consensus of ancient Middle American highland biogeography. **Journal of Biogeography 36**: 88-103.doi: 10.1111/j.1365-2699.2008.01991.x

Cerqueira R (1982) South American landscapes and their mammals, p. 539-539. In: Mares MA, Genoways HH (Eds.). **Mammalian Biology in South American.** Pittsburg, University of Pittsburg, Special Publication Pymatuning Laboratory of Ecology.

CERQUEIRA R, LEMOS B (2000) Morphometric differentiation between Neotropical black-eared opossums, *Didelphis marsupialis* and *D. aurita* (Didelphimorphia, Didelphidae). **Mammalia 64**: 319-327. doi: 10.1515/mamm.2000.64.3.319

COSTA LP (2003) The historical bridge between the Amazon and the Atlantic Forest of Brazil: a study of molecular phylogeography with small mammals. **Journal of Biogeography 30**(1): 71-86. doi: 10.2307/827350

DE OLIVEIRA PE, BEHLING H, LEDRU M-P, BARBERI M, BUSH M, SALGADO-LABOURIAU ML, MEDEANIC S, BARTH OM, BARROS MA, SCHELL-YHERT R (2005) Paleovegetação e paleoclimas do Quaternário do Brasil, p. 52-74. In: SOUZA CRG, SUGUIO K, OLIVEIRA AMS, DE OLIVEIRA PE (Ed.). **Quaternário do Brasil**. São Paulo, Holos, Associação Brasileira de Estudos do Quaternário.

EMMONS L (1984) Geographic variation in densities and diversities of non-flying mammals in Amazonia. **Biotropica 16**(3): 210-222. doi: 10.2307/2388054

FALCONER DS, MACKAY TFC (1996) **Introduction to Quantitative Genetics**. Harlow, Longman.

GARDNER AL (2005) Order Didelphimorphia, p. 3-18. In: WILSON DE, REEDER DM (Ed.). **Mammal species of the world: a taxonomic and geographic reference**. Baltimore, Johns Hopkins University Press, 3rd ed.

GARDNER AL (2008) **Mammals of South America**. Chicago, The University of Chicago Press.

GOODWIN GG (1942) Mammals of Honduras. **Bulletin American Museum of Natural History 79**: 107-195.

GOULD SJ, JOHNSTON RF (1972) Geographic variation. **Annual Review of Ecology and Systematics 3**: 457-499. doi: 10.1146/annurev.es.03.110172.002325

HAFFER J (1969) Speciation in Amazonian forest birds. **Science 165**: 131-137. doi: 10.1126/science.165.3889.131

HALLGRÍMSSON B, HALL BK (2005) **Variation. A Central Concept in Biology.** Elsevier Academic Press.

HIMES CMT, GALLARDO MH, KENAGY GJ (2008) Historical biogeography and post-glacial recolonization of South American temperate rain forest by the relictual marsupial *Dromiciops gliroides* **Journal of Biogeography 35**: 1415-1424. doi: 10.1111/j.1365-2699.2008.01895.x

HOLLISTER N (1914) Four new mammals from tropical America. **Proceedings of the Biological Society of Washington 27**: 103-106.

JOHNSON MA, SARAIVA PM, COELHO D (1999) The role of gallery forests in the ditribution of Cerrado mammals. **Revista Brasileira de Biologia 59**(3): 421-427. doi: 10.1590/S0034-71081999000300006

LEDRU M-P (1993) Late Quaternary environment and climatic changes in Central Brazil. **Quaternary Research 39**: 90-98. doi: 10.1006/qres.1993.1011

LEDRU M-P, SALGADO-LABOURIAUB ML, LORSCHEITTERC ML (1998) Vegetation dynamics in southern and central Brazil during the last 10,000yr. B.P. **Review of Palaeobotany and Palynology 99**: 131-142. doi: 10.1016/S0034-6667(97)00049-3

LEDRU MP (2002) Late Quaternary history and evolution of the Cerrados as revealed by palynological records, p. 33-50. In: OLIVEIRA PS, MARQUIS RJ (Eds.). **The Cerrados of Brazil: ecology and natural history of a Neotropical Savanna.** New York, Columbia University Press.

LEMOS B, CERQUEIRA R (2002) Morphological differentiation in the white-eared opossum group (Didelphidae: *Didelphis*). **Journal of Mammalogy 83**(2): 354-369. Available online at: http://jmammal.oxfordjournals.org/content/83/2/354 [Accessed: 20 April 2015]

LESSA G, GONÇALVES PR, PESSÔA LM (2005) Variação geográfica em caracteres cranianos quantitativos de *Kerodon rupestris* (Wied, 1820) (Rodentia, Caviidae). **Arquivos do Museu Nacional 63**(1): 75-88.

LIRA PK, FERNANDEZ FAS, CARLOS HSA, CURZIO PL (2007) Use of fragmented landscape by three species of opossum in southeastern Brazil. **Journal of Tropical Ecology 23**: 427-435. doi: 10.1017/S0266467407004142

LOPEZ-FUSTER MJ, PEREZ-HERNANDEZ R, VENTURA J (2008) Morphometrics of genus *Caluromys* (Didelphimorphia: Didelphidae) in northern South America. **Orsis 23**: 97-114.

LÓPEZ-FUSTER MJ, PÉREZ-HERNANDEZ R, VENTURA J, SALAZAR M (2000) Effect of environment on skull-size variation in *Marmosa robinsoni* in Venezuela. **Journal of Mammalogy 81**(3): 829. doi: 10.1093/jmammal/81.3.829

LÓPEZ-FUSTER MJ, SALAZAR M, PÉREZ-HERNÁNDEZ R, VENTURA J (2002) Craniometrics of the orange mouse opossum *Marmosa xerophila* (Didelphimorphia: Didelphidae) in Venezuela. **Acta Theriologica 47**(2): 201-209. doi: 10.1007/BF03192460

LÓSS S, COSTA LP, LEITE YLR (2011) Geographic variation, phylogeny and systematic status of *Gracilinanus microtarsus* (Mammalia: Didelphimorphia: Didelphidae). **Zootaxa 2761**: 1-33.

MACÊDO RH, MARES MA (1987) Geographic variation in the South American cricetine rodent *Bolomys lasiurus*. **Journal of Mammalogy 68**(3): 578-594. doi: 10.2307/1381594

MARCO D, PÁEZW SA (2002) Phenology and phylogeny of animal-dispersed plants in a Dry Chaco forest (Argentina). **Journal of Arid Environments 52**: 1-16. doi: 10.1006/jare.2002.0976

MAYR E (1977) **Populações, espécies e evolução.** São Paulo, Companhia Editora Nacional, Editora da Universidade de São Paulo.

OLSON DM, DINERSTEIN E, WIKRAMANAYAKE ED, BURGESS ND, POWELL GVN, D'AMICO JA, ITOUA I, STRAND HE, MORRISON JC, LOUCKS CJ, ALLNUTT TF, RICKETTS TH, KURA Y, LAMOREUX JF, WETTENGEL WW, HEDAO P, KASSEM KR (2001) Terrestrial ecoregions of the World: a new map of life on Earth. **BioScience 51**(11): 933-938. doi: http://dx.doi.org/10.1641/0006-3568(2001)051[0933:TEOTWA]2.0.CO;2

OLIFIERS N, VIEIRA MV, GRELLE CEV (2004) Geographic range and body size in Neotropical marsupials. **Global Ecology and Biogeography 13**(5): 439-444. doi: 10.1111/j.1466-822X.2004.00115.x

Patton JL, Costa LP (2003) Molecular phylogeography and species limits in rainforest didelphid marsupials of South America, p. 63-81. In: Jones ME, Dickman CR, Archer M (Eds.). **Predators with pouches: The Biology of Carnivorous Marsupials.** Collingwood, CSIRO Publishing.

Patton JL, Da Silva MNF, Malcom JR (2000) Mammals of the Rio Juruá and the evolutionary and ecological diversification of Amazonia. **Bulletin of the American Museum of Natural History 244:** 1-306.

Patton JL, Da Silva MNF (1997) Definition of species of pouched four-eyed opossums (Didelphidae, *Philander*). **Journal of Mammalogy 78**(1): 90. doi: 10.2307/1382642

Paynter Jr RA (1982) **Ornithological Gazetteer of Venezuela.** Cambridge, Museum of Comparative Biology, Harvard University.

Paynter Jr RA (1989) **Ornithological Gazetteer of Paraguay.** Cambridge, Museum of Comparative Biology, Harvard University.

Paynter Jr RA (1992) **Ornithological Gazetteer of Bolivia.** Cambridge, Museum of Comparative Biology, Harvard University.

Paynter Jr RA (1993) **Ornithological Gazetteer of Ecuador.** Cambridge, Museum of Comparative Biology, Harvard University.

Paynter Jr RA (1995) **Ornithological Gazetteer of Argentina.** Cambridge, Museum of Comparative Biology, Harvard University.

Paynter Jr RA (1997) **Ornithological Gazetteer of Colombia.** Cambridge, Museum of Comparative Biology, Harvard University.

Pennington RT, Prado DE, Pendry CA (2000) Neotropical Seasonally Dry Forests and Quaternary vegetation changes. **Journal of Biogeography 27**(2): 261-273. doi: 10.1046/j.1365-2699.2000.00397.x

Pérez-Emán JL (2005) Molecular phylogenetics and biogeography of the Neotropical redstarts (*Myoborus*; Aves, Parulinae). **Molecular Phylogenetics and Evolution 37:** 511-528. doi: 10.1016/j.ympev.2005.04.013

Pires AS, Lira PK, Fernandez FAS, Schittini GM, Oliveira LC (2002) Frequency of movements of small mammals among Atlantic Coastal Forest fragments in Brazil. **Biological Conservation 108**(2): 229-237. doi: 10.1016/S0006-3207(02)00109-X

Rohlf FJ (2006) **TpsDig.** Stony Brook, Department of Ecology and Evolution, State University of New York.

Rohlf FJ (2011) **TpsRegr.** Stony Brook, Department of Ecology and Evolution, State University of New York.

Rohlf FJ, Slice D (1990) Extensions of the Procrustes method for the optimal superimposition of landmarks. **Systematic Zoology 39:** 40-59. doi: 10.2307/2992207

Salgado-Labouriau ML, Barberi M, Ferraz-Vicentini KR, Parizzi MG (1998) A dry climatic event during the late Quaternary of tropical Brazil. **Review of Palaeobotany and Palynology 99:** 115-129. doi: 10.1016/S0034-6667(97)00045-6

Salgado-Labouriau ML, Casseti V, Ferraz-Vicentini KR, Martin L, Soubiés F, Suguio K, Turcq B (1997) Late Quaternary vegetacional and climatic changes in Cerrado and palm swamp from Central Brazil. **Palaeogeography, Palaeoclimatology, Palaeoecology 128:** 215-226. doi: 10.1016/S0031-0182(96)00018-1

Savage JM (1987) Systematics and distribution of the Mexican and Central American rainfrogs of the *Eleutherodactyllus gollmeri* group (Amphibia: Leptodactylidae). **Fieldiana, Zoology, 33:** 1-57.

Steiner C, Catzeflis FM (2004) Genetic variation and geographical structure of five mouse-sized opossums (Marsupialia, Didelphidae) throughout the Guiana Region. **Journal of Biogeography 31**(6): 959-973. doi: 10.1111/j.1365-2699.2004.01102.x

Storz JF, Balansingh J, Bhat HR, Nathan PT, Doss DPS, Prakash AA, Kunz TH (2001) Clinal variation in body size and sexual dimorphism in an India fruit bat, *Cynopterus sphinx*, (Chiroptera, Pteropodidae). **Biological Journal of Linnean Society 72:** 17-31. doi: 10.1111/j.1095-8312.2001.tb01298.x

Thomas O (1899) Descriptions of new Neotropical mammals. **Annals and Magazine of Natural History 4**(7): 278-288. doi: 10.1080/00222939908678198

Thomas O (1913) The geographical races of the woolly opossum (*Philander laniger*). **Annals and Magazine of Natural History 12**(8): 358-361. doi: 10.1080/00222931308693409

Thorpe RS (1987) Geographic variation: a synthesis of cause, data, pattern and congruence in relation to subspecies, multivariate analysis and phylogenesis. **Bolletino di Zoologia 54:** 3-11. doi: 10.1080/11250008709355549

Tribe CJ (1990) Dental age classes in *Marmosa incana* and other didelphoids. **Journal of Mammalogy 71**(4): 566-569. doi: 10.2307/1381795

Van Der Hammen T, Hooghiemstra H (2000) Neogene and Quaternary history of vegetation, climate, and plant diversity in Amazonia. **Quaternary Science Reviews 19:** 725-742. doi: 10.1016/S0277-3791(99)00024-4

Voss R, Jansa S (2009) Phylogenetic relationships and classification of didelphid marsupials, an extant radiation of New World metatherian mammals. **Bulletin American Museum of Natural History** (322): 1-177. doi: 10.1206/322.1

Voss RS, Tarifa T, Yensen E (2004) An introduction to *Marmosops* (Marsupialia: Didelphidae), with the description of a new species from Bolivia and notes on the taxonomy and distribution of other Bolivian forms. **American Museum Novitates 3466:** 1-40. doi: 10.1206/0003-0082(2004)466<0001:AITMMD>2.0.CO;2

Webster M, Sheets HD (2010) A practical introduction to landmark-based geometric morphometrics. p. 163-188. In: Alroy J, Hunt G (Eds). **Quantitative methods in Paleobiology.** Boulder, The Paleontological Society, Paleontological Society Papers, vol. 16.

Zelditch ML, Swiderski DL, Sheets HD, Fink WL (2012) **Geometric morphometrics for biologists: a primer.** Boston, Elsevier Academic Press.

22

New morphological data on *Solariella obscura* (Trochoidea: Solariellidae) from New Jersey, USA

Ana Paula S. Dornellas[1,2] & Luiz R.L. Simone[1]

[1]*Museu de Zoologia, Universidade de São Paulo. Caixa Postal 42494, 04218-970 São Paulo, SP, Brazil.*
[2]*Corresponding author. E-mail: dornellas.anapaula@usp.br*

ABSTRACT. Anatomical data on *Solariella obscura* (Couthouy, 1838) are presented and analyzed. The main features of this species, when compared with other known trochoids, are: ctenidium with thick lamellae; enlarged ureter (that may indicate sexual dimorphism) instead of a modified urogenital papilla; odontophore very different from other trochoids such as *Calliostoma*, *Agathistoma*, *Monodonta*, and *Gaza*, with short m6, large mj and m4 pairs and absent m8 pair and posterior cartilages; esophageal valve surrounding the odontophore ventrally; anterior and mid-esophagus composed of several thin folds and a very wide cerebral ganglion. *Solariella obscura* differs from *Solariella varicosa* (Mighels & Adams, 1842) by having lower spire, spiral cords weaker on the base and axial rib oblique. There are no differences between *S. obscura* and *S. varicosa* in the external morphology and radula. These internal anatomical data are described for the first time for a solariellid and might improve our understanding of the relationships within this taxon.

KEY WORDS. Anatomy; comparative data; North Atlantic; redescription.

Living Solariellidae is distributed in offshore waters worldwide and the fossil record is primarily in low latitudes during the Paleogene to higher latitudes in the Neogene (Hickman & McLean 1990). It comprises *Solariella* Wood, 1842, *Archiminolia* Iredale, 1929, *Bathymophila* Dall, 1881, *Hazuregyra* Shikama, 1962, *Ilanga* Herbert, 1987, *Microgaza* Dall, 1881, *Minolia* Adams, 1860, *Minolops* Iredale, 1929, *Spectamen* Iredale, 1924, *Zetela* Finlay, 1927 (Hickman & McLean 1990, Williams 2012).

Species belonging to Solariellidae are characterized by small, nacreous shells, short and straight radula, absent cephalic lappets, oral surface of snout with elongated papillae, anterior end of foot with well-developed lateral horns, and neck lobes bearing one or two tentacles (Herbert 1987, Hickman & McLean 1990, Williams et al. 2008). However, some shell characters in this family are convergent in both trochid subfamilies, Umboniinae and Margaritinae. Therefore, it has been inferred that shell characters in Solariellidae are misleading, and cannot be distinguished without knowledge of the external anatomy or radula (Hickman & McLean 1990).

The generic name *Solariella* was used in the past in reference to a wide variety of trochids with a round aperture and broad umbilicus. The type species, *Solariella maculata* Wood, 1842, is a fossil from the Pliocene, which makes it impossible to ascertain the state of its radular and soft parts. Nevertheless, recent species that resemble it and which are similar to the North Atlantic *Solariella amabilis* (Jeffreys, 1865) have been placed in *Solariella* by Herbert (1987), Hickman & McLean (1990), and Warén (1993). Thus, *Solariella* sensu stricto has three pairs of epipodial tentacles, no prominent epipodial lobes and a radula with nascent lateromarginal plates (Herbert 1987, Marshall 1999).

Herein we describe and analyze the external morphology and certain anatomical structures of the North Atlantic *Solariella obscura* (Couthouy, 1838). In order to achieve new insights that might prove useful in future taxonomic studies.

MATERIAL AND METHODS

Specimens preserved in 70% ethanol were extracted from their shells and subsequently dissected and photographed under the stereomicroscope. The terminology of the odontophore muscles follow Dornellas & Simone (2013). All drawings were made with the aid of a camera lucida. Samples examined with the SEM (radulae and protoconchs) were mounted on stubs and coated with gold-palladium alloy. The specimens were analyzed and photographed under a stereomicroscope.

Anatomical abbreviations: af, afferent vein; ai, intestinal loop; an, anus; ac, anterior cartilage of odontophore; ax, axis; bc, subesophageal connective; cb, cerebrobuccal connective; cc, cerebral commissure; cd, cerebropedal connective; cg, cerebral ganglion; cm: collumelar muscle; cp, cerebropleural connective; ct, ctenidium; cv, ctenidial vein; df, dorsal fold; dg, digestive gland; ef, esophageal fold; ep, epipodium; es, esophagus; et, epipodial tentacle; ev, esophageal valve; ft, foot; go, gonad; hg, hypobranchial gland; ho, horn; jw, jaws; la, left auricle; lg, labial ganglion; mb, mantle border; mj, jugal muscles; mt, mantle; nl, neck lobe; om, ommatophore; oa, opercular pad; os, osphadia; pc, pericardium; pe, pedal ganglion; pl, pleural ganglion; ps, papillary sac; pt, postoptic tentacle; ra, radula; ru, right auricle; rk, right kidney; ro, rod; sc, spiral caecum; sg, salivary gland; sk, skeletal rod; sm, stomach; sn, snout; st, statocysts; te, cephalic tentacle; ur, ureter; ve, ventricle.

Institutional acronyms. (USNM/NMNH) National Museum of Natural History, Smithsonian Institution, Washington, DC.

Material analyzed. Types: United States, between Marblehead and Nahant, Massachusetts Bay, 7 specimens, MCZ 154825. United States, off New Jersey, North Atlantic Ocean, 39°02'54"N 73°47'06"W, 6 specimens, USNM 828340; 3919'18"N 73°10'06"W, 2 specimens, USNM 828343.

TAXONOMY

Solariella obscura (Couthouy, 1838)
Figs. 1-34

Turbo obscurus Couthouy, 1838: 100 (pl. 3, fig. 12).

Solariella obscura: Tryon, 1889: 308 (pl. 57, figs. 44, 45); Locard, 1903: 43; Cushman, 1906: 16; Odhner, 1912: 70; Johnson, 1915: 89; Smith, 1951: 79 (pl. 31, fig. 19; pl. 71, fig. 16); Lopes & Cardoso, 1958: 62; MacGinite, 1959: 80; Talmadge, 1967: 236; Abbott, 1974: 40 (fig. 271); Procter, 1993: 172.

Margarites albula Gould, 1861: 36.

Margarita obscura: Gould, 1870: 283 (fig. 545).

Margarita bella Verkrüzen, 1875: 236.

Margarita obscura var *cinereaeformis* Leche, 1878: 45 (pl. 2, fig. 24).

Margarita obscura var *intermedia* Leche, 1878: 45 (pl. 2, fig. 25).

Solariella laevis Friele, 1886: 14.

Solariella obscura var *bella* (Verkrüzen): Tryon, 1889: 310 (pl. 64, figs. 57, 58); Blaney, 1906: 111; Johnson, 1915: 89; Lopes & Cardoso, 1958: 62.

Machaeroplax obscura var *planula* Verrill, 1882: 532; Johnson, 1915: 89 (pl. 17, fig. 6); Lopes & Cardoso, 1958: 63.

Machaeroplax obscura var *carinata* Verrill, 1882: 532; Johnson, 1915: 90; Lopes & Cardoso, 1958: 62.

Solariella obscura var *multilirata* Odhner, 1912: 79; Lopes & Cardoso, 1958: 62.

Type. Cotypes, 7 shells, MCZ 154825.

Type locality. Between Marblehead and Nahant, Massachusetts Bay, USA. In fish stomach.

Distribution. Arctic circumpolar; south of New England; eastern and western Greenland; Iceland; Canada, Labrador; USA, Maine, Massachusetts, Connecticut (Warén 1993, Rosenberg 2009).

Description. Shell (Figs. 1-10, 22, 23): up to 5½ whorls, 8 mm in height and 9 mm in diameter; deeply umbilicate; shape trochoid, whorls rounded or angular, suture impressed. Color grayish to pinkish tan, peristome thin and nearly complete, often worn, revealing iridescent color. Protoconch (Figs. 22, 23) of 1 whorl lighter-colored, smooth; about 250 μm diameter. Spire sculptured by weakly beaded spiral cords; about 12 cords on last

whorl. Some specimens with strong cord on middle whorl (Figs. 3, 6), forming carinate shoulder. Base weakly convex, with about 20 cords, thinner than those on spire; smooth. Strong beaded cord surrounding umbilical area; umbilical area with 7-10 cords. Umbilicus wide and deep, about 20% maximum shell width, funnel-shaped. Aperture rounded, inner surface iridescent; ~75% of shell length, ~55% of shell width.

Head-Foot (Figs. 24, 25, 34). Head bulging approximately in middle region of head-foot. Snout wide, cylindrical; distal end wider than base; distal surface papillated; papillae long, thin, cylindrical, with rounded tip. Outer lips mid-ventrally incomplete, mouth located in middle portion of ventral surface of snout. Pair of cephalic lappets absent. Cephalic tentacles (Fig. 24: te) ~20-30% longer than snout, sometimes asymmetrical in relation to one another, covered by small papillae, dorso-ventrally flattened, grooved, narrowing gradually up to lightly pointed tip. Ommatophore on outer base of cephalic tentacles, length 1/5 of tentacle length. Eyes dark, rounded, occupying anterior edge of ommatophore.

Foot thick, occupying half of total head-foot length; whitish, non papillated; anterior end truncated and drawn out laterally into two long processes (horns). Epipodium (Fig. 25: ep) surrounding latero-dorsal region of mesopodium, equidistant from sole and base of ommatophores. Right neck lobe with two tentacles, anterior tentacle postoptic (Fig. 25: pt); left neck lobe with one tentacle. Three pairs of epipodial tentacles (Figs. 24, 25) symmetrical on both sides; epipodial sense organs at base of epipodial tentacles. Opercular pad (Figs. 25, 34: oa) rounded, located in median dorsal region, with free edge in posterior area; posterior end with several chevron furrows, apex pointed posteriorly and two pairs of longitudinal furrows on median line. Furrow of pedal glands along entire anterior edge.

Operculum (Figs. 11, 12). Up to ~2.5 mm in diameter and ~7 whorls, closing entire aperture, corneous, thin, with central nucleus. Inner side convex, outer side concave. Color yellowish gold.

Mantle organs (Figs. 13, 26-28). Pallial cavity of 3/4 whorl. Mantle border (Fig. 26: mb) thick, white; anterior end papillated, occupying 1/3 of mantle border. Gill located on left side of pallial cavity; less than half of width and height of pallial cavity; projecting anteriorly, sustained by gill rod, lacking suspensory membrane (Fig. 26: sk). Anterior end of gill narrow, with acuminate tip. Ventral lamella larger than dorsal lamella (Figs. 27). Afferent gill vessel ~3/4 of gill's length, originating in transverse pallial vessel, running along distal region of central axis of gill. Transverse pallial vessel ~1/5 of afferent vessel length, originating in left nephrostome and discharging in afferent gill vessel. Ctenidial vein (efferent gill vessel) length more than twice afferent vessel length, running along basal region of central axis of gill; posterior end of vein (half) free from gill filaments, lying parallel to afferent vessel up to pericardium. Osphradium rounded, whitish, located at base of gill rod. Hypobranchial glands (Figs. 26: hg) on both sides of rectum; more developed on left side. Rectum occupying 1/4 of pallial cavity

Figures 1-17. *Solariella obscura*. (1-12) Shell and operculum: (1-4) Cotypes MCZ 154825, apertural and umbical views: (1-2) 7.2 mm height x 8.5 mm width, (3-4) 7.2 mm x 8.6 mm. (5-12) USNM 828340, shell and operculum; (5-10) shell, apertural, lateral and umbilical views: (5-7) 5.8 mm x 5.5 mm, (8-10) 8.9 mm x 9.1 mm; (11-12) operculum, external and internal views, 3 mm diameter. (13-17) Anatomy: (13) pallial cavity, ventral view; (14) jaws, ventral view; (15) foregut, anterior and mid-esophagus opened longitudinally, ventral inner view, odontophore removed; (16) pedal and pleural ganglia *in situ*, dorsal view; (17) anterior odontophore cartilages, muscles removed, ventral view. Scale bars: 14, 16 = 0.2 mm; 13, 15, 17 = 0.5 mm.

width, sigmoid; posterior region under kidneys. Anus siphoned, preceded by pleated and short free end, located on posterior right side of pallial cavity. Kidneys length more than half of rectum length, located on posterior region of pallial cavity.

Visceral mass (Fig. 34). Pericardium and posterior portion of right kidney exposed on pallial cavity roof. Stomach and spiral caecum (sc) located 1/3 whorl posterior to pallial cavity. Digestive gland (dg) located on left side and gonad (go) on right side, both posterior to right kidney.

Circulatory and excretory systems (Figs. 26, 28). Pericardium located between pallial cavity and visceral mass (Fig. 26: pc), close to median line and immediately posterior to kidneys. Left side of pericardium receiving ctenidial vein and right side receiving right pallial vein. Ventricle volume 1/3 of pericardium volume; surrounding rectum and flanked anteriorly by left auricle and posteriorly by right auricle; left auricle ventral, triangular, occupying about half of pericardium volume; right auricle weak, smaller than left one (Fig. 28). Papillary sac (or left kidney) base oval, wide, gradually narrowing towards

anterior portion, ending at left nephrostome; inner wall covered by numerous thin, long papillae. Right kidney (Fig. 28: ur, rk) divided into two regions; anterior region as hollow tube (ureter), right nephrostome located in anterior region; ureter might be as large as papillary sac (probably males) (Fig. 26) or twice papillary sac width (probably females) (Fig. 28); no mucus observed in females ureter. Posterior region spreading around visceral mass immediately beneath mantle, encircling inner surface of columellar muscle. Kidney expanding ventrally, covering half of right surface of adjacent visceral hump.

Digestive system (Figs. 14, 15, 17, 18-21, 29-33). Oral tube length ~½ odontophore length; walls with circular muscles (Figs. 29, 30); basal region with thick oblique fibers (Fig. 31: mj), originating gradually from dorsal surface, close to median line; fibers running posteriorly towards both sides, inserting in ventral surface of odontophore. Jaw plates (Fig. 14) thin, light brown, rounded, occupying half of odontophore length. Pair of dorsal folds starting posteriorly to jaws (Fig. 15: df), with each fold bending and forming two partially overlapping

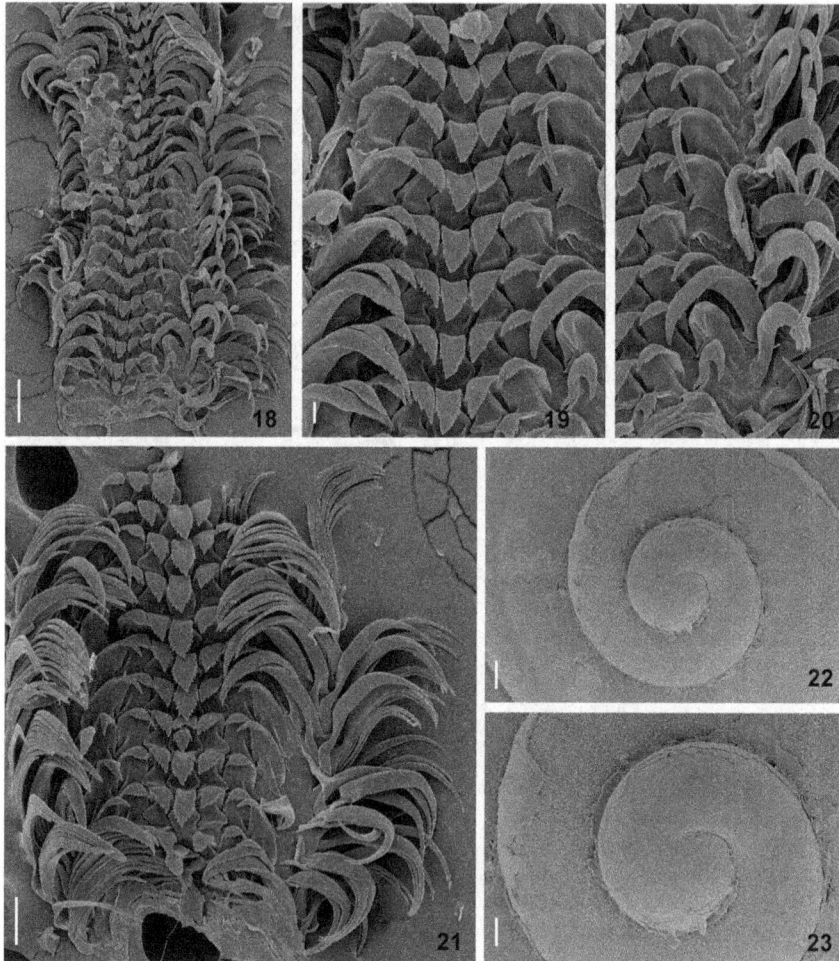

Figures 18-23. *Solariella obscura*, structures under SEM. (18-21) Radular ribbon: (18) middle region, whole view; (19) detail of central area, rachidian and lateral teeth; (20) detail of lateral and marginal teeth; (21) middle region, whole view; (22-23) Protoconch, apical views. Scale bars: 18 = 100 μm, 19 = 20 μm, 20 = 40 μm, 21-23 = 60 μm.

slits. Series of transverse muscles separating outer surface of esophagus from odontophore. Odontophore about 1/3 longer than snout. Odontophore muscles (Figs. 29-33): m1 (Fig. 29): series of small jugal muscles connecting buccal mass with adjacent inner surface of snout and haemocoel; m4 broad and long pair of dorsal tensor muscles of radula and subradular membrane, originating on ventral and lateral surface of anterior cartilages, at some distance from median line, inserting along dorsal region of subradular membrane (exposed inside buccal cavity), with portion in radular ribbon preceding buccal cavity; m5: pair of large accessory dorsal tensor muscles of radula, originating on posterior surface of anterior cartilages, running firstly towards dorsal and median regions and subsequently anteriorly, with insertion in radular ribbon region; m6: horizontal muscle, uniting over half of ventral edges of both anterior cartilages; m7: very small, thin pair of muscles, origi-

nating in middle region of inner ventral surface of radular sac, running anteriorly (insertion not observed); m10: broad pair of ventral protractor muscles of odontophore, originating in ventral region of mouth and buccal sphincter, running posteriorly, inserting in posterior region of anterior cartilage; m11: pair of thin ventral tensor muscles of radula, originating in middle region of ventral surface of anterior cartilage, running anteriorly and covering m6 and ventral surface of anterior cartilage, inserting on distal edge of subradular membrane; m11a; very long and thin pair of oblique ventral tensor muscles of radula, originating on anterior haemocoelic surface near pleural ganglia, running dorsally between anterior edge of anterior cartilages, inserting on distal edge of subradular membrane. Non-muscular structures of odontophore: ac: pair of anterior cartilages (Figs. 17, 33), antero-posteriorly elongated, flat, with anterior and posterior ends rounded, same length as

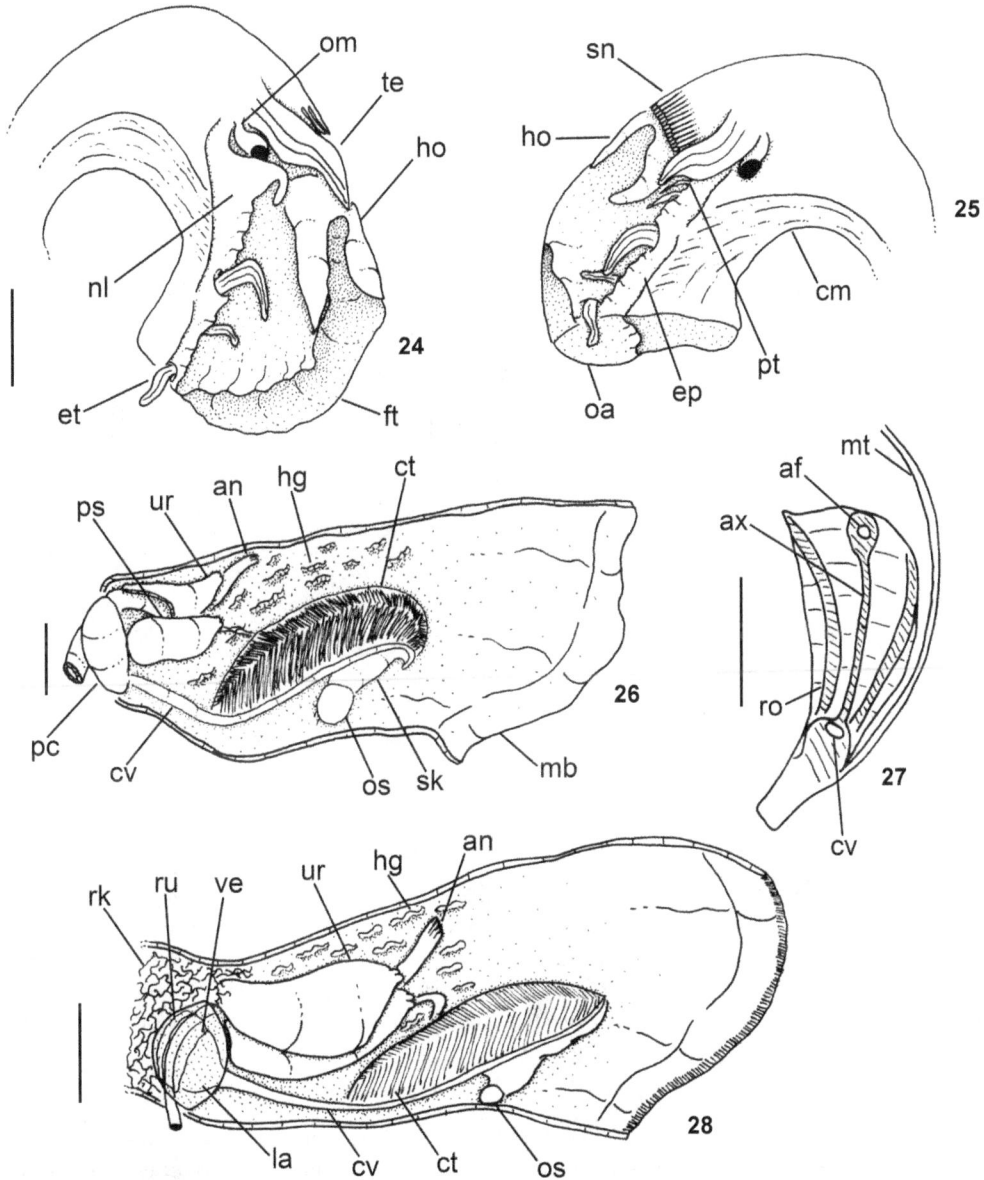

Figures 24-28. Anatomy of *Solariella obscura*: (24-25) head-foot, right and left views; (26) pallial cavity roof, male, ventral view; (27) transverse section in middle region of ctenidium; (28) pallial cavity roof, female, ventral view. Scale bars: 24-26, 28 = 1 mm; 27 = 0.5 mm.

odontophore. Pair of posterior odontophore cartilages absent. Odontophore cartilages whitish, rough; br: subradular membrane covering most of the exposed surface of odontophore in buccal cavity, where most of intrinsic odontophore muscles insert; sc: subradular cartilage maintaining radular ribbon.

Radular sac short (Fig. 31: ra), as long as odontophore. Radular nucleus (Fig. 32) located on ventral side of odontophore. Central complex (rachidian and laterals) well-developed, with interlocking process and correspondent sockets (Fig. 18); shafts expanding laterally, hood-shaped. Rachidian large (Fig.

19), triangular, cutting edge with projection turned posteriorly (almost 90°) and covering posterior end of preceding tooth; tip narrowly tapered, serrated; base and cusp with within-column interaction. Four lateral teeth (Figs. 20, 21); cusps oriented toward midline of radula, most strongly serrate along their outer margins; three inner lateral teeth similar to rachidian in shape; outermost lateral teeth broad, large, length twice of inner teeth length. Lateromarginal plate not observed. Marginal teeth (Fig. 18) as long as outermost lateral teeth, slender, serrate, ~10-12 teeth pairs. Anterior esophagus with esophageal

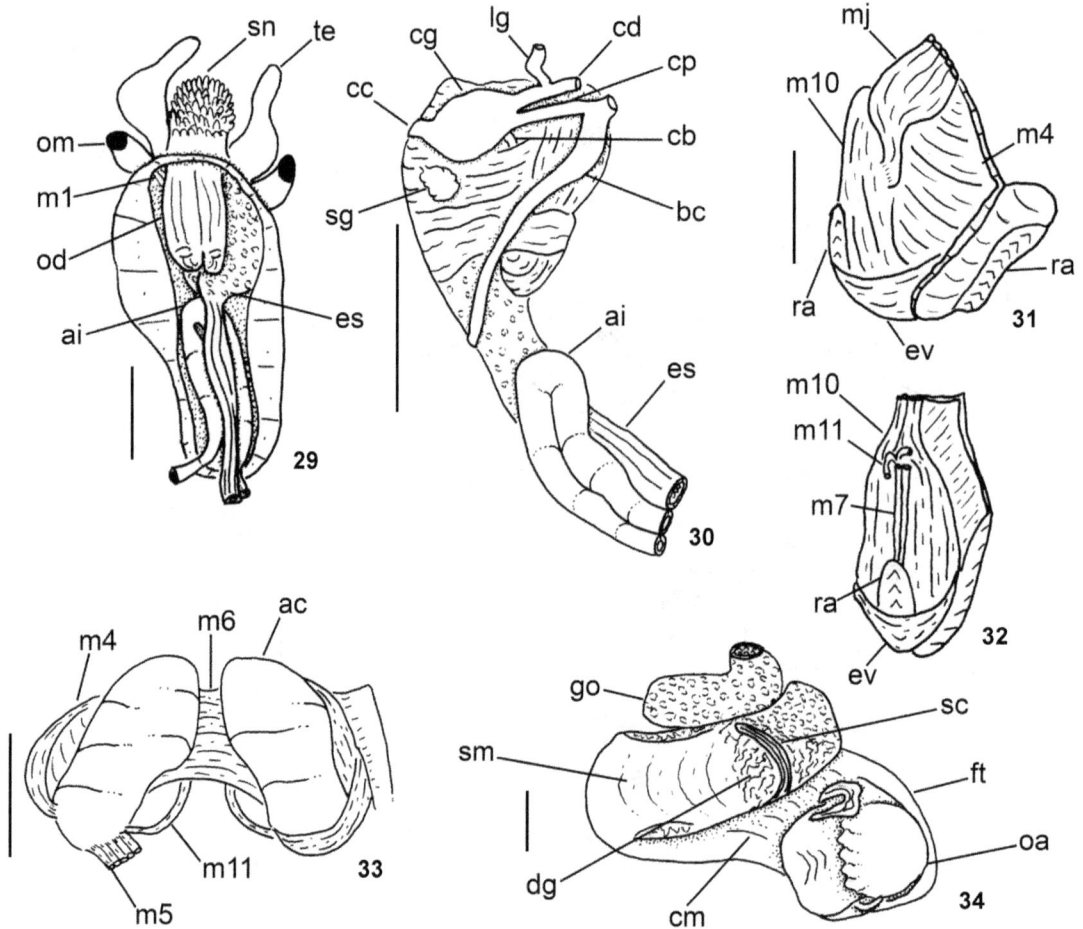

Figures 29-34. Anatomy of *Solariella obscura*: (29) head-foot haemocoel, ventral view, foot and columellar muscle removed; (30) buccal mass and central nervous system, right view; (31-32) odontophore, left, and ventral views; (33) odontophore, dorsal view, radular ribbon removed, left m5 extracted and reflected, m10 extracted; (34) head-foot and visceral mass, whole apertural view. Scale bars: 29-30, 34 = 1 mm; 31-33 = 0.5 mm.

valve covering ventral surface of odontophore. Anterior and mid esophagus (Fig. 15) with folds forming shallow chambers; epithelium entirely covered by villous papillae. Posterior esophagus narrow, with some thin longitudinal folds on inner surface. Stomach not observed. Spiral caecum with ½ counter clockwise (in dorsal view) whorls. Intestine (Figs. 26, 28-30) very wide, running anteriorly forward inside to cephalic haemocoel, bending abruptly, forming wide loop (Fig. 30: ai); anterior region of visceral mass with small loop surrounding kidney and pericardium, exiting in right-posterior corner of pallial cavity. Rectum and anus described above (pallial organs).

Central nervous system (Figs. 16, 30). Nerve ring surrounding anterior half of buccal mass. Cerebral ganglia rounded, located in lateral region of buccal mass (Fig. 30: cg), size ~1/3 of odontophore size; commissure thick, long, dorso-ventrally flattened; cerebropleural and cerebropedal (Fig. 30: cp, cd)

connectives long, thin, originating in anterior region of cerebral ganglia and running ventrally and back to pedal and pleural ganglia. Labial ganglia (Figs. 30: lg) 1/6 of cerebral ganglia, located in ventro-lateral region of buccal mass, anteriorly to cerebral ganglia; connected to cerebral ganglia by short cerebrolabial connective. Buccal ganglia posterior to cerebral ganglia; connected to cerebral ganglia by a buccolabial connective. Pleural and pedal ganglia (Fig. 16: pl, pe) close to each other, located inside pedal musculature immediately below ventral surface of haemocoel; both of about half size of cerebral ganglion. Pedal commissure thick, very short. Pedal nerve running forward from each pedal ganglion, surrounding medial pedal blood sinus. Supra-esophageal connective emerging from right pleural ganglia. Subesophageal connective (Fig. 30) emerging from left pleural ganglia. Statocysts (Fig. 16: st) rounded, bright, located very close to posterior side of pedal ganglia.

DISCUSSION

The organs and systems of *S. obscura* are congruent with the features of solariellid mentioned by previous authors (Herbert 1987, Hickman & McLean 1990, Williams 2012), such as: small and nacreous shells; presence of a ring of digitate papillae around the snout; short radula with 20-30 transverse teeth rows; anterior end of foot bilobed, forming the horn; long and thick cephalic tentacles, and an eye-stalk much shorter than the cephalic tentacle. Some of these features are not exclusive to solariellids, however, especially when compared with other trochoids. The presence of digitate papillae around the snout can also be found in *Gaza* Watson, 1879 (Simone & Cunha 2006), and the Umboniinae also have a bilobed foot (Hickman & McLean 1990, pers. obs.). The radula, on the other hand, is the main structure for characterizing this family and its genera (Herbert 1987, Marshall 1999), being short (20-30 rows of teeth), with reduced number of marginal teeth (~10 per half row).

The Artic species *Solariella varicosa* (Mighels & Adams, 1842) is similar in shape to *S. obscura*, but differs by having a taller spire, strong oblique axial rounded ribs and stronger spiral cords on the base. The distributions of both species overlaps in the Artic, south of Labrador and northern Canada (Warén 1993). The neck lobe shows a variety of shapes among trochoids and might be used to diagnose subfamilies, genera and even species (Hickman & McLean 1990, Dornellas & Simone 2013). They are usually digitate, fringed or smooth, as well as symmetrical or asymmetrical to each other according to the taxa. The neck lobes of solariellids show some inter-generic variation in shape (Herbert 1987, Marshall 1999) and are characterized by the presence of one or two short tentacles, the right neck lobe bearing the postoptic tentacle located at its anterior edge (Fig. 25: pt). The neck lobe of *S. obscura* was reported as being virtually identical to that of *S. varicosa* (Warén 1993: 161, figs. 4a, b).

Regarding the pallial cavity, the ctenidium of *S. obscura* has a thick lamella (Fig. 13), as is the case in *Solariella carvalhoi* Lopes & Cardoso, 1958 (pers. obs.), when compared with other trochoids such as *Calliostoma* Swainson, 1840, *Monodonta* Lamarck, 1789, *Lithopoma* Gray, 1850, *Agathistoma* Olsson & Harbison, 1953, *Gaza* (Fretter & Graham 1962, Righi 1965, Monteiro & Coelho 2002, Simone & Cunha 2006, Dornellas & Simone 2013). The enlarged ureter in some specimens may indicate sexual dimorphism in *Solariella* (Fig. 28), differently from other vetigastropods in which the females have modified urogenital papillae (Woodward 1901, Fretter & Graham 1962, Monteiro & Coelho 2002, Dornellas 2012, Dornellas & Simone 2013). However, this structure differs in shape among vetigastropods (see Dornellas & Simone 2013).

The odontophore of *S. obscura* is different from that of other trochoids such as *Calliostoma*, *Agathistoma*, *Monodonta* and *Gaza* (Fretter & Graham 1962, Righi 1965, Simone & Cunha 2006, Dornellas 2012, Dornellas & Simone 2013): the m6 is

shorter (occupying only half of the cartilages' length); the mj and m4 pairs are larger (more than twice) than the current size; the m8 pair and the posterior cartilages are lacking. The buccal cavity differs from that of other trochoids by the esophageal valve surrounding the odontophore ventrally (Figs. 31, 32: ev), also observed in *S. carvalhoi* (personal observation). The salivary gland, located in latero-dorsal area of the buccal mass, is rounded and concentrated, similar to that observed in *Tegula viridula* (Gmelin, 1791), *Monodonta labio* (Linnaeus, 1758) and *Lithopoma olfersii* (Philippi, 1846) (Righi 1965, pers. obs.).

Usually, the anterior and mid-esophagus of vetigastropods is composed of four folds that compartmentalize it (Fretter & Graham 1962, Fretter 1964, Haszprunar 1988, Sasaki 1998, Dornellas & Simone 2013). In *S. obscura*, on the other hand, the anterior and mid-esophagus are composed of several thin folds (Fig. 15). The presence of papillate glands covering the inner wall of that region, which is also present in *S. obscura*, is a morphological synapomorphy of the clade Vetigastropoda (Haszprunar 1988, Salvini-Plawen 1988, Sasaki 1998).

The radula of *S. obscura* is a typical solariellid radula, comprising a straight and short radular ribbon (as long as the odontophore), triangular rachidian, with the outermost lateral tooth being larger than the innermost teeth, and reduced number of marginal teeth (~10 teeth along the same row). In *Solariella*, the radula is characterized by the presence of well-developed, elongate, cuspless lateromarginal plates (Herbert 1987). *Solariella obscura* and *S. varicosa* lack lateromarginal plates (Warén 1993, Fretter & Graham 1977), whereas all southern *Solariella* bear lateromarginal plates (Marshall 1999).

Despite the gastric spiral caecum being a variable structure, a large spiral caecum is considered derived within Vetigastropoda. This structure opens ventrally in the posterior end of the stomach, more or less as continuous extensions of the typhlosoles (Strong 2003). The 0.5 whorl long caecum observed in *S. obscura* (Fig. 31: sc) can also be found in *Calliostoma depictum* Dall, 1927, but the number of spiral caecum whorls seems to be an inter-specific feature rather than a generic one, because it may vary among congeners such as in *Calliostoma* and *Lithopoma* (Monteiro & Coelho 2002, Dornellas & Simone 2013).

The central nervous system of *S. obscura* demonstrates a trochoid pattern (Sasaki 1998) but the cerebral ganglion is proportionally wider (Fig. 30) when compared to those described in other vetigastropods (Fretter & Graham 1962, Sasaki 1998, Simone & Cunha 2006, Simone 2008, Dornellas 2012, Dornellas & Simone 2013).

Solariellidae has been recently recognized as a family (Bouchet et al. 2005), with molecular studies supporting this rank (Williams et al. 2008, Williams 2012). As discussed above, several features seem to be exclusive of Solariellidae such as the above-mentioned radular pattern and external morphology. These features have also been further used to trace patterns between solariellid genera (Herbert 1987). On the other hand, there is no described data about the internal anatomy for any solariellid, and the new data described herein for *S. obscura* might improve our understanding about the relationships within this taxon, at generic and even suprageneric levels.

ACKNOWLEDGEMENTS

We thank Jerry Harasewych (NMNH) for lending us specimens; FAPESP (Fundação de Amparo à Pesquisa do Estado de São Paulo) process numbers 2010/18864-3; 2012/25173-2 for supporting our research; Lara Guimarães (MZSP) for help with the SEM; Diogo Couto for taking photos of the type, and Daniel Cavallari for helping with grammar revision.

REFERENCES

ABBOTT RT (1974) **American Seashells.** New York, Van Nostrand Rheinhold, 2nd ed., 663p.

BLANEY D (1906) Shell-bearing Mollusca of Frenchman's Bay, Maine. **The Nautilus 19**(10): 110-111.

BOUCHET P, ROCROI J, FRÝDA J, HAUSDORF B, PONDER W, VALDÉS A, WARÉN A (2005) Classification and nomenclator of gastropod families. **Malacologia 47**(1-2): 1-397.

COUTHOUY JP (1838) Descriptions of new species of Mollusca and shells, and remarks on several polypi found in Massachusetts Bay. **Boston Journal of Natural History 2:** 53-111.

CUSHMAN JA (1906) The Pleistocene deposits of Sankoty Head, Nantucket, and their fossils. **Publications of the Nantucket Maria Mitchell Association 1**(1): 1-21.

DORNELLAS APS (2012) Description of a new species of *Calliostoma* (Gastropoda, Calliostomatidae) from southeastern Brazil. **ZooKeys 224:** 89-106. doi: 10.3897/zookeys.224.3684

DORNELLAS APS, SIMONE LRL (2013) Comparative morphology and redescription of three species of *Calliostoma* (Gastropoda, Trochoidea) from Brazilian coast. **Malacologia 56**(1-2): 267-293.

FRIELE H (1886) Mollusca II. **The Norwegian North-Atlantic Expedition, 1876-1878 3:** 1-35.

FRETTER V (1964) Observations on the anatomy of *Mikadotrochus amabilis* Bayer. **Bulletin of Marine Science of the Gulf and Caribbean 14**(1): 172-184.

FRETTER V, GRAHAM A (1962) **British prosobranch molluscs. Their functional anatomy and ecology.** Ray Society Publications, London, XVI+755p.

FRETTER V, GRAHAM A (1977) The prosobranch molluscs of Britain and Denmark. Part 2. Trochacea. **Journal of Molluscan Studies 3**(Suppl.): 39-100.

GOULD AA (1861) Descriptions of shells collected by the North Pacific Exploring Expedition. **Proceedings of the Boston Society of Natural History 8:** 14-40.

GOULD AA (1870) **Report on the Invertebrata of Massachusetts.** Boston, 2nd ed., 524p.

HASZPRUNAR, G (1988) On the origin and evolution of major gastropod groups, with special reference to the *Streptoneura*. **Journal of Molluscan Studies 54**: 367-441.

HERBERT DG (1987) Revision of the Solariellinae (Mollusca: Prosobranchia: Trochidae) in southern Africa. **Annals of the Natal Museum 28**(2): 283-382.

HICKMAN CS, MCLEAN JH (1990) Systematic revision and suprageneric classification of trochacean gastropods. **Science Series, Natural History Museum of Los Angeles Country 35**: 1-169.

JOHNSON CW (1915) Fauna of New England 13. List of Mollusca. **Occasional Papers of the Boston Society of Natural History 7**: 1-231.

LECHE W (1878) Öfversigt öfver de af Svenska Expeditionerna till Novaja Semlja och Jenissej 1875 och 1876 Insamlade: Hafs-Mollusker. **Kongliga Svenska Vetenskaps-Akademiens Handlingar 16**(2): 1-86.

LOCARD A (1903) **Conquilles des Mers D'Europe. Turbinidae.** Lyon, Société d'Agriculture, Sciences et Industrie de Lyon, p. 1-66.

LOPES HS, CARDOSO PS (1958) Sobre um novo gastrópodo brasileiro do gênero *Solariella* Wood, 1842 (Trochidae). **Revista Brasileira de Biologia 18**(1): 59-64.

MACGINITE B (1959) Marine Mollusca of Point Barrow, Alaska. **Proceedings of the United States National Museum 10**(3412): 59-208.

MARSHALL BA (1999) A Revision of the Recent Solariellinae (Gastropoda: Trochoidea) of the New Zealand Region. **The Nautilus 113**(1): 4-42.

MONTEIRO JC, COELHO ACS (2002) Comparative morphology of *Astraea latispina* (Philippi, 1844) and *Astraea olfersii* (Philippi, 1846) (Mollusca, Gastropoda, Turbinidae). **Brazilian Journal of Biology 62**(1): 135-150. doi: 10.1590/S1519-69842002000100016

ODHNER NH (1912) Northern and Artic invertebrates in the collection of the Swedish state Museum. **Kungliga Svenska Vetenskapsakademiens Handlingar 48**(1): 1-93.

PROCTER W (1993) **Biological survey of the Mount Desert Region, part V – Marine Fauna.** Philadelphia, Wistar Institute of Anatomy and Biology, 402p.

RIGHI G (1965) Sobre *Tegula viridula* (Gmelin, 1971). **Boletim da Faculdade de Filosofia, Ciências e Letras da Universidade de São Paulo (Zoologia) 25**: 325-390.

ROSEMBERG G (2009) **Malacolog 4.1.1: A Database of Western Atlantic Marine Mollusca.** Avalaible online at: http://www.malacolog.org [Accessed: 22 December 2014]

SALVINI-PLAWEN L VON (1988) The structure and function of molluscan digestive systems, p. 301-379. In: Trueman ER, CLARKE MR (Eds) **The Mollusca: Form and Function.** San Diego, Academic Press, vol. 11, 504p.

SASAKI T (1998). Comparative anatomy and phylogeny of the recent Archaeogastropoda (Mollusca: Gastropoda). **The University of Tokyo Bulletin 38**: 1-224.

SIMONE LRL (2008) A new species of *Fissurella* from São Pedro e São Paulo Archipelago, Brazil (Vetigastropoda, Fissurellidae). **The Veliger 50**(4): 292-304.

SIMONE LRL, CUNHA CM (2006) Revision of genera *Gaza* and *Callogaza* (Vetigastropoda, Trochidae), with description of a new Brazilian species. **Zootaxa 1318**: 1-40.

SMITH M (1951) **East coast marine shells.** Ann Arbor, Edwards Brothers, 4th ed., 314p.

STRONG EE (2003) Refining molluscan characters: morphology, character coding and a phylogeny of the Caenogastropoda. **Zoological Journal of the Linnean Society 137**: 447-554.

TALMADGE RR (1967) Notes on the Mollusca of Prince William sound, Alaska. Part II. **The Veliger 9**(2): 235-238.

TRYON GW (1889) **Manual of Conchology, with illustrations of the species. Trochidae, Stomatiidae, Pleutotomariidae, Haliotidae.** Philadelphia, Published by the Author, vol. 1, #11, 519p.

VERKRÜZEN TA (1875) Bericht über einen Schabe – Ausflug in Sommer 1874. **Jahrbücher der Deutschen Malakozoologischen Gesellschaft 2**: 229-240.

VERRILL AE (1822) Catalogue of marine Mollusca added to the fauna of the New England region, during the past ten years. **Transactions of the Connecticut Academy of Arts and Sciences 5**: 451-587.

WARÉN A (1993) New and little known Mollusca from Iceland and Scandinava. Part 2. **Sarsia 78**: 159-201.

WILLIAMS ST (2012) Advances in molecular systematics of the vetigastropod superfamily Trochoidea. **Zoologica Scripta 41**(6): 571-595. doi: 10.1111/j.1463-6409.2012.00552.x

WILLIAMS ST, KARUBE S, OZAWA T (2008) Molecular systematics of Vetigastropoda: Trochidae, Turbinidae and Trochoidea redefined. **Zoologica Scripta 37**(5): 483-506. doi: 10.1111/j.1463-6409.2008.00341.x

WOODWARD MF (1901) The anatomy of *Pleurotomaria beyrichii*, Hilg. **Bulletin of the Museum of Comparative Zoology 8**: 215-268.

23

Reproduction and diet of *Imantodes cenchoa* (Dipsadidae: Dipsadinae) from the Brazilian Amazon

Kellen R. M. de Sousa[1,3], Ana Lúcia C. Prudente[2] & Gleomar F. Maschio[1]

[1] Instituto de Ciências Biológicas, Universidade Federal do Pará. Rua Augusto Corrêa 01, Guamá, Caixa Postal 479, 66075-110 Belém, PA, Brazil.
[2] Laboratório de Herpetologia, Coordenação de Zoologia, Museu Paraense Emílio Goeldi. Avenida Perimetral 1901, Terra Firme, Caixa Postal 399, 66017-970 Belém, PA, Brazil.
[3] Corresponding author. E-mail: kellenrayanne@gmail.com

ABSTRACT. *Imantodes cenchoa* (Linnaeus, 1758) is distributed from the east coast of Mexico to Argentina. In Brazil, it occurs in the north, central-west and northeast regions. We present information on the reproductive biology and diet of *I. cenchoa* from analysis of 314 specimens deposited in the Herpetological Collection of the Museu Paraense Emílio Goeldi (MPEG). *Imantodes cenchoa* displays sexual dimorphism in the snout-vent length, where sexually mature females are larger than mature males (t = 4.02, p < 0.01; N males = 150, N females = 71), head length ($f_{1.218}$ = 98.29, p < 0.01; N males = 150, N females = 71), and head width ($f_{1.218}$ = 112.77, p < 0.01, N males = 150; N females = 71). Bi-sexual maturity is observed, with males becoming sexually mature earlier than females. Females with eggs were recorded from November to January (rainy season) and from April to July (dry season), suggesting two reproductive peaks throughout the year, with recruitment occurring mainly during the rainy season, when there is a greater supply of food. *Imantodes cenchoa* is a nocturnal active forager, capturing prey that are asleep on the vegetation. In 32.80% of the analyzed specimens, food contents were present, of which 84.11% were lizards of the genera *Norops* (69.16%, N = 74) and *Gonatodes* (14.95%, N = 16). The other 15.89% of the contents were made up of items in an advanced state of digestion, preventing their identification. Some specimens had more than one food item in their digestive tract, accounting for 107 prey items in total. There was no ontogenetic variation in the diet of *I. cenchoa*, and the predominant direction of prey ingestion was antero-posterior (71.96%). Larger snakes tended to feed on larger prey, although these did not exclude small prey from their diet.

KEY WORDS. Ecology; natural history; Neotropical; snakes.

Results obtained from natural history studies, and which investigate reproductive modes and strategies, habitat use, food habits, behavior and defensive tactics (according to Greene 1986, 1994), can increase our knowledge of the local fauna (Cunha & Nascimento 1978).

Reproductive biology is widely discussed within the context of the natural history of snakes, mainly due to the great diversity of reproductive strategies in the group (Seigel & Ford 1987, Shine 2003). These studies also include descriptions of reproductive mode, period of vitellogenesis, ovulation and gestation, period of spermatogenesis and mating, fecundity, maturation size and sexual dimorphism (Parker & Plummer 1987, Seigel & Ford 1987, Shine 1993). The trophic biology of snakes requires more detailed studies than those available in the literature, which consist mainly of taxonomic lists of ingested items. Since the 1990's, more detailed information has begun to emerge on predator-prey relationships, absolute and relative frequencies of food items in the diet, importance of food items in the diet, ontogenetic variation, volume and calorific value (e.g., Arnold 1993, Henderson 1993, Cundall 1995, Marques & Sazima 1997, Maschio et al. 2010).

Despite the increasing number of ecological studies involving Neotropical snakes, mostly during the 21st century, (e.g., Albuquerque et al. 2007, Leite et al. 2007, Marques & Muriel 2007, Maschio et al. 2007, 2010, Prudente et al. 2007, Rivas et al. 2007a, b, Pizzatto & Marques 2007, Martins et al. 2008, Parpinelli & Marques 2008, Pinto et al. 2008, Pizzatto et al. 2008a, b, 2009, Sawaya et al. 2008; Sturaro & Gomes 2008, Tozetti & Martins 2008, Hartmann et al. 2009a, b, Marques et al. 2009, Scartozzoni et al. 2009, Turci et al. 2009, Albarelli & Santos-Costa 2010, Ávila et al. 2010, Bernarde & Abe 2010, Orofino et al. 2010, Araújo & Almeida-Santos 2011, Bernarde et al. 2011a, Rodrigues & Prudente 2011, Bernarde & Gomes 2012, Maria-Carneiro et al. 2012) and considering the great diversity and complexity of the snake fauna in the Amazon Biome (Bernarde et al. 2011a, b, Mendes-Pinto et al. 2011, Araújo et al. 2012), the reproductive and dietary patterns of many species of snakes are still poorly known (Martins & Oliveira 1999, Balestrin & Di-Bernardo 2005, Pizzatto et al. 2006, Sturaro & Gomes 2008, Maschio et al. 2010).

Popularly known as vine-snakes, or tree-snakes (Cunha & Nascimento 1993), Imantodes (Duméril, 1853) includes six species with arboreal habits: I. cenchoa (Linnaeus, 1758), I. gemmistratus (Cope, 1861), I. inornatus (Boulenger, 1896), I. lentiferus (Cope, 1894), I. phantasma (Myers, 1982), I. tenuissimus (Cope, 1866). They are distributed between the Tropic of Cancer and the Tropic of Capricorn, from Mexico to Argentina, inhabiting very different environments, such as tropical forests and savannas (Zug et al. 1979, Myers 1982, Costa et al. 2010).

Imantodes cenchoa is distributed from the tropical region of the island of Trinidad and Panama, through Mexico, Venezuela, Colombia, Guyana, Suriname, French Guiana, Bolivia and Brazil (north, central-west and northeast regions), reaching Paraguay and Argentina (Cunha & Nascimento 1978). It is an oviparous species and feeds on small frogs and lizards, mainly of Norops (Daudin, 1802) (Robinson 1977, Zug et al. 1979, Aveiro-Lins et al. 2006). Individuals can be seen foraging during the night on bromeliads, shrubs, in the underbrush, or on palm bracts, which during the day are primarily used for sleeping (Henderson & Nickerson 1976, Bartlett & Bartlett 2003). Despite the arboreal habits of I. cenchoa, this snake has been observed on the ground of forests (Henderson & Nickerson 1976, Marques & Sazima 2004) and dead on paved roads, killed by cars, indicating possible terrestrial foraging.

In this study, we present data to increase our understanding of the reproductive and dietary strategies of I. cenchoa from the Brazilian Amazon region.

MATERIAL AND METHODS

We analyzed 314 specimens (200 males and 114 females) of I. cenchoa, from forested environments of the Brazilian Amazon. They are deposited in the Herpetological Collection of the Museu Paraense Emílio Goeldi (MPEG), state of Pará, Brazil (Appendix).

The Amazon region is formed mostly of constantly humid environments, with precipitation close to 3000 mm/year. The driest month never has less than 60 mm of rainfall and humidity is around 80%, with an average temperature of 25.9°C. The climate of this region is classified as type Af (according to Köppen's classification). The vegetation varies from a low-altitude forest (the Andean portion) to the Amazon rainforest, with predominance of the "terra firme" forests, flooded forests, "várzeas" and "igapós" (PNRH 2006).

For each specimen, the snout-vent length (SVL), tail length (TL), head width (HW), head length (HL), head height (HH) and distance between the eyes (DBE) were measured, as well as the mass.

Through a longitudinal incision in the abdominal region, the gonads of both males and females were analyzed

macroscopically, in order to assess the state of the vas deferens of the males and the oviduct of the females, and to infer the number and length of ovarian follicles. Thus, females with SVL equal to or larger than females having follicles in secondary vitellogenesis (diameter 10.0 mm), and/or oviductal eggs, and/or oviducts with evidence of eggs, were considered sexually mature. Males with SVL equal to or larger than the smallest male with turgid testes, convoluted and opaque efferent ducts, were considered sexually mature (Shine 1977b, c, 1988, Balestrin & Di-Bernardo 2005). Fecundity was inferred from the ratio of the number of follicles in secondary vitellogenesis and the number of oviductal eggs with snout-vent length of the female (Aldridge 1979, DeNardo 1996, Thompson & Speake 2002, Santos & Llorente 2004).

Analysis of the reproductive cycle was performed using adult (sexually mature) specimens, observing the temporal distribution of follicles in secondary vitellogenesis or the presence of oviductal eggs (Shine 1977b, 1988).

Stomach contents were observed directly in the digestive tract of each specimen. The quantitative analysis of the diet was performed on the number of prey items observed in the stomach or intestine; in order to perform the qualitative analysis, we identified each prey item to the lowest possible taxonomic level, using the help of experts and the literature. Partially digested prey were measured (SVL and TL) and their mass was inferred from comparison with intact conspecific specimens of similar size (according to Rodriguez-Robles & Greene 1999) from nearby locations, and preserved in the Herpetological Collection of MPEG.

The direction of ingestion was classified according to the alignment of the head of the prey in relation to the body of the snake. Thus, the direction of items ingested head first was considered antero-posterior, and the direction of items ingested tail first was considered postero-anterior.

To verify the existence of ontogenetic variation in the diet, the types and size of prey found in the digestive tract of sexually immature and mature snakes were compared.

For data analyses, the program Statistica 7.1 was used. The significance level (α) used for all tests was 0.05. A Student t-test was performed to test for the presence of sexual dimorphism in the SVL (for data with normality and homogeneity of variance). To compare the sexual dimorphism in relation to the TL (excluding specimens with broken tails), a one-way ANCOVA was used, with sex as a factor and SVL as a covariate. To test the sexual dimorphism of the HL, a one-way ANCOVA was used with sex as a factor and HL subtracted from SVL as a covariate. In the case of HW, a one-way ANCOVA was used with sex as a factor and SVL as a covariate, with log-transformed data. All data submitted to ANCOVA were tested for normality and homogeneity of variance. Pearson correlation analysis was performed to find the relationships between SVL and HL, and SVL and HW. In all tests of sexual dimorphism, only data from sexually mature females and males were used. Pearson correlation analysis was performed to find the ratio between the SVL of females and the number of eggs and vitellogenic follicles. Pearson correlation analysis was also used to find the relationship between predator-prey SVL, snake HL and prey SVL, and between mass of prey and mass of snake.

RESULTS

Reproductive aspects

The SVL of sexually mature females was significantly greater than that of sexually mature males (t = 4.02, p < 0.01; N males = 150, N females = 71). There was no significant difference between the TL of males and females ($f_{1.196}$ = 2.40, p = 0.12; N males = 136, N females = 63). Sexual dimorphism of the HL was observed ($f_{1.218}$ = 98.29, p < 0.01; N males = 150, N females = 71), with sexually mature females having a longer head than mature males (females = 11.60 to 15.91 mm; males = 10.20 to 15.50 mm). Sexual dimorphism was also present in the HW ($f_{1.218}$ = 112.77, p < 0.01; N males = 150, N females = 71), with females having wider heads than males (females = 2.02 to 2.42 mm; males = 1.80 to 2.28 mm) (Table I).

Table I. Morphometry of sexually mature females and males of *Imantodes cenchoa* from the Brazilian Amazon, showing the number of specimens examined (N), mean, standard deviation and range. (SVL) Snout-vent length (mm), (TL) tail length (mm), (HL) head length (mm), (HW) head width (mm).

	Number of specimens	Mean ± Standard Deviation	Range
Female			
SVL	71	731.53 ± 67.84	560.00-900.00
TL	63	333.60 ± 35.97	247.00-403.00
HL	71	14.04 ± 1.00	11.60-15.91
HW	71	2.21 ± 0.09	2.02-2.42
Male			
SVL	150	690.60 ± 71.86	498.00-829.00
TL	136	311.23 ± 38.74	220.00-395.00
HL	150	12.78 ± 0.99	10.20-15.50
HW	150	2.07 ± 0.08	1.80-2.28

A correlation was found between SVL and HL (r^2 = 0.68, p < 0.01) and between SVL and HW (r^2 = 0.44, p < 0.01), where individuals with larger snout-vent length had larger head length and width.

The SVL of sexually mature specimens ranged from 438 to 829 mm in males (Fig. 1) and between 560 and 900 mm in females (Fig. 2), emphasizing the sexual bi-maturity, with males reaching sexual maturity at a smaller size than females.

One to two follicles in secondary vitellogenesis (N = 8, mean = 1.87) and one to three eggs (N = 14, mean = 2.07) were found in mature females. The length of eggs ranged from 22.85 to 45.10 mm (N = 14, mean = 30.06 mm). Larger females had a

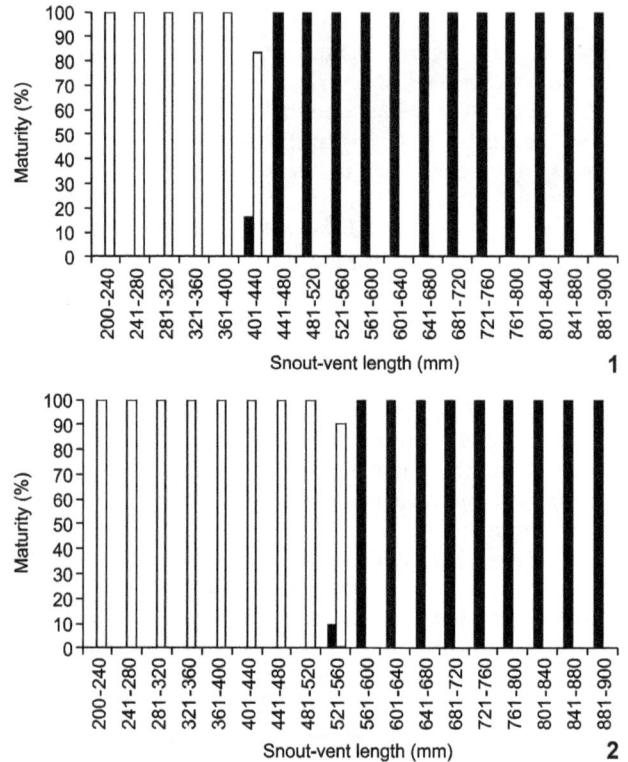

Figures 1-2. Percentage of sexually immature and mature males (1) and females (2) of *Imantodes cenchoa* from the Brazilian Amazon for the different size classes. (□) Immature, (■) mature.

greater number of follicles over 10 mm and a greater number of eggs (r^2 = 0.22, p = 0.03) (Fig. 3).

Females of *I. cenchoa* with eggs were found between November and January, during the rainiest season, and from April to July, during the dry season (Fig. 4).

Diet

Of the 314 specimens of *I. cenchoa* analyzed, 32.80% (N = 103) had some type of content in the stomach and/or intestines. Among the females with secondary follicles and/or eggs (N = 21), six (28.57%) presented food contents.

Among the food items, 84.11% were *Norops* (69.16%, N = 74) and *Gonatodes* (Fitzinger, 1843) (14.95%, N = 16) lizards, and the other 15.89% was made up of insect wings, eggs, scales, tails, lizard skulls, and other items in an advanced state of digestion, preventing their identification. Seven specimens had two food items in their digestive tract, yielding a total of 107 preys (Table II).

The predominant direction of ingestion was antero-posterior (71.96%, N = 77), and prey ingested postero-anteriorly accounted for 1.87% (N = 2) of the total. The direction of ingestion could not be determined for 26.17% (N = 28) of the items (Table II).

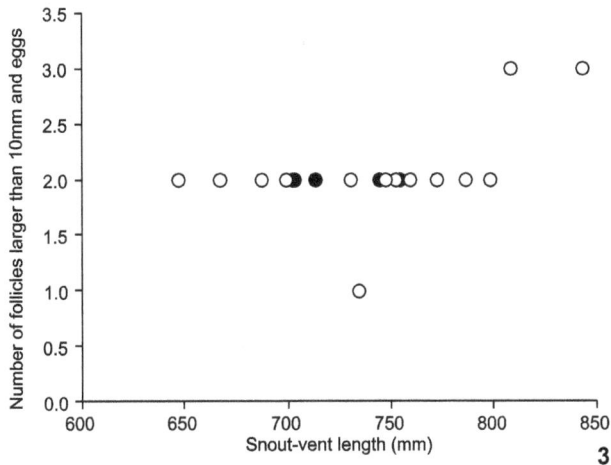

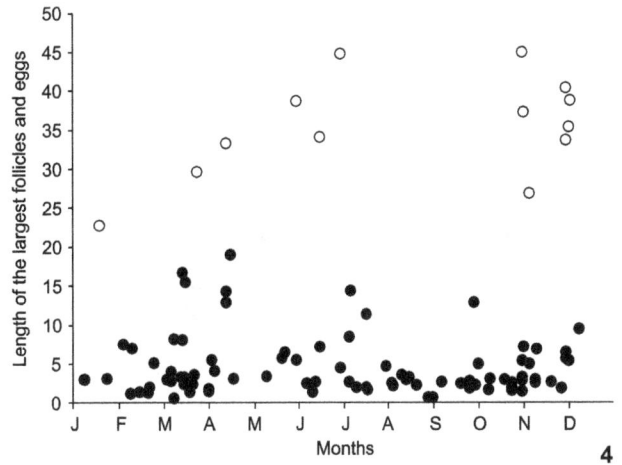

Figures 3-4. (3) Fecundity of *Imantodes cenchoa*, from the Brazilian Amazon, indicating the relationship between the number of follicles in secondary vitellogenesis (●) and eggs (○) with female SVL. (4) Seasonal distribution of the largest follicles (●) and eggs (○) of *Imantodes cenchoa*, from the Brazilian Amazon.

Table II. Contents of the digestive tract of *Imantodes cenchoa* from the Brazilian Amazon. (N) Number of prey; Undetermined = food content in an advanced state of digestion; (Di) Direction of ingestion, (AP) antero-posterior, (PA) postero-anterior; Unidentified = food content for which the direction of ingestion was not verified due to its advanced state of digestion.

Prey/Item	N	(%)	DI
Reptilia, Squamata			
Norops sp. (Daudin 1802)	17	15.89	AP
Norops sp. (Daudin 1802)	1	0.93	PA
Norops sp. (Daudin 1802)	8	7.48	Unidentified
Norops chrysolepis (Duméril & Bibron, 1837)	3	2.80	AP
Norops fuscoauratus (D'Orbigny, 1837)	34	31.78	AP
Norops fuscoauratus (D'Orbigny, 1837)	2	1.87	Unidentified
Norops punctatus (Daudin, 1802)	2	1.87	AP
Norops ortonii (Cope, 1868)	6	5.61	AP
Norops ortonii (Cope, 1868)	1	0.93	PA
Gonatodes sp. (Fitzinger, 1843)	3	2.80	AP
Gonatodes humeralis (Guichenot, 1855)	12	11.21	AP
Gonatodes humeralis (Guichenot, 1855)	1	0.93	Unidentified
Undetermined	17	15.89	Unidentified
Total	107	100.00	

Both immature and mature females and males frequently preyed on *Norops fuscoauratus* and no evidence of ontogenetic variation was observed (Figs 5 and 6).

The results show that larger snakes tend to eat larger prey items with larger mass, although smaller prey with smaller mass were not eliminated from their diet, (r^2 = 0.13, p = 0.01; r^2 = 0.22; p < 0.01, respectively) (Figs 7 and 8).

DISCUSSION

Sexual dimorphism in snakes can occur in characters such as size and/or body shape, color, position and/or size of organs, and behavior (KING 1989a, SHINE 1993, BONNET et al. 1998, KEOGH & WALLACH 1999, PIZZATTO & MARQUES 2006), as well as the relative size of the head (CAMILLERI & SHINE 1990). Most frequently sexually mature females have greater snout-vent length than their sexually mature male counterparts, as observed in this study, is (e.g., MARQUES & PUORTO 1998). This is related to the fact that there is a positive relationship between body size and the ability to produce and maintain a larger amount of eggs/embryos (SHINE 1994, PIZZATTO et al. 2006).

The absence of sexual dimorphism in the length of the tail of *Imantodes cenchoa*, observed in this study, had been previously reported by MYERS (1982), as was the exponential increase in length and width of the head of females relative to that of males. Adult males of *I. cenchoa* in Honduras and Costa Rica had a slightly longer tail than females. However, there was an absence of sexual dimorphism in tail length in specimens from Panama, Ecuador and Peru (ZUG et al. 1979). Sexual dimorphism in tail length occurs in most colubrids, in which males have larger tails than females. The larger tail of males is necessary to hold the copulatory organ and associated muscles. Conversely, females need a larger body in order to produce more offspring (KING 1989). The absence of sexual dimorphism in the tail of *I. cenchoa* can be related to its arboreal habit, since selection favors longer tails, which provide better balance and movement in trees (PIZZATTO et al. 2007).

Sexual dimorphism in head size, as found in *I. cenchoa*, has evolved, according to SHINE (1986), to enable males and females to feed on prey of different sizes, maximizing foraging efficiency and minimizing competition between the sexes. Simi-

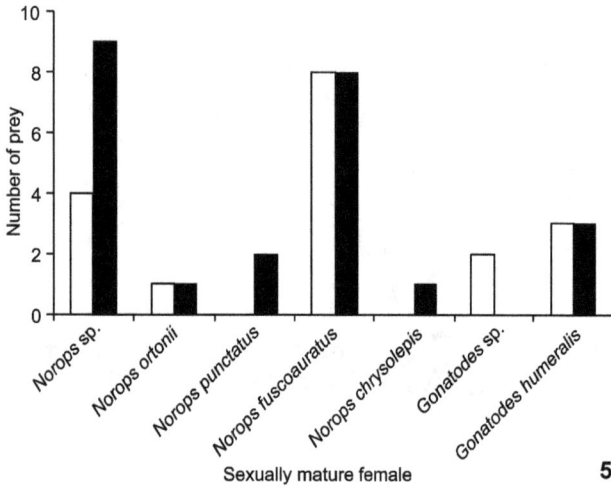

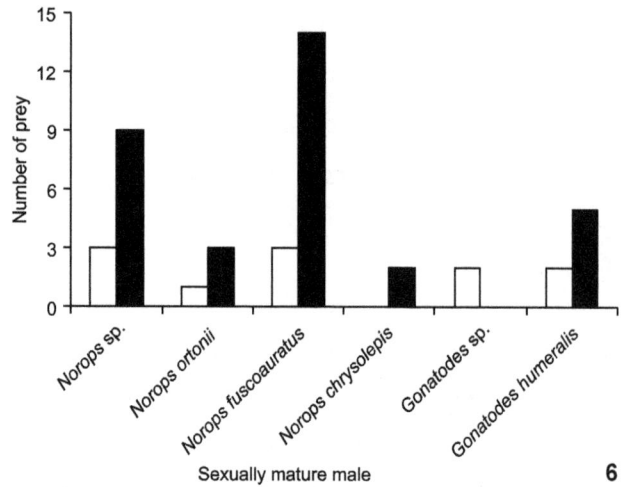

Figures 5-6. Relationship between the number and type of prey consumed by sexually immature and mature females (5) and male (6) of *Imantodes cenchoa* from the Brazilian Amazon. (□) Immature, (■) mature.

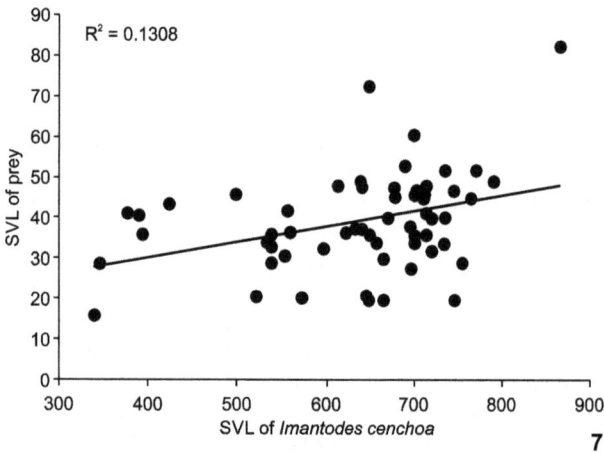

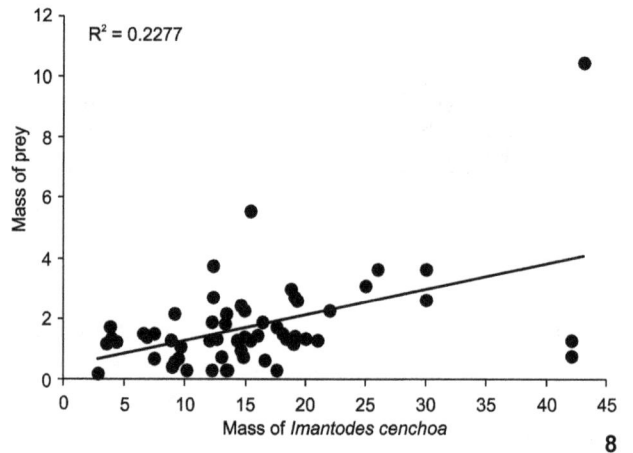

Figures 7-8. (7) Relationship between snout-vent length (SVL, in mm) of *Imantodes cenchoa*, and snout-vent length (SVL, in mm) of their prey, from the Brazilian Amazon. (8) Relationship between the mass (g) of *Imantodes cenchoa* and mass (g) of their prey, from the Brazilian Amazon.

larly, several authors (e.g., SHINE 1989, 1994, HOUSTON & SHINE 1993, PEARSON *et al.* 2002, SHETTY & SHINE 2002, SHINE *et al.* 2002) have suggested that head length dimorphism is related to intersexual resource competition instead of sexual selection, since head length is not crucial in mate choice or male fighting (SHINE 1991).

The fact that males reach sexual maturity earlier than females is probably related to energy cost, which is higher for females than for males of different sizes (PARKER & PLUMMER 1987, SANTOS-COSTA *et al.* 2006), but similar between large females and large males (MADSEN & SHINE 1993). Natural selection would favor of a more delayed sexual maturity in females. This results in larger females that are consequently more fertile (FITCH 1970,

1982, VITT & VANGILDER 1983, SHINE 1994) and could explain the bi-maturity in some species of snakes, such as *Erythrolamprus aesculapii* (Linnaeus, 1766) (MARQUES 1996), *Tantilla melanocephala* (Linnaeus, 1758) (MARQUES & PUORTO 1998, SANTOS-COSTA *et al.* 2006); *Helicops infrataeniatus* (Jan, 1865) (AGUIAR & DI-BERNARDO 2005), *Atractus reticulatus* (Boulenger, 1885) (BALESTRIN & DI-BERNARDO 2005), and as observed in *Imantodes cenchoa*.

The results on the fecundity of *Imantodes cenchoa*, presented in this study, are consistent with a trend found for most species of snakes, where larger females produce more eggs (FITCH 1970, SEIGEL & FORD 1987). The abdominal space may limit the reproductive investment relative to the size or the total mass of offspring, resulting, in many species, a strong relationship between the body size of the mother and the size of the litter (FITCH 1981), emphasizing why larger females are favored. LILLYWHITE and HENDERSON (1993) speculated that the slender body shape of arboreal snakes may constrain the female's capacity to carry eggs. This hypothesis was strongly supported by the results of PIZZATTO et al. (2007), and would explain the small amount of eggs produced by I. cenchoa. Some previous studies (e.g., JAMES & JOHNSTON 1998, PLAUT 2002, GHALAMBOR et al. 2004) have demonstrated that the movement and escape ability of females is related to the habitat type and the offspring/egg weight. These results suggest that the weight of eggs has a significant cost. Likewise, the use of arboreal habitats by snakes could have provided selective advantages, such as a better ability to cradle and camouflage in between tree branches (POUGH et al. 1988, LILLYWHITE & HENDERSON 1993), along with the laterally compressed body shape, as in I. cenchoa.

Snakes have different reproductive cycles, and males and females of the same species may also have distinct cycles (SEIGEL & FORD 1987). PIZZATTO et al. (2008a) suggested that the cycle of I. cenchoa is continuous for about eight months, differing from that observed in this study, where females containing eggs were found in the beginning of the rainy season (November to January) and in the first months of the dry season (April to July), suggesting two reproductive peaks throughout the year, but with recruitment occurring mainly in the period (rainy season) when there seems to be more food available. A third pattern was observed by ZUG et al. (1979). In his data, females of I. cenchoa with eggs were found throughout the year, indicating a prolonged breeding season, which is correlated with the rainy season in the region.

In temperate regions, the period when the temperatures are higher influence the timing of the reproductive cycle of snakes. Higher temperatures are important not only for egg development, but are also associated with greater availability of food during recruitment (SHINE 1977a). In Iquitos, Peru, FITCH (1982) found no evidence that snake species reproduce only in certain seasons. DUELLMAN (1978), based on work carried out in Santa Cecilia, Ecuador, concluded that non-seasonal reproduction is a trend among Amazonian snakes. However, according to data obtained by MARTINS & OLIVEIRA (1999), recruitment for most species occurs during the rainy season, when there is a greater supply of food than in the dry season.

In the present study, only 32.8% of the specimens analyzed had food contents in their stomachs. These results were expected, since the frequency of specimens containing food items is usually low, ranging between 14 and 30% (MASCHIO et al. 2010). According to GREGORY & ISAAC (2004), this low frequency may be related to the period in which the specimen was collected. SHINE (1987) suggested that female snakes generally tend to reduce food consumption during their gestation period. This was observed in females of Natrix natrix (Linnaeus, 1758) and Anilius scytale (Linnaeus, 1758) (READING & DAVIES 1996, MASCHIO et al. 2010, respectively), which began to feed

again after the reproductive period. The small number of females of I. cenchoa containing eggs and/or follicles in this study does not allow us to draw conclusions about the influence of gestation on stomach contents.

Analyses of stomach contents of I. cenchoa showed that Norops lizards are the most frequent items in their diet, demonstrating that these are their main prey (see also STUART 1948, 1958, WEHEKIND 1955, LANDY et al. 1966, HENDERSON & NICKERSON 1976). This result is in agreement with the findings of ZUG et al. (1979), who stated that I. cenchoa forages actively during the night, feeding on small diurnal arboreal lizards (mainly Norops) that sleep on the vegetation (see ÁVILA-PIRES 1995).

In this study, no food item of I. cenchoa was identified as an anuran. Despite this result, it is possible that these amphibians may occasionally serve as food for this species, as noted by different authors, as follows: MARTINS & OLIVEIRA (1999) found Pristimantis fenestratus (Steindachner, 1864) in the digestive tract of an adult male and a gravid female; TEST et al. (1966), observed I. cenchoa under captivity feeding on Pristimantis sp. (Duméril & Bibron, 1841); and BEEBE (1946) observed an individual of I. cenchoa stalking an individual of Ololygon rubra (Daudin, 1803).

According to MARTINS & GORDO (1993) the occurrence of arthropods in the digestive tract of snakes is the result of their anuran-based diets. In our study, arthropods were characterized as secondary stomach contents because they were always associated with lizard vestiges. This may be explained by the fact that Norops sp. and/or Gonatodes sp. are insectivorous (ARAÚJO 1991).

There is no record of I. cenchoa eating reptile eggs. Even though eggs have been found in the stomachs of these snakes by some studies (e.g. LANDY et al. 1966), these findings are associated with the ingestion of a gravid female lizard, where most of the prey was digested, except the eggs (HENDERSON & NICKERSON 1976).

The anteroposterior direction of ingestion, prevalent in this study, follows the pattern found in most species of snakes (PALMUTI et al. 2009), for example Helicops infrataeniatus (Jan, 1865) (AGUIAR & DI-BERNARDO 2004) and Thamnodynastes strigatus (Günther, 1858) (RUFFATO et al. 2003). According to GREENE (1976), this direction prevents the limbs of the prey and the disposition of their scales from offering resistance to swallowing, and minimizes the potential risk of injuries caused by the claws or teeth of the prey. Consequently, intake by the head results in decreased time, effort and energy expenditure during the process of prey capture and ingestion (PINTO & LEMA 2002). Furthermore, the chances of prey escaping are minimized by ingesting the lizards head-first, since some species have caudal autotomy. Caudal autotomy is a successful defensive strategy used by many lizards (CHAPPLE & SWAIN 2002), including those of Norops (SCHOENER & SCHOENER 1980) and geckos (CONGDON et. al. 1974), for instance species of Gonatodes, the main prey of I. cenchoa.

The specimens containing *N. fuscoauratus* and *G. humeralis* were recorded in almost every month of the year except May, September and November for *N. fuscoauratus* and April, August and December for *G. humeralis*. *Norops fuscoauratus*, the most consumed item by *I. cenchoa* in this study, can produce at least three clutches per year, laying one egg per clutch (VITT *et al.* 2008), indicating that this species has a continuous reproductive cycle (ÁVILA-PIRES 1995), which would justify the occurrence of *N. fuscoauratus* in the digestive tract of *I. cenchoa* throughout the year (see FITCH 1970, HOOGMOED 1973, DIXON & SOINI 1975, 1986, DUELLMAN 1978). Females of *Gonatodes humeralis* lay one egg at a time and may return a few days later to the same place to lay another egg (HOOGMOED 1973). Its reproduction also seems to occur throughout the year (DIXON & SOINI 1975, 1986) and it can lay up to twelve eggs (ÁVILA-PIRES 1995).

The selection of prey with specific sizes can be influenced by prey density and availability (PLUMMER & GOY 1984, SHINE 1987, 1991, MASCHIO *et al.* 2010). Although species normally have continuous reproductive cycles in the tropics, reproduction of some species may be seasonal (SHINE 2003. ALVES *et al.* 2005), with recruitment occurring mainly during the rainy season, when prey availability appears to be higher (MARTINS & OLIVEIRA 1999). This does not seem to be the case of *I. cenchoa*, for which both immature and mature males and females feed on common lizard species, as seen previously, in all seasons, in the studied area.

ACKNOWLEDGMENTS

We are very grateful to Maria C. dos S. Costa for helpful comments on the manuscript. We thank Teresa C.S. Avila-Pires, Annelise B. D'Angiolella and Marcelo J. Sturaro for identifying prey. We also thank the Museu Paraense Emílio Goeldi for logistic support, the Conselho Nacional de Desenvolvimento Científico e Tecnológico (CNPq) for productivity scholarship (ALCP, process 308950/2011-9) and financial support (ALCP, PROTAX project 562171/2010-0), and Fundação Amazônia Paraense de Amparo à Pesquisa (FAPESPA) for scholarship (KRMS).

REFERENCES

AGUIAR, L.F.S. & M. DI-BERNARDO 2004. Diet and feeding behavior of *Helicops infrataeniatus* (Serpentes: Colubridae: Xenodontinae) in southern Brazil. **Studies on Neotropical Fauna and Environment 39** (1): 7-14. doi: 10.1080/01650520412331270927.

AGUIAR, L.F.S. & M. DI-BERNARDO. 2005. Reproduction of the water snake *Helicops infrataeniatus* (Colubridae) in southern Brazil. **Amphibia-Reptilia 26** (2005): 527-533. doi:10.1163/156853805774806205.

ALBARELLI, P.P. & M.C. SANTOS-COSTA. 2010. Feeding ecology of *Liophis reginae semilineatus* (Serpentes: Colubridae: Xenodontinae) in eastern Amazon, Brazil. **Zoologia 27** (1): 87-91. doi: 10.1590/S1984-46702010000100013.

ALBUQUERQUE, N.R.; U. GALATTI & M. DI-BERNARDO. 2007. Diet and feeding behaviour of the Neotropical parrot snake (*Leptophis ahaetulla*) in the Northern Brazil. **Journal of Natural History 41**: 1237-1243. doi: 10.1080/00222930701400954.

ALDRIDGE, R.D. 1979. Female reproductive cycles of the *Arizona elegans* and *Crotalus viridis*. **Herpetologica 38**: 5-16.

ALVES, F.Q.; A.J.S. ARGOLO & J. JIM. 2005. Biologia reprodutiva de *Dipsas neivai* Amaral e *D. catesbyi* (Sentzen) (Serpentes, Colubridae) no sudeste da Bahia, Brasil. **Revista Brasileira de Zoologia 22** (3): 573-579. doi: 10.1590/S0101-81752005000300008.

ARAÚJO, A.F.B. 1991. Structure of a white sand-dune lizard community of coastal Brazil. **Revista Brasileira de Biologia 51** (4): 857-865.

ARAUJO, C.O. & S.M. ALMEIDA-SANTOS. 2011. Herpetofauna de um remanescente de Cerrado no Estado de São Paulo, sudeste do Brasil. **Biota Neotropica 11** (3): 47-62. doi: 10.1590/S1676-06032011000300003.

ARAÚJO, J.S.; M.B. SOUZA; T.A. FARIAS; D.P. SILVA; N.M. VENÂNCIO; J.M.L. MACIEL & P.R. MELO-SAMPAIO. 2012. *Liophis dorsocorallinus* Esqueda, Natera, La Marca and Ilija-Fistar, 2007 (Squamata: Dipsadidae): Distribution extension in southwestern Amazonia, state of Acre, Brazil. **Check List 8** (3): 518-519.

ARNOLD, S. 1993. Foraging theory and prey-size-predator-size relations in snakes, p. 87-115. *In*: R.A. SEIGEL & J. T. COLLINS (Eds.) **Snakes: Ecology and Behaviour.** New York, McGraw-Hill.

AVEIRO-LINS, G.; O. ROCHA-BARBOSA; M.G. SALOMÃO; G. PUORTO & M.F.C. LOGUERCIO. 2006. Topographical Anatomy of the Blunthead Treesnake, *Imantodes cenchoa* (Linnaeus, 1758) (Colubridae: Xenodontinae). **International Journal of Morphology 24** (1): 43-48. doi: 10.4067/S0717-95022006000100009.

ÁVILA, R.W.; R.A. KAWAHITA-RIBEIRO; V.L. FERREIRA & C. STRÜSMANN. 2010. Natural history of the coral snake *Micrurus pyrrhocryptus* Cope 1862 (Elapidae) from Semideciduous Forests of Western Brazil. **South American Journal of Herpetology 5** (2): 97-101. doi: 10.2994/057.005.0204.

ÁVILA-PIRES, T.C.S. 1995. Lizards of Brazilian Amazonia. **Zoologische Verhandelingen 299** (1): 1-706.

BALESTRIN, R.L. & M. DI-BERNARDO. 2005. Reproductive biology of *Atractus reticulatus* (Boulenger, 1885) (Serpentes, Colubridae) in southern Brazil. **Herpetological Journal 15** (3): 195-199.

BARTLETT, R.D. & P.P. BARTLETT. 2003. **Reptiles and Amphibians of the Amazon.** Gainesville, University Press of Florida, XVIII+291p.

BEEBE, W. 1946. Field Notes on the Snakes of Kartabo, British Guiana, and Caripito, Venezuela. **Zoologica 31** (1): 11-52.

BERNARDE, P.S. & A.S. ABE. 2010. Hábitos alimentares de serpentes em Espigão do Oeste, Rondônia, Brasil. **Biota Neotropical 10** (1): 167-173. doi: 10.1590/S1670-06032010000100017.

BERNARDE, P.S. & J.O. GOMES. 2012. Serpentes peçonhentas e ofidismo em Cruzeiro do Sul, Alto Juruá, Estado do Acre, Brasil. **Acta Amazonica 42** (1): 65-72. doi: 10.1590/S0044-59672012000100008.

BERNARDE, P.S.; H.C. COSTA; R.A. MACHADO & V.A. SÃO-PEDRO. 2011a. *Bothriopsis bilineata bilineata* (Wied, 1821) (Serpentes: Viperidae): New records in the states of Amazonas, Mato Grosso and Rondônia, northern Brazil. **Check List 7** (3): 343-347.

BERNARDE, P.S.; R.A. MACHADO & L.C.B. TURCI. 2011b. Herpetofauna da área do Igarapé Esperança na Reserva Extrativista Riozinho da Liberdade, Acre – Brasil. **Biota Neotropica 11** (3): 117-144. doi: 10.1590/S1676-06032011000300010.

BONNET, X.; R. SHINE; G. NAULLEAU & M. VACHER-VALLAS. 1998. Sexual dimorphism in snakes: different reproductive roles favour different body plans. **Proceedings of the Royal Society of London, Series B 265** (1392): 179-183. doi: 10.1098/rspb.1998.0280.

CAMILLERI, C. & R. SHINE. 1990. Sexual dimorphism and dietary divergence: differences in trophic morphology between male and female snakes. **Copeia 1990** (3): 649-658.

CHAPPLE, G.C. & SWAIN, R. 2002. Distribution of energy reserves in a viviparous skink: Does tail autotomy involve the loss of lipid stores? **Austral Ecology 27**: 565-572.

CONGDON, J.D.; L.J. VITT; & W.W. KING. 1974. Geckos: adaptive significance and energetics of tail autotomy. **Science 184**: 1379-1380.

COSTA, H.C.; A.B. BARROS; L.R. SUEIRO & R.N. FEIO. 2010. The blunt-headed vine snake, *Imantodes cenchoa* (Linnaeus, 1758), in Minas Gerais, southeastern Brazil. **Biotemas 23** (4): 173-176. doi: 10.5007/2175-7925.2010v23n4p173.

CUNDALL, D. 1995. Feeding behaviour in *Cylindrophis* and its bearing on the evolution of alethinophidia snakes. **Journal of Zoology: Proceedings of the Zoological Society of London 237**: 353-376. doi: 10.1111/j.1469-7998.1995.tb02767.x.

CUNHA, O.R. & F.P. NASCIMENTO. 1978. Ofídios da Amazônia. X. As cobras da região leste do Pará. **Publicações Avulsas do Museu Paraense Emilio Goeldi 32**: 1-218.

CUNHA, O.R. & F.P. NASCIMENTO. 1993. Ofídios da Amazônia. As cobras da região leste do Pará. **Boletim do Museu Paraense Emilio Goeldi, Série Zoologia, 9** (1): 1-191.

DENARDO, D. 1996. Reproductive biology, p. 212-223. *In*: D.R. MADER (Ed.). **Reptile medicine and surgery**. Philadelphia, W.B. Saunders Co, 512p.

DIXON, J.R. & P. SOINI. 1975. The reptiles of the upper Amazon Basin, Iquitos region, Peru. Part I. Lizards and Amphisbaenians. **Contributions in biology and geology/Milwaukee Public Museum 4**: 1-58.

DIXON, J.R. & P. SOINI. 1986. **The reptiles of the upper Amazon Basin, Iquitos region, Peru**. Milwaukee, Contributions in biology and geology, Milwaukee Public Museum, 154p.

DUELLMAN, W.E. 1978. **The biology of an equatorial herpetofauna in Amazonian Ecuador**. Lawrence, University of Kansas, Museum of Natural History, Miscellaneous Publication 65, 352p.

FITCH, H. 1970. Reproductive cycles of lizard and snakes. **Miscellaneous Publication (University of Kansas, Natural History) 52**: 1-247.

FITCH, H. 1981. Sexual size differences in reptiles. **Miscellaneous Publication (University of Kansas, Natural History) 70**: 1-72.

FITCH, H. 1982. Reproductive cycles in tropical reptiles. **Miscellaneous Publication (University of Kansas, Natural History) 96**: 1-53.

GHALAMBOR, C.K.; D.N. REZNICK & J.A. WALKER. 2004. Constraints on adaptative evolution: the functional trade-off between reproduction and fast-start swimming performance in the Trinidadian guppy (Poecilia reticulata). **The American Naturalist 164**: 38-50.

GREENE, H.W. 1976. Scale overlap a directional sign stimulus for prey ingestion by ophiophagous snakes. **Zeitschrift für Tierpsychologie 41** (2): 113-120. doi: 10.1111/j.1439-0310.1976.tb00473.x.

GREENE, H.W. 1986. Natural history and evolutionary biology, p. 99-108. *In*: M.E. FEDER & G.V. LAUDER (Eds). **Predator-prey relationships: Perspectives and approaches from the study of lower vertebrates**. Chicago, University of Chicago Press, 208p.

GREENE, H.W. 1994. Systematics and natural history, foundations for understanding and conserving biodiversity. **American Zoologist 34**: 48-56. 10.1093/icb/34.1.48.

GREGORY, P.T. & L.A. ISAAC. 2004. Food habits of the Grass Snake in southeastern England: is *Natrix natrix* a generalist predator? **Journal of Herpetology 38** (1): 88-95. doi: 10.1670/87-03A.

HARTMANN, P.A.; M.T. HARTMANN & M. MARTINS. 2009a. Ecology of snakes assemblage in the Atlantic Forest of Southeastern Brazil. **Papéis Avulsos de Zoologia 49** (27): 343-360. doi: 10.1590/S0031-10492009002700001.

HARTMANN, P.A.; M.T. HARTMANN & M. MARTINS. 2009b. Ecologia e história natural de uma taxocenose de serpentes do Núcleo Santa Virgínia do Parque Estadual da Serra do Mar, no sudeste do Brasil. **Biota Neotropica 9** (3): 173-184. doi: 10.1590/S1676-06032009000300018.

HENDERSON, R.W. 1993. On the diets of some arboreal Boids. **Herpetological Natural History 1**: 91-96.

HENDERSON, R.W. & M.A. NICKERSON. 1976. Observations on the behavioral ecology of three species of *Imantodes* (Reptilia, Serpentes, Colubridae). **Journal of Herpetology 10** (3): 205-210.

HOOGMOED, M.S. 1973. Notes on the herpetofauna of Surinam IV: The lizards and amphisbaenians of Surinam. **Biogeographica 4**: 1-425.

JAMES, R.S. & I.A. JOHNSTON. 1998. Influence of spawning on swimming performance and muscle contractile properties in the short-horn sculpin. **Journal of Fish Biology 53**: 485-501.

KEOGH, J.S. & V. WALLACH. 1999. Allometry and sexual dimorphism in the lung morphology of prairie rattlesnakes, *Crotalus viridis viridis*. **Amphibia-Reptilia 20**: 377-389. doi: 10.1163/156853899X00420.

KING, R.B. 1989a. Body size variation among island and mainland snake populations. **Herpetologica 45**: 84-88.

KING, R.B. 1989b. Sexual dimorphism in tail lenght: sexual selection, natural selection, or morphological constraint? **Biological Journal of Linnean Society 38**: 133-154. doi: 10.1111/j.1095-8312.1989.tb01570.x.

KLAUBER, L.M. 1972. **Rattlesnakes their habits, life-histories, and influence of mankind.** Berkeley, University of California Press, 740p.

LANDY, M.J.; A.L. DAVID; E.O. MOLL & H.M. SMITH. 1966. A Collection of Snakes from Volcan Tacana, Chiapas, Mexico. **Journal of the Ohio Herpetological Society 5** (3): 93-101.

LEITE, P.T.; S.F. NUNES & S.Z. CECHIN. 2007. Dieta e uso de hábitat da jararaca-do-brejo. *Mastigodryas bifossatus* Raddi (Serpentes, Colubridae) em domínio subtropical do Brasil. **Revista Brasileira de Zoologia 24** (3): 729-734. doi: 10.1590/S0101-81752007000300025.

LILLYWHITE, H.B. & R.W. HENDERSON. 1993. Behavioral and functional ecology of arboreal snakes, p. 1-48. *In*: R.A. SEIGEL & J.T. COLLINS (Eds). **Snakes: ecology and behavior.** New York, McGraw-Hill.

MADSEN, T. & R. SHINE. 1993. Costs of reproduction in a population of European adders. **Oecologia 94** (4): 488-495. doi: 10.1007/BF00566963.

MARIA-CARNEIRO, T.; M. WACHLEVSKI & C.F.D. ROCHA. 2012. What to do to defend themselves: description of three defensive strategies displayed by a serpent *Dipsas alternans* (Fischer, 1885) (Serpentes, Dipsadidae). **Biotemas 25** (1): 207-210. doi: 10.5007/2175-7925.2012v25n1p207.

MARQUES, O.A.V. 1996. Biologia reprodutiva de *Erythrolamprus aesculapii* Linnaeus (Colubridae), no sudeste do Brasil. **Revista Brasileira de Zoologia 13**: 747-753. doi: 10.1590/S0101-81751996000300022.

MARQUES, O.A.V. & G. PUORTO. 1998. Feeding, reproduction and growth in the crowned snake *Tantilla melanocephala* (Colubridae), from southeastern Brazil. **Amphibia-Reptilia 19** (3): 311-318. doi: 10.1163/156853898X00214.

MARQUES, O.A.V. & A.P. MURIEL. 2007. Reproductive biology and food habits of the swamp racer *Mastigodryas bifossatus* (Colubridae) from Southeastern South America. **Herpetological Journal 17**: 104-109.

MARQUES, O.A.V. & I. SAZIMA. 1997. Diet and feeding behavior of the coral snakes, *Micrurus corallinus*, from the Atlantic Forest of Brazil. **Herpetological Natural History 5** (1): 253-259.

MARQUES, O.A.V. & I. SAZIMA. 2004. História natural dos répteis da estação ecológica Juréia-Itatins, p. 257-277. *In*: O.A.V. MARQUES (Ed.). **Estação Ecológica Juréia-Itatins: Ambiente Físico, Flora e Fauna.** Ribeirão Preto, Holos, 384p.

MARQUES, O.A.V.; S.M. ALMEIDA-SANTOS; M. RODRIGUES & R. CAMARGO. 2009. Mating and Reproductive Cycle in the Neotropical Colubrid snake *Chironius bicarinatus*. **South American Journal of Herpetology 4** (1): 76-80. doi: 10.2994/057.004.0110.

MARTINS, M. & M. GORDO. 1993. *Bothrops atrox* (Common Lancehead). Diet. **Herpetolological Review 24**: 151-152.

MARTINS, M. & M.E. OLIVEIRA. 1999 [1998]. Natural history of snakes in forests of the Manaus region, Central Amazonia, Brazil. **Herpetological Natural History 6** (2): 78-150.

MARTINS, M.; O.A.V. MARQUES & I. SAZIMA. 2008. How to be arboreal and diurnal and still stay alive: microhabitat use, time of activity, and defense in Neotropical forest snake. **South American Journal of Herpetology 3** (1): 58-67. doi: 10.2994/1808-9798(2008)3[58:HTBAAD]2.0.CO;2.

MASCHIO, G.F.; A.L.C. PRUDENTE & D.T. FEITOSA. 2007. Reproductive biology of *Anilius scytale* (Linnaeus, 1758) (Serpentes, Aniliidae) from Eastern Amazonia, Brazil. **South American Journal of Herpetology 2** (3): 179-183. doi: 10.2994/1808-9798(2007)2[179:RBOASL]2.0.CO;2.

MASCHIO, G.F.; A.L.C. PRUDENTE; F.S. RODRIGUES & M.S. HOOGMOED. 2010. Food habits of *Anilius scytale* (Serpentes: Aniliidae) in the Brazilian Amazonia. **Zoologia 27** (2): 184-190. doi: 10.1590/S1984-46702010000200005.

MENDES-PINTO, T.J.; M.A. MAGALHÃES & J.C.R. TELLO. 2011. Predação de *Scinax ruber* (Laurenti, 1768) (Anura: Hylidae) por *Liophis typhlus* (Linnaeus, 1758) (Serpente: Dipsadidae) na Amazônia Central, Brasil. **BioFar 5** (2): 122-126.

MYERS, C.W. 1982. Blunt-Headed Vine Snakes (*Imantodes*) in Panama, Including a New Species and Other Revisionary Notes. **American Museum of Natural History** (2738): 1-50.

OROFINO, R.P.; L. PIZZATTO & O.A.V. MARQUES. 2010. Reproductive biology and food habits of *Pseudoboa nigra* (Serpentes: Dipsadidae) from the Brazilian Cerrado. **Phyllomedusa 9** (1): 53-61.

PALMUTI, C.F.S.; J. CASSIMIRO & J. BERTOLUCI. 2009. Food habits of snakes from the RPPN Feliciano Miguel Abdala, an Atlantic Forest fragment of southeastern Brazil. **Biota Neotropica 9** (1): 263-269. doi: 10.1590/S1676-06032009000100028.

PARKER, W.S. & M.V. PLUMMER. 1987. Population ecology, p. 253-301. *In*: R.A. SEIGEL; J.T. COLLINS & S.S. NOVAK (Eds). **Snakes: ecology and evolutionary biology.** New York, MacMillan Publish Co.

PARPINELLI, L. & O.A.V. MARQUES. 2008. Seasonal and daily activity in the Pale-headed Blindsnake *Liotyphlops beui* (Serpentes: Anomalepididae) in Southeastern Brazil. **South American Journal of Herpetology 3** (3): 207-212. doi: 10.2994/1808-9798-3.3.207.

PINTO, C.C. & T. LEMA. 2002. Comportamento alimentar e dieta de serpentes, gêneros *Boiruna* e *Clelia* (Serpentes, Colubridae). **Iheringia, Série Zoologia, 92** (2): 9-19. doi: 10.1590/S0073-47212002000200002.

PINTO, R.R.; R. FERNANDES & O.A.V. MARQUES. 2008. Morphology and diet of two sympatric colubrid snakes, *Chironius flavolineatus* and *Chironius quadricarinatus* (Serpentes: Colubridae). **Amphibia-Reptilia 29**: 149-160.

PIZZATTO, L.; M. CANTOR; J.L. OLIVEIRA; O.A.V. MARQUES; V. CAPOVILLA & M. MARTINS. 2008a. Reproductive ecology of Dipsadine snakes, with emphasis on South American species. Herpetologica 64 (2): 168-179. doi: 10.1655/07-031.1.

Pizzatto, L. & O.A.V. Marques. 2006. Interpopulational variation in sexual dimorphism, reproductive output, and parasitism of Liophis miliaris (Colubridae) in the Atlantic forest of Brazil. Amphibia-Reptilia 27 (1): 37-46.

Pizzatto, L. & O.A.V. Marques. 2007. Reproductive ecology of boine snakes with emphasis on Brazilian species and comparison to pythons. South American Journal of Herpetology 2: 107-122. doi: 10.2994/1808-9798(2007)2[107:REOBSW]2.0.CO;2.

Pizzatto, L.; O.A.V. Marques & K. Facure. 2009. Food habits of Brazilian boid snakes: overview and new data, with special reference to Corallus hortulanus. Amphibia-Reptilia 30: 533-544. doi: 10.1163/156853809789647121.

Pizzatto, L.; R. Jordão & O.A.V. Marques. 2008b. Overview of reproductive strategies in Xenodontini (Serpentes: Colubridae: Xenodontinae) with new data for Xenodon neuwiedii and Waglerophis merremii. Journal of Herpetology 42: 153-162. doi: 10.1670/06-150R2.1.

Pizzatto, L.; S.M. Almeida-Santos & O.A.V. Marques. 2006. Biologia reprodutiva de serpentes brasileiras, p. 201-221. In: L.B. Nascimento & M.E. Oliveira (Eds). Herpetologia no Brasil II. Belo Horizonte, Sociedade Brasileira de Herpetologia, 354p.

Pizzatto, L.; S.M. Almeida-Santos & R. Shine. 2007. Life-history adaptations to arboreality in snakes. Ecology 88 (2): 359-366.

Plano Nacional de Recursos Hídricos (PNRH). 2006. Caderno da Região Hidrográfica Amazônica/Ministério do Meio Ambiente. Available online at: http://www.mma.gov.br/estruturas/161/_publicacao/161_publicacao03032011024915.pdf [Accessed: July 8th 2013].

Plaut, I. 2002. Does pregnancy affect swimming performance of female mosquitofish, Gambusia affinis? Functional Ecology 16: 290-295.

Plummer, M.V. & J.M. Goy. 1984. Ontogenetic dietary shift of water snakes (Nerodia rhombifera) in a fish hatchery. Copeia (2): 550-552p. doi: 10.2307/1445218.

Pough, F.H.; R.M. Andrews; J.E. Cadle; M.L. Crump; A.H. Savitsky & K.D. Wells. 1988. Herpetology. Upper Saddle River, Prentice Hall.

Prudente, A.L.C.; G.F. Maschio; C.E. Yamashina & M.C. Santos-Costa. 2007. Morphology, reproductive biology and diet of Dendrophidion dendrophis (Schlegel, 1837) (Serpentes, Colubridae) in Brazilian Amazon. South American Journal of Herpetology 2 (1): 53-58. doi: 10.2994/1808-9798(2007)2 [53:MRBADO]2.0.CO;2.

Reading, C.J. & J.L. Davies. 1996. Predation by Grass Snakes (Natrix natrix) at a site in Southern England. Journal of Zoology 239 (1): 73-82. doi: 10.1111/j.1469-7998.1996.tb05437.x.

Rivas, J.A.; M.C. Muños; G. Burghardt & J.B. Thorbjarnarson. 2007a. Sexual size dimorphism and mating system of the

Green Anaconda (Eunectes murinus), p. 312-325. In: R.W. Henderson & G.W. Powell (Eds). Biology of Boas, Pythons, and Related Taxa. Eagle Mountain, Eagle Mountain Publishing Company.

Rivas, J.A.; M.C. Muños; J.B. Thorbjarnarson; G. Burghardt; W. Holmstrom & P. Cadle. 2007b. Natural History of the Green Anaconda (Eunectes murinus) in the Venezuelan Ilanos, p. 128-138. In: R.W. Henderson & G.W. Powell (Eds). Biology of Boas, Pythons, and Related Taxa. Eagle Mountain, Eagle Mountain Publishing Company.

Robinson, D.C. 1977. Herpetofauna bromelícola costarricense y renacuajos de Hyla picadoi Dunn, Historia Natural de Costa Rica, p. 31-43. In: L.D. Gómez (Ed.). Biologia de las Bromelias. San José, Museo Nacional de Costa Rica.

Rodrigues, F.S. & A.L.C. Prudente. 2011. The snake assemblage (Squamata: Serpentes) of a Cerrado-Caatinga transition area in Castelo do Piauí, state of Piauí, Brazil. Zoologia 28 (4): 440-448. doi: 10.1590/S1984-46702011000400005.

Rodriguez-Robles, J. & H.W. Greene. 1999. Food habits of the long-nosed snake (Rhinocheilus lecontei), a "specialist" predator? Journal Zoology 248 (4): 489-499. doi: 10.1111/j.1469-7998.1999.tb01048.x.

Ruffato, R.; M. Di-Bernarno & G.F. Maschio. 2003. Dieta de Thamnodynastes strigatus (Serpentes, Colubridae) no Sul do Brasil. Phyllomedusa 2 (1): 27-43.

Santos-Costa, M.V. & A.L.C. Prudente. 2005. Imantodes cenchoa (Chunk-headed Snake). Mating. Herpetological Review 36 (3): 324.

Santos-Costa, M.C.; A.L.C. Prudente & M. Di-Bernardo. 2006. Reproductive biology of Tantilla melanocephala (Linnaeus, 1758) (Serpentes, Colubridae) from Eastern Amazonia, Brazil. Journal of Herpetology 40 (4): 553-556. doi: 10.1670/0022-1511(2006)40[553:RBOTML]2.0.CO;2.

Santos, X. & G.A. Llorente. 2004. Lipid dynamics in the viperine snake, Natrix maura, from the Ebro Delta (NE Spain). Oikos 105 (1): 132-140. doi: 10.1111/j.0030-1299.2004.12222.x.

Sawaya, R.J.; O.A.V. Marques & M. Martins. 2008. Composição e história natural das serpentes de Cerrado de Itirapina, São Paulo, sudeste do Brasil. Biota Neotropica 8 (2): 153-175. doi: 10.1590/S1676-06032008000200015.

Scartozzoni, R.R.; M.G. Salomão & S.M. Almeida-Santos. 2009. Natural history of the vine snake Oxybelis fulgidus (Serpentes, Colubridae) from Brazil. South America Journal of Herpetology 4 (1): 81-89. doi: 10.2994/057.004.0111.

Seigel, R. & N. Ford. 1987. Snakes: Ecology and Evolutionary Biology, p. 210-252. In: R. Seigel; J. Collins & S. Novak (Eds) Reproductive ecology. New York, MacMillan Publish Co.

Shine, R. 1977a. Habitats, diet and sympatry in snakes: a study from Australia. Canadian Journal of Zoology 55: 1118-1128. doi: 10.1139/z77-144.

Shine, R. 1977b. Reproduction in Australian elapid snakes II. Female Reproductive Cycles. Australian Journal of Zoology 25: 655-666. doi: 10.1071/ZO9770655.

SHINE, R. 1977c. Reproduction in Australian elapid snakes I. Testicular cycles and mating seasons. **Australian Journal of Zoology 25**: 647-653. doi: 10.1071/ZO9770647.

SHINE, R. 1986. Sexual differences in morphology and niche utilization in an aquatic snake, *Acrochordus arafurae*. **Oecologia 69** (2): 260-267. doi: 10.1007/BF00377632.

SHINE, R. 1987. Ecological ramifications of prey size: food habits and reproductive biology of Australian copperhead snakes (*Austrelaps*, Elapidae). **Journal of Herpetology 21** (1): 21-28.

SHINE, R. 1988. Constraints on reproductive investment in six species of Australian elapid snakes. **Herpetologica 34**: 73-79.

SHINE, R. 1988. Food habits and reproductive biology of small australian snakes of the genera Unechis and Suta (Elapidae). **Journal of Herpetology 22** (3): 307-315.

SHINE, R. 1991. Intersexual dietary divergence and the evolution of sexual dimophism in snakes. **The American Naturalist 138**: 103-122.

SHINE, R. 1991. Why do larger snakes eat larger prey items? **Functional Ecology 5** (4): 493-502.

SHINE, R. 1993. Sexual dimorphism in snakes, p. 49-86. *In*: R.A. SEIGEL & J.T. COLLINS (Eds). **Snakes: ecology and behavior.** New York, McGraw-Hill.

SHINE, R. 1994. Sexual size dimorphism in snakes revisited. **Copeia** (2): 326-346.

SHINE, R. 2003. Reproductive strategies in snakes. **Proceedings of the Royal Society of London, Series B 270** (1519): 995-1004. doi: 10.1098/rspb.2002.2307.

SCHOENER, T.W. & A. SCHOENER. 1980. Ecological and demographic correlates of injury rates in some Bahamian Anolis lizards. **Copeia 1980** (4): 839-850.

STUART, L. 1948. The Amphibians and Reptiles of Alta Verapaz, Guatemala. **Miscellaneous Publications (Museum of Zoology, University of Michigan) 69**: 1-109.

STUART, L.C. 1958. Study of the Herpetofauna of the Uaxactun-Tikal Area of Northern El Peten, Guatemala. **Contributions from the Laboratory of Vertebrate Biology 75**: 1-30.

STURARO, M.J. & J.O. GOMES. 2008. Feeding behavior of the Amazonian Water Snake *Helicops hagmanni* Roux, 1910 (Reptilia: Squamata: Colubridae: Hydropsini). **Boletim do Museu Paraense Emílio Goeldi 3** (3): 225-228.

TEST, F.H.; O.J. SEXTON & H. HEATWOLE. 1966. Reptiles of Rancho Grande and Vicinity, Estato Aragua, Venezuela. **Miscellaneous Publications (Museum of Zoology, University of Michigan) 128**: 1-63. doi: 2027.42/56372.

THOMPSON, M.B. & B.K. SPEAKE. 2002. Energy and nutrient utilization by embryonic reptiles. **Comparative Biochemistry and Physiology A 133** (3): 529-538.

TOZETTI, A.M. & M. MARTINS. 2008. Habitat use by the South-American rattlesnakes (*Crotalus durissus*) in South-easthern Brazil. **Journal of Natural History 42** (19-20): 1435-1444. doi: 10.1080/00222930802007823.

TURCI, L.C.B.; S. ALBUQUERQUE; P.S. BERNARDE & D.B. MIRANDA. 2009. Uso do hábitat, atividade e comportamento de *Bothriopsis bilineata* e de *Bothrops atrox* (Serpentes, Viperidae) na floresta do Rio Moa, Acre, Brasil. **Biota Neotropica 9** (3): 197-206. doi: 10.1590/S1676-06032009000300020.

VITT, L.J.; W.E. MAGNUSSON; T.C. ÁVILA-PIRES & A.P. LIMA. 2008. **Guia de Lagartos da Reserva Adolpho Ducke, Amazônia Central (Guide to the Lizards of Reserva Adolpho Ducke, Central Amazonia).** Manaus, Áttema Design Editorial, 176p.

VITT, L.J. & L.D. VANGILDER. 1983. Ecology of snake community in the northeastern Brazil. **Amphibia-Reptilia 4**: 273-296. doi: 10.1163/156853883X00148.

ZUG, G.R.; S.B. HEDGES & S. SUNKEL. 1979. Variation in reproductive parameters of three neotropical snakes, *Coniophanes jissidens*, *Dipsas catesbyi*, and *Imantodes cenchoa*. Smithsonian Contributions to Zoology. **Smithsonian Institution Press** (300): 11-16.

PERMISSIONS

LIST OF CONTRIBUTORS

Sheila M. Simão
Laboratório de Bioacústica e Ecologia de Cetáceos, Departamento de Ciências Ambientais, Instituto de Florestas, Universidade Federal Rural do Rio de Janeiro. Rodovia BR 465 km 7, 23890-000 Seropédica, RJ, Brazil

Rodrigo H. Tardin
Laboratório de Bioacústica e Ecologia de Cetáceos, Departamento de Ciências Ambientais, Instituto de Florestas, Universidade Federal Rural do Rio de Janeiro. Rodovia BR 465 km 7, 23890-000 Seropédica, RJ, Brazil
Departamento de Ecologia, IBRAG, Universidade do Estado do Rio de Janeiro. Rua São Francisco Xavier 524, Maracanã, 20550-011 Rio de Janeiro, RJ, Brazil

Maria Alice S. Alves
Departamento de Ecologia, IBRAG, Universidade do Estado do Rio de Janeiro. Rua São Francisco Xavier 524, Maracanã, 20550-011 Rio de Janeiro, RJ, Brazil

Míriam P. Pinto
Departamento de Ecologia, IBRAG, Universidade do Estado do Rio de Janeiro. Rua São Francisco Xavier 524, Maracanã, 20550-011 Rio de Janeiro, RJ, Brazil
Departamento de Botânica, Ecologia e Zoologia, Centro de Biociências, Universidade Federal do Rio Grande do Norte. Rodovia BR-101, Campus Universitário, Lagoa Nova, 59072-970 Natal, RN, Brasil

Beria Falakali Mutaf and Ahmet Balci
Akdeniz University, Faculty of Aquatic Sciences and Fisheries, Antalya, Turkey

Deniz Aksit
Akdeniz University, Faculty of Aquatic Sciences and Fisheries, Antalya, Turkey

Rina Ramírez
Departamento de Malacología y Carcinología, Museo de Historia Natural, Universidad Nacional Mayor de San Marcos, Apartado 14-0434, Lima-14, Perú

Jorge L. Ramirez
Departamento de Malacología y Carcinología, Museo de Historia Natural, Universidad Nacional Mayor de San Marcos, Apartado 14-0434, Lima-14, Perú

Marcus Vinícius Vieira
Departamento de Ecologia, Universidade Federal do Rio de Janeiro. Avenida Carlos Chagas Filho 373, Cidade Universitária, 21941-902 Rio de Janeiro, RJ, Brazil

Carlos Frederico Duarte Rocha
Departamento de Ecologia, Universidade do Estado do Rio de Janeiro. Rua São Francisco Xavier 524, 20550-900 Rio de Janeiro, RJ, Brazil

Mauricio Almeida-Gomes
Departamento de Ecologia, Universidade Federal do Rio de Janeiro. Avenida Carlos Chagas Filho 373, Cidade Universitária, 21941-902 Rio de Janeiro, RJ, Brazil

Marcelo M. Dalosto, Alexandre V. Palaoro and Davi de Oliveira
Núcleo de Estudos em Biodiversidade Aquática, Programa de Pós-Graduação em Biodiversidade Animal, Centro de Ciências Naturais e Exatas, Universidade Federal de Santa Maria. Avenida Roraima 1000, 97105-900 Santa Maria, RS, Brazil.

Évelin Samuelsson
Programa de Pós-Graduação em Ecologia, Departamento de Ciências Biológicas, Universidade Regional Integrada do Alto Uruguai e das Missões. Avenida Sete de Setembro 1621, 99700-000 Erechim, RS, Brazil

Sandro Santos
Núcleo de Estudos em Biodiversidade Aquática, Programa de Pós-Graduação em Biodiversidade Animal, Centro de Ciências Naturais e Exatas, Universidade Federal de Santa Maria. Avenida Roraima 1000, 97105-900 Santa Maria, RS, Brazil

Leandro Lourenço Dumas
Universidade Federal do Rio de Janeiro, Laboratório de Entomologia, Departamento de Zoologia, Instituto de Biologia, Caixa Postal 68044, Cidade Universitária, 21941-971, Rio de Janeiro, RJ, Brazil

Ana Lucia Henriques-Oliveira
Universidade Federal do Rio de Janeiro, Laboratório de Entomologia, Departamento de Zoologia, Instituto de Biologia, Caixa Postal 68044, Cidade Universitária, 21941-971, Rio de Janeiro, RJ, Brazil

Lucas M.R.P. dos Santos and Daniele M.P. Nogueira
Laboratório de Anatomia Animal, Universidade Federal do Pampa. Rodovia BR-472, km 585, Caixa postal 118, 97500-970 Uruguaiana, RS, Brazil

Marcelo Abidu-Figueiredo
Departamento de Biologia Animal, Universidade Federal Rural do Rio de Janeiro. Rodovia BR-465, km 07, 23890-000 Seropédica, RJ, Brazil

André L.Q. Santos
Laboratório de Ensino e Pesquisa em Animais Silvestres, Universidade Federal de Uberlândia. Avenida Amazonas 2245, 38405-302 Uberlândia, MG, Brazil

Paulo de Souza Junior
Laboratório de Anatomia Animal, Universidade Federal do Pampa. Rodovia BR-472, km 585, Caixa postal 118, 97500-970 Uruguaiana, RS, Brazil

Andressa Paladini
Departamento de Zoologia, Universidade Federal do Paraná. Caixa Postal 19020, 81531-980 Curitiba, PR, Brazil

Rodney Ramiro Cavichioli
Departamento de Zoologia, Universidade Federal do Paraná. Caixa Postal 19020, 81531-980 Curitiba, PR, Brazil

Luiz F. Andrade
Programa de Pós-graduação em Biologia Animal, Universidade Federal Rural do Rio de Janeiro. Rodovia BR 465, km 7, 23890-000 Seropédica, RJ, Brazil

Rodrigo Johnsson
Laboratório de Invertebrados Marinhos: Crustacea, Cnidaria & Fauna Associada, Instituto de Biologia, Universidade Federal da Bahia. Rua Barão de Jeremoabo 147, Ondina, 40170-290 Salvador, BA, Brazil

Mariana S. Zanon
Programa de Pós-graduação em Ecologia e Evolução, Instituto de Biologia Roberto Alcantara Gomes, Universidade do Estado do Rio de Janeiro. Rua São Francisco Xavier 524, 20550-011 Rio de Janeiro, RJ, Brazil
Departamento de Ecologia, Instituto de Biologia Roberto Alcantara Gomes, Universidade do Estado do Rio de Janeiro. Rua São Francisco Xavier 524, 20550-011 Rio de Janeiro, RJ, Brazil

Mariana M. Vale
Departamento de Ecologia, Instituto de Biologia, Centro de Ciências da Saúde, Universidade Federal do Rio de Janeiro. Avenida Carlos Chagas 373, Ilha do Fundão, Cidade Universitária, 21941-902 Rio de Janeiro, RJ, Brazil

Maria Alice S. Alves
Departamento de Ecologia, Instituto de Biologia Roberto Alcantara Gomes, Universidade do Estado do Rio de Janeiro. Rua São Francisco Xavier 524, 20550-011 Rio de Janeiro, RJ, Brazil

Alice A. Notini, Talita O. Farias, Sônia A. Talamoni and Hugo P. Godinho
Programa de Pós-graduação em Zoologia de Vertebrados, Departamento de Ciências Biológicas, Pontifícia Universidade Católica de Minas Gerais, Avenida Dom José Gaspar, 500, 30535-610, Belo Horizonte, Minas Gerais, Brazil

Carla C. Siqueira
Programa de Pós-Graduação em Ecologia, Instituto de Biologia, Universidade Federal do Rio de Janeiro. Avenida Carlos Chagas Filho 373, Bloco A, Cidade Universitária, 21941-902 Rio de Janeiro, RJ, Brazil
Departamento de Ecologia, Universidade do Estado do Rio de Janeiro. Rua São Francisco Xavier 524, Maracanã, 20550-013 Rio de Janeiro, RJ, Brazil

Leonardo Dantas, Vagner L. R. Gomes, Helena G. Bergallo and Carlos Frederico D. Rocha
Departamento de Ecologia, Universidade do Estado do Rio de Janeiro. Rua São Francisco Xavier 524, Maracanã, 20550-013 Rio de Janeiro, RJ, Brazil

Davor Vrcibradic
Departamento de Zoologia, Universidade Federal do Estado do Rio de Janeiro. Avenida Pasteur 458, Urca, 22290-240 Rio de Janeiro, RJ, Brazil

Paulo Nogueira-Costa and Angele R. Martins
Departamento de Vertebrados, Museu Nacional, Universidade Federal do Rio de Janeiro. Quinta da Boa Vista, 20940-040 Rio de Janeiro, RJ, Brazil

Vivian Trevine and Maurício C. Forlani
Museu de Zoologia, Universidade de São Paulo. Avenida Nazaré 481, 04263-000 São Paulo, SP, Brazil

Célio F. B. Haddad
Departamento de Zoologia, Instituto de Biociências, Universidade Estadual Paulista. Avenida 24-A, 1515, Bela Vista, 13506-900 Rio Claro, SP, Brazil

Hussam Zaher
Museu de Zoologia, Universidade de São Paulo. Avenida Nazaré 481, 04263-000 São Paulo, SP, Brazil

João P. Vieira and Michelle N. Lopes
Laboratório de Ictiologia, Universidade Federal do Rio Grande. Avenida Itália km 8, 96201-900 Rio Grande, RS, Brazil

Fábio Lameiro Rodrigues and João Paes Vieira
Laboratório de Ictiologia, Instituto de Oceanografia, Universidade Federal do Rio Grande. Avenida Itália km 8, 96203-900 Rio Grande, RS, Brazil

Pryscilla Moura Lombardi
Laboratório de Ictiologia, Instituto de Oceanografia, Universidade Federal do Rio Grande. Avenida Itália km 8, 96203-900 Rio Grande, RS, Brazil

Daniel M. A. Pessoa
Departamento de Fisiologia, Centro de Biociências, Universidade Federal do Rio Grande do Norte. Campus Universitário Lagoa Nova, 59078-970 Natal, RN, Brazil

Ana Maria Rui
Departamento de Ecologia, Zoologia e Genética, Instituto de Biologia, Universidade Federal de Pelotas. Campus Universitário Capão do Leão, Caixa Postal 354, 96001-970 Pelotas, RS, Brazil

Marília A. S. Barros
Departamento de Fisiologia, Centro de Biociências, Universidade Federal do Rio Grande do Norte. Campus Universitário Lagoa Nova, 59078-970 Natal, RN, Brazil

Eraldo Medeiros Costa Neto
Departamento de Ciências Biológicas, Universidade Estadual de Feira de Santana. Avenida Transnordestina, Novo Horizonte, 44036-900 Feira de Santana, BA, Brazil

Sérgio Schwarz da Rocha
Laboratório de Bioecologia de Crustáceos, Centro de Ciências Agrárias, Ambientais e Biológicas, Universidade Federal do Recôncavo da Bahia. Rua Rui Barbosa 710, 44380-000 Cruz das Almas, BA, Brazil

Tiago Rozário da Silva
Departamento de Ciências Biológicas, Universidade Estadual de Feira de Santana. Avenida Transnordestina, Novo Horizonte, 44036-900 Feira de Santana, BA, Brazil

Mingming Zhang
College of Wildlife Resources, Northeast Forestry University, No.26 Hexing Road, Xiangfang District, Harbin 150040, P.R. China

Zhensheng Liu
College of Wildlife Resources, Northeast Forestry University, No.26 Hexing Road, Xiangfang District, Harbin 150040, P.R. China
Key Laboratory of Conservation Biology, State Forestry Administration, No.26 Hexing Road, Xiangfang District, Harbin 150040, P.R. China

Liwei Teng
College of Wildlife Resources, Northeast Forestry University, No.26 Hexing Road, Xiangfang District, Harbin 150040, P.R. China
Key Laboratory of Conservation Biology, State Forestry Administration, No.26 Hexing Road, Xiangfang District, Harbin 150040, P.R. China

Rosa Sá Gomes Hutchings
Coordenação de Biodiversidade, Instituto Nacional de Pesquisas da Amazônia. Caixa Postal 2223 AC Andre Araujo, 69080-971 Manaus, AM, Brazil

Gilmar Perbiche-Neves and Carlos E.F. da Rocha
Departamento de Zoologia, Instituto de Biociências, Universidade de São Paulo. Rua do Matão, travessa 14, 321, 05508-900 São Paulo, SP, Brazil

Marcos G. Nogueira
Departamento de Zoologia, Instituto de Biociências, Universidade Estadual Paulista. Distrito de Rubião Júnior, 18618-970 Botucatu, SP, Brazil

Raul Fonseca
Departamento de Zoologia, Universidade Federal de Pernambuco. Avenida Professor Moraes Rêgo, s/n, Cidade Universitária, 50670-901 Recife, PE, Brazil
Departamento de Zoologia, Universidade do Estado do Rio de Janeiro. 20550-013 Rio de Janeiro, RJ, Brazil

Diego Astúa
Departamento de Zoologia, Universidade Federal de Pernambuco. Avenida Professor Moraes Rêgo, s/n, Cidade Universitária, 50670-901 Recife, PE, Brazil

Luiz R.L. Simone
Museu de Zoologia, Universidade de São Paulo. Caixa Postal 42494, 04218-970 São Paulo, SP, Brazil

Ana Paula S. Dornellas
Museu de Zoologia, Universidade de São Paulo. Caixa Postal 42494, 04218-970 São Paulo, SP, Brazil

Gleomar F. Maschio
Instituto de Ciências Biológicas, Universidade Federal do Pará. Rua Augusto Corrêa 01, Guamá, Caixa Postal 479, 66075-110 Belém, PA, Brazil

Ana Lúcia C. Prudente
Laboratório de Herpetologia, Coordenação de Zoologia, Museu Paraense Emílio Goeldi. Avenida Perimetral 1901, Terra Firme, Caixa Postal 399, 66017-970 Belém, PA, Brazil

Kellen R. M. de Sousa
Instituto de Ciências Biológicas, Universidade Federal do Pará. Rua Augusto Corrêa 01, Guamá, Caixa Postal 479, 66075-110 Belém, PA, Brazil

Index